T0231522

# WATER RESOURCES MANAGEMENT AND THE ENVIRONMENT

Respectfully dedicated to the memory of the savant,

Abdus Salam

who will forever be remembered
for his efforts to promote science in the Third World.

# Water Resources Management and the Environment

U. ASWATHANARAYANA

*Adviser on Environment & Technology, c/o Ministry of Environment, Maputo, Mozambique*

*Former Commonwealth Visiting Professor, Universidade Eduardo Mondlane, Maputo, Mozambique*

Taylor & Francis
Taylor & Francis Group

LONDON AND NEW YORK

*Library of Congress Cataloging-in-Publication Data*

*Applied for*

Cover design: Studio Jan de Boer, Amsterdam, The Netherlands.
Typesetting: Composition & Design Services, Minsk, Belarus.

Published by
Taylor & Francis
2 Park Square, Milton Park, Abingdon, Oxon, OX14 4RN
270 Madison Ave, New York NY 10016

Transferred to Digital Printing 2007

www.routledge.com

ISBN 90 5809 322 0 (HB)
ISBN 90 5809 339 5 (PB)

**Publisher's Note**
The publisher has gone to great lengths to ensure the quality of this reprint
but points out that some imperfections in the original may be apparent

# Contents

# Foreword

Professor Aswathanarayana starts his book with the words '*All life is critically dependent upon water*' and ends with the advice that '*There is little doubt that in the twenty-first Century, water issues are going to emerge as the most serious problem facing humankind*'.

In 1977 the United States Nations Water Conference was held in Mar del Plata, Argentina. It was the first full intergovernmental meeting devoted solely to freshwater and its final report opens with the words:

'The accelerated development and orderly administration of water resources constitute a key factor in efforts to improve the economic and social conditions of mankind, especially in the developing countries, and … it will not be possible to ensure a better quality of life and promote human dignity and happiness unless specific and concentrated action is taken to find solutions and to apply them at the national, regional and international levels'.

Fifteen years later, in preparation for the UN Conference on Environment and Development (Rio de Janeiro 1992), the World Meteorological Organization (WMO) convened the International Conference on Water and the Environment which, with the support of all UN agencies and many non-governmental organizations involved in water matters, was held in Dublin, Ireland. It adopted the famous 'Dublin Principles', which read as follows:

'Principle No. 1 – Fresh water is a finite and vulnerable resource, essential to sustain life, development and the environment,

Principle No. 2 – Water development and management should be based on a participatory approach, involving users, planners and policy-makers at all levels,

Principle No. 3 – Women play a central part in the provision, management and safeguarding of water,

Principle No. 4 – Water has an economic value in all its competing uses and should be recognized as an economic good.'

Over all these years, therefore the message has remained the same: water is a precious resource that must be assessed and managed with care for the benefit of mankind and the natural environment.

These wise words can only achieve results, however, if they are acted upon by those who have the power to do so. They were addressed primarily to Governments, but they were also intended as a call for institutions, agencies and individuals to take action in the face of what was seen as an evolving crisis in the availability and use of freshwater.

Eight years later and the crisis has deepened and again the attention of the international community is being drawn to the serious consequences that will result if adequate action is not taken. This time the call has come from the Ministerial Conference held in March 2000 in the Hague, The Netherlands in conjunction with the Second World Water Forum. The Ministers declared that 'Water is vital for the life and health of people and ecosystems and a basic requirement for the development of countries, but around the world women, men and children lack access to adequate and safe water to meet their most basic needs. Water resources, and the related ecosystems that provide and sustain them, are under threat from pollution, unsustainable use, land-use changes, climate change and many other forces. The link between these threats and proverty is clear, for it is the poor who are hit first and hardest. This leads to one simple conclusion: business as usual is not an option. There is, of course, a huge diversity of needs and situations around the globe, but together we have one common goal: to provide water security in the 21st Century.'

In 2002 the United Nations will be reviewing progress with the implementation of the recommendation on environment and development adopted to Riò de Janeiro ten years before. It is likely to find that progress is far from satisfactory and once more there will be expressions of concern and a call for renewed action.

In all of this, Professor Aswathanarayana and WMO share an important concern for what the Dublin Conference called the 'Knowledge base'. As part of the Action Agenda adopted in Dublin, we read:

'Measurement of components of the water cycle, in quantity and quality, and of other characteristics of the environment affecting water are an essential basis for undertaking effective water management. Research and analysis techniques applied on an interdisciplinary basis, permit the understanding of these data and their application to many uses.

With the threat of global warming due to increasing greenhouse gas concentrations in the atmosphere, the need for measurements and data exchange on the hydrological cycle on a global scale is evident. The data are required to understand both the world's climate system and the potential impacts on water resources of climate change and sea level rise. All countries must participate and, where necessary, be assisted to take part in the global monitoring, the study of the effects and the development of appropriate response strategies.'

This call for action assumes that staff have been trained to undertake this important work and that they have available to them the materials that they need. Basic among these are reference books that introduce students and research workers to the subjects they are studying and offer them the information and advice they need to continue their work.

In this Prof. Aswathanarayana has done a real service by providing us with a book that covers an unusually wide range of topics within the water field. He takes a refreshingly open approach to many subjects, often challenging established practices, and provides valuable references to a wealth of literature.

In this way, he can be said to have made a valuable contribution to the important efforts to develop the knowledge base that is so vital if we are to overcome the water crisis that now threatens the world.

G.O.P. Obasi
Secretary-General
World Meteorological Organization

# Preface

This book is not about *finding* water – it is about *managing* the multiple uses and reuses of water optimally, sustainably and equitably, through an understanding of the linkages and dynamics of interactions involving water. In keeping with this ethos, the book adopts an integrated techno-socio-economic approach for water resources management.

Every human being on earth is a stakeholder in water resources management. And so, for that matter, is every animal, domesticated and wild – only, they do not have a constituency. We need to protect other living things not for any altruistic reasons, but in our own self-interest. We should realize that that our well-being is inseparable from the well-being of the ecosystem. If frogs are dying, we would surely be next in the line!

In his insightful analysis of the linkages between the geosphere and biosphere in the context of global change, Berrien Moore of IGBP (1999) explains how water couples the terrestrial biogeochemical system to the climate system. As is well known, the plant productivity and the sustainability of the natural ecosystems are critically dependent upon the availability of water. In their turn, the terrestrial ecosystems recycle water vapor and other trace gases at the land surface/atmosphere boundary. Soil moisture plays two inter-related key roles: on one hand, it determines the fertility and productivity of the soil, and constitutes the means by which climate regulates (and is partly regulated by) the ecosystem. On the other hand, soil moisture feeds the rivers and groundwater, depending upon the land-use and land-cover. The rivers deliver not only water but also nutrients to the coastal oceans, and are thus linked to coastal fisheries, and so on. We should understand how these linkages operate at a watershed level, in order to integrate the biophysical and socioeconomic approaches of the use and reuse of water.

The International Conference on Water and Environment, Dublin 1992, enunciated two crucially important guiding principles, namely, that all human beings have a basic right to have access to clean water and sanitation at an *affordable price*, and that water has an economic value in all its competing uses and should be recognized as economic good.

In his keynote address to the UNESCO-organized World Conference on Science, Budapest, Hungary, June 26-July 1, 1999 Dr M.S. Swaminathan of India gave his vision of a hunger-free world. He envisages three approaches:

1. Improving food production through the optimization of various inputs, via people-participatory technologies in an ecologically-sustainable manner,
2. Increasing access to food by promoting job-led economic growth, and
3. Facilitating the availability of safe drinking water and environmental hygiene, in order to ensure the biological absorption of food in the human body. It may be noted that water resources management happens to be at the core of all the three approaches. Plants need tremendous amounts of water to grow – about 30-40 cm of water is needed to produce a food crop, such as wheat. Thus irrigation will continue to be the principal user of water, and also the key input to enhance agricultural productivity. Water-related activities (e.g. rainwater harvesting, storage, distribution, use and reuse, etc.) have considerable potential to create jobs. About 5000 children (mostly in the Developing countries) die *every day* due to water-related diseases. These deaths are entirely preventable if only adequate quantities of clean water is available for drinking and sanitation.

The techno-socio-economic problems involved in water resources management are indeed formidable, but not insoluble (both figuratively and literally!). As Larry Kohler of ICSU points out that we should never underestimate the ingenuity of humans and their societies to adapt to change. For instance, according to some earlier scenarios of environmental degradation, London should now be about five metres deep in horse manure. Also, Stone Age did not end because of a global shortage of stones !

There is little doubt all development processes in the 21st century will be knowledge-driven. In order to be relevant to the needs of the society, geological science should be focused on ecologically-sustainable use of water, soils and minerals (in that order), rather than being taught in the form of traditional subject disciplines. In his own small way, the author has embarked on providing teaching texts with this perspective. The present volume is the third member of the trilogy, the other two being, '*Geoenvironment: An Introduction*' (A.A. Balkema, Rotterdam, 1995), '*Soil Resources and the Environment*' (Science Publishers Inc., Enfield, NH, USA/Oxford-IBH, New Delhi, 1999).

In keeping with its interdisciplinary character, the book carries a Foreword and a Prologue which complement each other. Prof. G.O.P. Obasi, Secretary General, World Meteorological Organization, Geneva, wrote the Foreword. Apart from being an eminent atmospheric scientist, Dr Obasi has steered WMO during the last two decades to achieve high levels of public usefulness through systems such as GOOS (Global Ocean Observing System) which could give early warning of extreme weather and climate events. The Prologue has been contributed by Mr Derish M. Wolff, President of Louis Berger International Incorporated, (LBII), New Jersey, USA. LBII is one of the world's leading engineering consultancy companies in the area of water use (irrigation, water supply, sanitation, navigation, recreation, etc.) whose hallmarks are Innovation, Integration, and Lower costs. I am grateful to Dr Obasi and Wolff for their kind contributions.

The book will be useful to university students and professionals in the areas of earth, environmental and water sciences and technologies, resource management, agriculture, ecology, civil engineering, etc.

When I was a visiting fellow at Caltech, Pasadena, California in 1957, I was a neighbor for a few months, of the great Bubridges, the distinguished astrophysicist couple. They had a baby daughter at that time, and they named a star after her! When our daughter, Indira, was born in 1972, she brought us much joy (and later, much honor). In Sanskrit, indira variously connotes water, lotus and Lakshmi (goddess of wealth). The thought occurred to me later that though I cannot do what Bubridges have done so spectacularly, I could still do something at a more mundane level, such as writing a book on *water*.

The author is grateful to Michel Meybeck (France), Asa Sjöblom (Sweden), I. Radhakrishna (India), S.V. Durvasula (Kuwait), Karl Roehl (Germany), Thomas Puempel (Austria), Herman Bouwer (USA), Willard Moore (USA), Arne Tollan (Norway) and Adáo Matonse (Mozambique) for the provision of literature and kind suggestions to improve the text, and to Eng. Luis de Macedo, Mahomed Essak, and Sandesh Hegde  for practical help.

Maputo, Mozambique
U. Aswathanarayana
June, 2000

# Prologue

The management of water, a resource which is becoming an increasingly scarce in many parts of the world, requires an understanding of the science of hydrology as well as an appreciation of socioeconomic values of communities. In this book, the author has effectively and elegantly integrated, the two strands of science of water and the management of water, to produce a book which is both scientifically authoritative and eminently readable.

Water resource planning and management is not a new field. There is extensive theoretical and applied experience in the field to guide practitioners to successfully design and implement viable water resource systems, facilities and structures, as Prof. Aswathanarayana so ably points out. Advances have often gone in lock step with humankind's social and economic progress, since much of our early development occurred in the alluvial valleys of the Nile, Euphrates, Tigris, Indus, Yellow and Yangtze Rivers.

Today's water planners and designers are heirs to a technology as old as recorded history with the earliest reference to water resource management occurring in Egypt during the 5th millennium BC. Modern water resource managers, however, face increasing challenges in the light of rapidly growing populations and competing demands for an increasingly scarce resource. In many developing nations, rapid population growth mandates an increasingly intensive use of scarce agricultural lands, accompanied by improved water resource management to provide multiple harvests where only one grew before.

The related practice of sanitary engineering, which dates at least to the Romans, is one of the fastest growing infrastructure fields. Water and sewage facilities directly impact our health as few other engineering activities do. But as Professor Aswathanarayana points out, facilities have all to frequently trailed population and economic growth, a particularly acute issue in the developing world. According to a recent World Bank publication, 20% of the world's population lacks access to potable water and nearly half to decent sanitary services. An estimated 25% of rural water supply facilities in developing nations are inoperative. In Latin America, a fourth of the urban population is not connected to public water, and a half to public sewerage, and the region must invest USD 12 billion annually for the next 10 years to reach acceptable water and sanitation levels. In Africa, 65% of those living in rural areas lack access to adequate water and 73%

lack access to sanitation. While the percentages drop in urban areas to 25 and 43%, respectively, the startling statistic remains that over 250 million Africans-half of the continent's population-lack access to portable water. On average, 80% in the United States and United Kingdom are connected to modern sewerage facilities. But other advanced nations are worse off, including Greece, Spain and Japan, where only a third of the population is connected to modern sewerage, as well as France and Portugal, where only half of wastewater is collected, and a fourth of treated water is below internationally acceptable standards.

Furthermore, adequate water and sanitation are becoming increasingly critical in urban areas, as rapid urbanization and industrialization strain already limited resources. This is of particular concern in Asia, which holds more than 50% of the world's population and is home to number of the world's megacities. But such explosive growth is a global problem. Well over 300 cities have populations over 1 million, double the total in 1975. By 2025, 60% of the global population will live in urban areas, a daunting number given the health hazards already present in many of our cities.

The author points out, with the world's population likely to grow by 3.7 billion from 1990 to 2030, and with water use increasing twice as fast as population during this century, water is likely to become an increasingly scarce and volatile resource. By the year 2025, as much as two-thirds of the world's population will face moderate to severe water scarcities.

Given the author's concern for the growing water crises we are likely to face, he wisely addresses the important issue of wastewater reuse. The author comes up with some spectacular estimates of the volume of water currently disposed of as wastewater and, quite rightfully, presents this as both a loss and an opportunity. Recycling of wastewaters is obviously important and yet, as the author points out, historic problems with wastewater quality including concerns over water borne disease and confusion with sewage has discouraged its use. The author suggests such innovative approaches as marshing, geopurification, bio-ponds and eco-ponds, as cost effective solutions to wastewater recycling, while stressing the need to carefully monitor the health implications of such reuse.

He also point out both the need for water in irrigation as well as the short falls of many current systems and suggest techniques to improve utilization. He especially warns about the frequent error of over-irrigating, accompanied by informative case studies, as well as a wide- ranging discussion of the industrial and municipal uses of water.

The book includes a careful analysis of opportunities for improved utilization and conservation of water resources and presents a number of innovative techniques and approaches for groundwater recharge and protection of existing resources. The author also analyzes the opportunities and problems in desalinization for both sea and brackish waters, an increasingly popular, though expensive approach.

Water has long been a free good. Rarely, except in times of crisis have we viewed it as a scarce resource and our pricing systems have done little to encourage conservation. The author provides a useful discussion of the various methodologies for establishing the economic cost of water, dealing with such issues as measuring water's utility, exchange value and scarcity or marginal utility, and describing some of the solutions recommended by various experts. The author addresses the secondary uses of water such as navigation, hydro-power generation, recreation, needs for flood control, etc. to develop a mechanism for properly pricing water in order to establish guidelines for optimization of this increasingly scarce resource.

The author describes the best water resource management practices supported by some excellent case studies. He projects the likely impact of climatic changes on current water resources, and suggests ways and means of coping with them.

Thus, the author in this important work has tried to describe and define the entire water resources environment, its dynamics of interaction, the impact of various usages and changing technologies on water quality, accompanied by appropriate measures for evaluating the needs of various users.

The book also includes some excellent guidelines for practical practitioners in the field and exercises for the practitioner in the latest techniques of modern water resource management.

Finally, the author uses the book to transcend the large technical scope of the water resource field to address such issues as the proper institutional framework for water resource management including regulatory measures, economic instruments and likely future technological innovations and the continued role the water sector can play in integrating and encouraging social and economic change and development in the Third World.

As a practical practitioner in the field, I found the book both informative and useful for the trained and experienced planners and managers as well as for the entry-level professionals who for the first time must address the issues, challenges and concerns in this field.

East Orange, New Jersey, USA      Derish M. Wolff
September 2000          President, Louis Berger Int. Inc.

# Copyright Acknowledgement

Grateful acknowledgement is made to the publishers, authors, editors of journals, books from which some figures that appear in this volume have been redrawn or adapted. The figures in the volume that are so drawn are shown against the publisher of the concerned book or journal.

*Ambio*, 6.2, 12.3

*Acta Geophysica Polonica*, 12.1

Amer. Geophy. Uni., *World Water Resources and their future* 1.1-1.6, 1.21-1.24, 12.18

AWWA (Amer. Water Works Assoc.) 5.3, 5.7

Balkema, *Geochemistry, Groundwater and Pollution* 2.14, 2.27, 3.1, 3.7, 3.14, 8.7, 8.8, 10.3, 10.4

Butterworths, Artificial Recharge of Groundwater 5.4

Clarendon Press, *Computer Modelling in Environmental Sciences* 1.8, 1.9

Elsevier, *Environmental Modelling* 8.10; *Topics in Environment & Health*, 10.2

Fizika 2.18

*Geol. Soc. Am. Bull* 3.12, 3.13

*Global & Planetary Change* 3.4, 3.5, 3.6, 3.8, 3.9

*Groundwater Monitoring Rev.* 1.12, 6.11

Harper & Row, *Solutions, Minerals and Equilibria* 2.12; $H_2O$ 2.25

IAHS (Int. Assn. Hydrol. Sci.), no. 150 (1984) 2.31; 169 (1987) 6.3, 6.4; 180 (1989) 12.7, 12.16; 225 (1995) 12.5; 231 (1995) 2.3, 2.4, 6.5-6.8, 12.9, 12.11, 12.17; 248 (1998) 2.5, 2.17, 2.22-2.24, 3.15-3.17

IIASA (Int. Inst. for Appl. Systems Analysis) 12.2

IAEA (Int. Atomic Energy Agency) 2.6-2.9, 2.16, 2.20, 2.32, 2.33, 6.9, 6.10; *Int. J. Remote Sensing* 12.10

IGBP (Int. Geosphere – Biosphere Program) 12.13

John Wiley, *Environmental Impact Assessment for Waste Treatment and Disposal Facilities* 3.2, 3.3, 8.4-8.6; *J. Geophy. Res.* 2.19; *J. Trace Elem. Exp. Med.* 10.1; KIWA 2.15

Lewis Publishers, *Constructed Wetlands for wastewater treatment* 4.8; LOICZ 3.10

Martinus Nijhoff, *Water Resources and Landuse Planning: A Systems Approach*
5.5, 8.1, 8.2, 9.4-9.10
McGraw-Hill, *Principles of chemical Sedimentology* 2.30
Mir Publishers, *General Hydrogeology* 1.10, 1.11, 1.13-1.20, 1.22-1.24
Nat. Inst. of Env. Studies (Tsukuba) 2.1; NOAH (USA) 2.2
Pergamon Press (*Water Sci. & Tech.*) 5.1, 5.2
*Science* 2.11a,b; *Soil Sci. Soc. Amer. Bull.* 5.6
Springer-Verlag (*Env. Geol.*) 2.35, 2.36; Swedish Univ. of Agr. Sci. 2.34
UNEP, *Rain and Stormwater Harvesting in Rural areas* 4.1, 4.2, 4.3a,b, 4.4a,b,
4.7: UNEP-UNESCO, *Water Management and Geoenvironment* 1.7, 8.9, 12.6
UNESCO, *Water: A Looming Crisis?* 4.5, 4.6, 9.1-9.3, 12.12, 12.14, 12.15;
*World Water Resources ...* 12.19, 12.20
US CSP (Country Study Program) 12.4; USDA 6.1; Univ. of Linköping 12.8;
US GS 2.29, *Water Resour. Res.* 2.21, 3.11; Wiley, *Hydrogeology* 2.10;
*Aquatic Chemistry* 2.13, *Water quality in Catchment Ecosystems* 2.28; World
Bank 1.25, 6.11; *Z.dt. Geol. Ges.* (Dutch) 8.3.

# Sleeping under the stars

Sherlock Holmes and Dr Watson went on a camping trip. As they lay down for the night, Holmes said, 'Watson, look up the sky and tell me what you see'.
Watson: 'I see millions and millions of stars'
Holmes: 'And what does that tell you?'
Watson: 'Astronomically, it tells me that there are millions of galaxies, and potentially billions of planets. Theologically, it tells me that God is great, and that we are small. Meteorologically, it tells me that we will have a beautiful day tomorrow. What does it tell you?
Holmes: '*Elementary, my dear Watson! Somebody stole our tent*'

Modified after Jack Zirkel, *Eos*, p. 3, Jan. 5, 99.

Moral of the story: Before developing elaborate models, check whether there might be a simple explanation for the phenomenon you have observed.

'The sea is our mother and father. All life comes from it. It is still far richer in major groupings of animals than on land: of 34 animal phyla, 29 occur in the sea, and 14 of them only in the sea. The complexity of its species and ecosystems is immense. Yet, as on land, we are engaged in the process of extinguishing populations and ecosystems at something like 1000 times the natural rate...'

Foreword of Sir Crispin Tickell to the volume       *Marine Biodiversity (1997)*

'It is a curious situation that the sea, from which life first arose, should now be threatened by the activities of one form of that life. But the sea, changed in a sinister way, will continue to exist; the threat is rather to life itself'
*Rachel Carson, 1950*

'Let me challenge you with two simple facts.
    First, every one of you sitting here today is carrying at least 500 measurable chemicals in your body that were not part of human chemistry before the 1920s.

We are walking experiments, differing from all generations of human ancestry in this regard.

Second, there is now incontrovertible scientific proof that a mother shares some of these man-made chemicals with her baby while it is in her womb. No baby has been born on the planet for at least two decades without exposure to novel chemicals in the womb. Some with little exposure. Some with a lot. But none with none.

In all likelihood, some, perhaps many of these compounds will turn out to be benign, with no impact. But some we know already cause problems ...'

*Excerpts from the speech of Dr J.P. Myers*
*in the Rio + 5 Forum on 14 March 1997*

The following is a hymn to the River Sarasvati in Rig Veda (older than 3000 BC): 'Coming together, glorious, loudly roaring Sarasvati, Mother of seven rivers, fulfilling our desires, strongly flowing, from their water swelling in volume'(7-36-6) 'Sarasvati, you are the best of mothers, you are the best of rivers, you are the best of the goddesses. Although we are of no repute, dear mother, grant us distinction'(2-41-16) (Source: River Saraswati in Rigveda by B.P. Radhakrishna, J. Geol. Soc. Ind., v. 51(6): 725-730, 1998).

Yajurveda (2500-2000 BC) carries the following hymn to Water:
'O Water, thou art the reservoir of welfare and prosperity, sustain us to become strong. We look up to thee to be blessed by thy kind nectar on this earth. O Water, we approach thee to get rid of our sins. May the water cleanse the earth and the earth cleanse me. May the holy waters make me devoid of sins. May the waters remove my bad deeds. The waters that kept the Agni (fire) inside them, bless us. The waters that generate all prosperity on earth and heaven and those which dwell in different forms in the atmosphere, those which irrigate the earth, may those waters be kind to us and bless us. O Water, kindly touch me with thy divine self and establish strength, radiance, intellect and wisdom in me.'
(source: *Water – the fulcrum of ancient Indian Socio-religious traditions* by K.N. Sharma, in the Proc. Int. Conf. on the World Water Resources at the Beginning of the 21st century, Unesco, June, 1998, p. 471-476).

Highest good is like water... because water excels in benefiting the myriad creatures without contending with them and settles when none would like to be, it comes close to the way ... In the world there is nothing more submissive and weak then water... Yet for attacking that which is hard and strong, nothing can surpass it ... Tasteless, it accepts all tastes, colourless, all colours, reflecting the sky, refracting the white stones of its bed, dissolving or suspending the soils and minerals over which it flows. The pulse of the bodies is liquid, as indeed all living pulses are...'

Sayings of Lao Tzu, as quoted and annotated by Vikram Seth, in his book, *From Heaven Lake* (1983, p. 165).

The beef in one hamburger involves: 0.79 kg of feed grain, which needs 795 l of water to grow it, clearance of 5.1 m$^2$ of tropical forest, and 5.44 kg of animal faeces.

*Time* (Nov. 8, 1999)

# Abbreviations

*Organizations*

| | |
|---|---|
| EC | European Community |
| FAO | Food and Agriculture Organization |
| ICRCL | Inter-departmental Committee on the Redevelopment of Contaminated Land, UK |
| OECD | Organization of Economic Cooperation and Development |
| UNEP | United Nations Environment Programme |
| UNESCO | United Nations Educational, Scientific and Cultural Organization |
| USEPA | United States Environmental Protection Agency |
| WHO | World Health Organization |
| WMO | World Meteorological Organization |

*Terms*

| | |
|---|---|
| ADI | Acceptable Daily Intake |
| ALARP | As Low As Reasonably Practicable |
| AQS | Air Quality Standard |
| AROI | Acceptable Range of Oral Intake |
| BAT | Best Available Technology |
| BATNEEC | Best Available Techniques Not Entailing Excessive Cost |
| BOD | Biochemical Oxygen Demand |
| BPEO | Best Practicable Environmental Option |
| BPM | Best Practicable Means |
| CBA | Cost – Benefit Analysis |
| CEC | Cation Exchange Capacity (meq/100 g or $mols_c$ $kg^{-1}$) |
| CFCs | Chlorofluorocarbons |
| COD | Chemical Oxygen Demand |
| DOC | Dissolved Organic Carbon |
| EIA | Environmental Impact Assessment |
| EMS | Environmental Management System |
| EQS | Environmental Quality Standard |
| ERL | Environmental Reference Level |
| ES | Environmental Statement |
| GIS | Geographic Information System |

| | |
|---|---|
| HI | Hazard Index |
| MCNC | Most Common Natural Concentrations |
| MEL | Maximum Exposure Level |
| NIMBY | 'Not In My Backyard' |
| NOEL | No Observed Effect Level |
| OEL | Occupational Exposure Limit |
| RDI | Recommended Daily Intake |
| SIA | Social Impact Assessment |
| SWQO | Statutory Water Quality Objective |
| TDS | Total Dissolved Salts (mg $l^{-1}$, mmol $l^{-1}$) |
| TIC | Total Inorganic Carbon (mg $l^{-1}$, mmol $l^{-1}$) |
| TOC | Total Organic carbon (mg $l^{-1}$, mmol $l^{-1}$) |

*Substances*

| | |
|---|---|
| PCBs | Polychlorinated biphenyls |
| PCDDs | Polychlorinated dibenzodioxins |
| PCDFs | Polychlorinated dibenzofurans |
| TCDD | Tetrachlorobenzodioxin |
| TCE | Trichloroethylene |
| VOCs | Volatile Organic Compounds |

*Measurements*

| | |
|---|---|
| Bq | becquerel (1 Bq = $2.7 \times 10^{-11}$ curies) |
| EC | Electrical conductivity of water sample ($\mu S$ $cm^{-1}$) |
| Eh | Redox potential relative to standard state $H_2/H^+$ reaction (V) |
| $EC_{50}$ | Effect-specific concentration lethal to 50% of exposed organisms |
| ESR | Exchangeable Sodium Ratio |
| $LC_{50}$ | Concentration lethal to 50% of those exposed |
| $LD_{50}$ | Dose lethal to 50% of those exposed |
| SAR | Sodium Absorption Ratio |

*Notations*

| | |
|---|---|
| $A_0$ | Total cation concentration in solution (eq/l) |
| $A_s$ | Specific surface ($m^2$/kg) |
| $C$ | Solute concentration (mol/l, mol/kg, mg/l) |
| $\Delta G_r^0$ | Gibbs free energy (enthalpy) of a reaction at standard state conditions (25°C, I atm., unit activity of the species) (J/mol or kcal/mol) |
| $I$ | Ionic strength (mol/l) |
| $K$ | Mass Action Constant or Solubility product |
| $K_d$ | Distribution coefficient (mg sorbed or solute/l of porewater) |
| $K_d'$ | Distribution coefficient (mg sorbed/g solid)/mg solute/l of porewater) (l $kg^{-1}$) |
| $K_{oc}$ | Distribution coefficient for the given chemical among 100% organic carbon and water (l $kg^{-1}$) |

$K_{ow}$      Distribution coefficient for the given chemical among octanol and water ( $kg^{-1}$)

$M$      Molality (mol/kg of $H_2O$)

$N$      Normality (eq/l)

$Q$      Discharge ($m^3$/s)

CHAPTER 1

# Introduction

## 1.1 WATER, CIVILIZATION AND QUALITY OF LIFE

All life is critically dependent upon water – the origin and evolution of life became possible on earth because of the presence of liquid water. Animals and plants consist mostly of water. Water (40 l) constitutes three-fourth of the human organism.

Since ancient times, man has settled along the rivers, has drunk the river water, has moved along the river to unknown parts, and has eaten the fish caught in the rivers. In terms of civilization, *Homo sapiens* who survived solely by hunting and food gathering, are indistinguishable from other animals. Modern civilization can be deemed to have dawned about 10,000 years ago, when man has learnt to cultivate land to grow food. It is not an accident that the fertile alluvial valleys of the Nile, Euphrates-Tigris and Indus constituted the early sites for significant human settlements and civilizations. The worship of gods and goddesses, the religious practices, and the whole tenor of life of the people living on the banks of the rivers are inextricably linked with the rhythm of the river.

Among the great existing religions in the world, Hinduism is unique in that it originated in a region blessed with abundant water. The Sarasvati was a mighty river in northwest India in Rigvedic time, and the vedic texts are replete with invocations to it. When it lost its Yamuna headwaters about 4000 years ago due to river piracy, its discharge dwindled to nothing. It is now just a dry stream bed (see, the evocative account of the disappearance of Sarasvati, by Valdiya, *Resonance*, v. 1(5), p. 19-28, 1996).

Traditionally, water had and continues to have, a central role in the rituals, worship and prayers of the Hindus (for instance, *abhishekam* involves the ritual anointing of gods using water or milk). In the course of worship, a Hindu is required to identify the perennial river (*jeeva nadi*) on the banks of which and nearest to which, he/she resides (the author's daughter who lives in St. Louis, USA, says in her worship that she lives on the banks of the Missouri River!). For a Hindu, the Ganga is just not any river – Ganga is the river goddess celebrated in myth and legend. In Sanskrit, Ganga means water. Hindus believe that all water in India – surface and underground – is ultimately connected to the Ganga. So much so, even when a Hindu is taking a bath with water from the well in his backyard, he ritually evokes the connection of that water with Ganga. Hindus generally cremate their dead, and ceremonially immerse the ashes in the waters of perennial rivers (the confluence of the Ganga and Yamuna rivers in Allahabad is considered to be the most sacred site for this purpose). *Life originated in water, and is returned to water.*

Egypt has rightly been called the 'Gift of Nile'. The Nile waters are so crucial to the life of the people in Egypt that that virtually all institutions – governments, religion, gods and goddesses and priesthood, rituals, etc. – arose from the need to control the utilization of the flood waters of the Nile. Egypt is thus a most extraordinary example of intertwining between the government and irrigation, for millennia. The earliest village at Marimda in the Nile Delta is said to be 5000 BC, i.e. 7000 years old. The earliest irrigation ditches were dug there around 3000 BC (presently, there are 16,000 km of irrigation canals in the Nile Delta). Observations on the water level on the Nile started to be made as long ago as 2000 BC Ancient Nilometers have been preserved in Aswan, and the Roda Island within the city limits of Cairo. On the basis of the measurement of the flood water in the Nile, priests would figure out the extent of the overflow of the flood water and hence the kind of harvest that could be expected. Higher flood water meant bigger harvests, and lower flood waters meant poor harvests (presently, the Nile water is used so completely that the river is not discharging any more water, nutrients or sediments to the coastal zone, but that is a different story).

The need to measure the dimensions of land under cultivation, led to the development of geometry. Domestication of animals, and building of canals made for a more productive agriculture. Larger populations could then be supported because of the availability of greater quantities of food. The use of human and animal muscle power, and burning of animal dung and agricultural refuse as fuel, resulted in the rise of energy consumption.

It is not easy to define, let alone quantify, the quality of life of a community or a family. There is, however, little doubt that quality of life is closely linked to the quality and quantity of water available for drinking, cooking, washing, bathing, sanitation, gardening, etc. The rate of consumption of water for domestic purposes (in terms of cubic meters per capita per annum) bears a general direct relationship to the quality of life. One way to quantify the quality of life is the Human Development Index (HDI is a parameter developed by UNDP, based on life expectancy, adult literacy, and standard of living, as measured by Real GDP in terms of Purchasing Power Parity dollars or PPP$; sources: 'Human Development Report' 1997; 'World Development Report' 1992) (see Table 1.1).

There does not appear to exist any similar relationship between the quantity of water used in industry and agriculture (in terms of $m^3$ per capita per annum) and GNP of a country. This is so because water is only one (not necessarily the most crucial) of the several parameters that have a bearing on the industrial and agricultural productivity of a country. A few examples are given to illustrate this point (see Table 1.2).

Table 1.1. Relationship between HDI and water consumption.

| Country | Domestic consumption of water (m³/capita/annum) | Human development index (HDI) |
|---|---|---|
| Canada | 193 | 0.960 |
| Brazil | 91 | 0.783 |
| South Africa | 65 | 0.716 |
| China | 28 | 0.626 |
| Mozambique | 13 | 0.281 |
| Ethiopia | 5 | 0.244 |

Table 1.2. Relationship between water consumption and GDP.

| Country | Consumption of water in industry and agriculture (m³) | Real GDP per capita (PPS$) |
|---|---|---|
| Pakistan | 2032 | 2,154 |
| Madagascar | 1658 | 694 |
| Chile | 1528 | 9,129 |
| Bulgaria | 1488 | 4,533 |
| Egypt | 1118 | 3,846 |
| Netherlands | 972 | 19,238 |

## 1.2   FROM RAINWATER TO TERRESTRIAL WATERS

The total amount of water in the hydrosphere: 1386 M m³. Out of this, saline waters constitute 97.5%, and fresh water 2.5%. The greater portion of freshwater (68.7%) is locked up in the form of ice and permanent snow cover in the Antarctic, and the Arctic, and high mountain chains, like the Himalayas and Alps. Groundwater constitutes 29.9% of freshwater, 0.9% of fresh water exists in the form of soil moisture, swamp water and permafrost. Surface water in the form rivers, reservoirs and lakes account for 0.26% of freshwater, which alone is renewable (Shiklomanov 1998, p. 4).

As water vapor in the atmosphere condenses as rain, it may acquire (or lose) various ions in the process of its movement through the atmosphere. Near the coast, the composition of rainwater is akin to that of diluted seawater. As we move inland from the coast, the concentration of ions derived from seawater (such as, $Cl^-$) decreases and the concentration of ions derived from terrestrial sources (such as, $Ca^{2+}$) increases. For instance, in the case of Netherlands, the $Cl^-$ concentration in the precipitation decreases from about 30 mg $l^{-1}$ at the coast to 2 mg $l^{-1}$, 200 km inland (Ridder 1978). On the other hand, the continental rain in India is characterized by a high content of $Ca^{2+}$ presumably derived from the calcite dust particles. $Na^+$ in the aerosols may be derived from the sea salt, or from dust particles in arid cles in arid climates (playa lakes have encrustations of salts). Burning of waste plastics in USA is known to contribute $Cl^-$ to the rainwater.

A convenient way to trace the modifications (addition or depletion of ions) that rainwater in an area has undergone vis-à-vis the seawater is to make use of the fractionation factor based on Cl/Na ratio. The fractionation factor, $F_{Na}$, is given by:

$$F_{Na} = \frac{(Cl/Na)_{rain}}{(Cl/Na)_{seawater}} \qquad (1.1)$$

The range (maximum and minimum) of fractionation factors (relative to $F_{Na}$) for seawater components in marine aerosols are as follows:

| | | | |
|---|---|---|---|
| $Mg^{2+}$ | = 1.07-0.98 | $Ca^{2+}$ | = 1.22-0.97 |
| $K^+$ | = 1.05-0.97 | $Sr^{2+}$ | = 0.89-0.84 |
| $Cl^-$ | = 1.0-0.93 | $SO_4^{2-}$ | = ?-1 |
| $Br^-$ | = 12-1 | Org. N | = $1.10^6$-$2.10^4$ |

An examination of the above data leads to the following conclusions:
- The minimum fractionation factor of 0.93 for $Cl^-$ is probably a consequence of some volatilization,
- The maximum fractionation factor of 1.22 for $Ca^{2+}$ relative to $Na^+$ is indicative of terrestrial source,
- The large enrichment of $Br^-$ and organic $-N$ in rainwater is attributable to the concentration of lipid containing organic matter at the interface between water and air.

That the fractionation factors for major elements ($Mg^{2+}$, $Ca^{2+}$, $K^+$, $Sr^{2+}$, $Cl^-$, $SO_4^{2-}$) are around 1, indicate that these elements have not undergone any significant fractionation. That is to say that the rainwater is essentially diluted seawater. As they move inland, the air masses pick up continental dust of natural or anthropogenic origin. Thus the chemistry of continental rainwater in a given area carries the 'signature' of the terrestrial sources.

The concentrations (in $\mu$ mol $l^{-1}$) of various ions in seawater are as follows:

| | | | |
|---|---|---|---|
| $Na^+$ | = 485 | $Cl^-$ | = 566 |
| $K^+$ | = 10.6 | $SO_4^{2-}$ | = 29.3 |
| $Mg^{2+}$ | = 55.1 | $NO_3^-$ | = $5 \times 10^{-6}$ |
| $Ca^{2+}$ | = 107 | $HCO_3^-$ | = 2.4 |
| $NH_4^+$ | = $2 \times 10^{-6}$ | | |

In order to trace the impact of terrestrial sources on the composition of rainwater, we have to take into consideration a conservative datum reflecting the seawater. The concentration of $Cl^-$ fulfills this requirement admirably. As seawater has $Na^+$ concentration of 485 $\mu$ mol $l^{-1}$ and $Cl^-$ concentration of 566 $\mu$ mol $l^{-1}$, the ratio of Na/Cl in seawater is therefore 0.86. The $Cl^-$ concentration of rainwater in (say) Delhi is 28 $\mu$ mol $l^{-1}$ and the contribution from seawater is therefore $28 \times 0.86 =$

Table 1.3. Terrestrial contribution to rainwater.

| Ion | Kiruna, Sweden | Hubbard Brook, USA | De Kooy, Netherlands | Delhi, India |
|---|---|---|---|---|
| $Ca^{2+}$ | 16 | 4 | 12 | 28 |
| $NH_4^+$ | 6 | 12 | 78 | |
| $SO_4^{2-}$ | 20 | 30 | 48 | 3 |
| $NO_3^-$ | 5 | 12 | 63 | |

24 $\mu$mol l$^{-1}$. The observed Na$^+$ concentration in the case of Delhi is 30 $\mu$ mol l$^{-1}$. Thus, $30 - 24 = 6$ $\mu$ mol l$^{-1}$ of Na$^+$ must have a non-marine origin. Similar calculations can be made for other ions.

The magnitude of non-marine (natural terrestrial or anthropogenic) contribution to rainwater could be judged from Table 1.3 (units are $\mu$ mol l$^{-1}$) (source: Appelo & Postma 1996, p. 24-25).

The high Ca$^{2+}$ in the case of Delhi is attributed to calcareous dust of natural origin. The markedly high SO$_4^{2-}$ and NO$_3^-$ contents in the case of Hubbard Brook (USA) and De Kooy (Netherlands) can be traced to the industrial and automobile emissions of NO$_x$ and SO$_2$, which get oxidized to HNO$_3$ and H$_2$SO$_4$ and cause the *acid rain*. The different aspects of acid rain are discussed in Chapter 8.


## 1.3 WATER CYCLE

Following the usage of L'vovich (1979), the term Water Cycle (rather than Hydrological Cycle) is used in this volume.

Water is different from other natural resources in being mobile. Most of the water on earth is involved in a continuous movement called the Water Cycle. Water evaporates from the surface of the oceans, seas, lakes, rivers, etc., forms clouds and moves inland. When the water vapor in the clouds condenses, it precipitates as rain or snow depending upon the ambient temperature. Part of the precipitation on land forms surface runoff, and part percolates downward forming groundwater. The force of gravity moves surface water and groundwater into depressions (river valleys) and forms rivers which feed the oceans. Part of the rainfall is retained in the soil as soil moisture, which undergoes evaporation (including transpiration). The vapors of the atmosphere are carried by air currents, condense and form precipitation, and a new cycle starts (Fig. 1.1; source: L'vovich 1979, p. 24).

The water cycle involves interaction with lithosphere, atmosphere and biosphere. It links all the forms of hydrosphere – namely, oceans, rivers, soil moisture, groundwater and atmospheric water. The water cycle is energized by solar energy and gravity. Evaporation and condensation of water vapor take place under the influence of solar heat, whereas gravity controls the falling of rain drops,

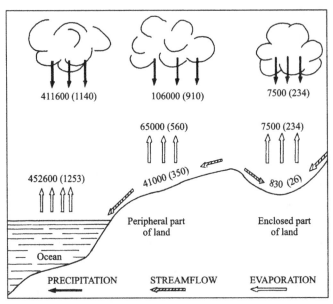

Figure 1.1. Water cycle (source: L'vovich 1979, p. 24). The numbers in parentheses are in millimeters, and the numbers without parentheses are in km$^3$. Thus, the evaporation from the oceans is 452,600 km$^3$ (1253 mm), and the precipitation on the oceans is 411,600 km$^3$ (1140 mm).

the flow of rivers, the movement of groundwater and soil moisture. Frequently, the two processes operate together – for instances, both thermal processes and gravity are involved in the circulation of the atmosphere.

In his monumental work, L'vovich (1979) distinguished the following links in the water cycle – atmospheric, oceanic, continental (including lithogenic), soil, river, glacial, biological and human or economic links. The system of the water cycle may be considered to be a closed circuit taking the earth as a whole, but strictly speaking, none of these links can be deemed to be clo sed circuits. However, for all practical purposes, an individual river basin or lake may be taken as a closed circuit.

### 1.3.1   *The atmospheric link*

The atmospheric link in the water cycle relates to the transport of moisture in the circulating air, and atmospheric precipitation. These processes lead to the redistribution of moisture over the globe. Atmospheric moisture carried from the ocean over the land and precipitating on land, amounts to 40,000 to 43,000 km$^3$ yr$^{-1}$. The precipitation on the continents exceeds the moisture received by the atmosphere due to evaporation from land. Since some moisture is carried from over land to the oceans, the actual amount of precipitation falling on land is more than this figure. Thus, the accrual of 40,000 to 43,000 km$^3$ yr$^{-1}$ is the net gain by land.

This process is of great importance since it augments the water resources of the continents.

The difference between the air moisture carried from the oceans over land and in the opposite direction, is manifested in the form of discharge of stream flow and groundwater from land to the oceans.

Marine transgressions and regressions, and the consequent redistribution between land and sea, have taken place throughout the geological history. But these changes take place very slowly. They have hardly any effect on the air moisture regime and the water cycle.

### 1.3.2 *The oceanic link*

Continuous replenishment of water vapor by evaporation from the sea is the most significant aspect of the oceanic link – evaporation from the oceans accounts for 86% of the total air moisture, with evaporation from land producing only 14%. Neither evaporation nor precipitation is uniform in the oceans, due to differences in temperature and cloud cover. In the tropical and subtropical zones of the oceans, where there is much heat and cloud cover is thin, the amount of moisture evaporated from the surface of the oceans exceeds that falling as precipitation. Under conditions of warm temperatures and high cloud cover in the equatorial zones, precipitation is more than evaporation. In the temperate latitudes, evaporation is less than precipitation because of lack of heat.

The volume of water involved in internal water exchange in different oceans, and the residence times involved, are summarized in Table 1.4 (source: L'vovich 1979, p. 27).

The average number of years it takes for the waters in a given ocean to be moved or replaced (residence time) is longest in the Pacific ocean (110 yr) and shortest in the Indian Ocean and North Arctic Ocean (about 40 yr). The residence time for the oceans as a whole is about 60 years. As should be expected, the turnover time for surface waters of the oceans is much smaller (of the order of a few years).

The exchange of seawater involved in ocean currents is about 50 times more intense than the water exchange from evaporation and precipitation.

Table 1.4. Dynamics of exchange of oceanic water.

|  | Ocean | | | | Oceans as a whole |
|---|---|---|---|---|---|
|  | Pacific | Atlantic | Indian | North Arctic |  |
| Area (in millions of km$^2$) | 180 | 93 | 75 | 13 | 363 |
| Volume (in millions of km$^3$) | 725 | 338 | 290 | 17 | 1370 |
| Annual flow of water transported ($10^6$ km$^3$) | 6.56 | 7.30 | 7.40 | 0.44 | 21.70 |
| Residence time (yrs) | 110 | 46 | 39 | 38 | 63 |

### 1.3.3   *The lithogenic link*

The lithogenic link is manifested in the form of participation of groundwater in the water cycle. Groundwater is of two types. Deep groundwaters are fossil groundwater or juvenile water. They have formed slowly over many million years and at great depths (1-2 km). They tend to be mineralized and stable, and have very little connection with the surface groundwater.

On the other hand, surface groundwater occurs in the active zone of water exchange in the upper parts of the crust, drained by streams. It tends to be fresh because of its passage through the phase of condensation of atmospheric moisture.

Groundwater plays an important role in ensuring that some rivers remain perennial. Rivers without such a feeding source, would have water only when it rains or when the snow melts. Subterranean flow into the rivers (including the surface runoff) for the entire globe has been estimated to be 12,000 km$^3$ yr$^{-1}$ (which is about one-third of the total stream flow). In arid regions, replenishment of groundwater does not take place, and hence the groundwater is unable to feed the rivers. For instance, the wadis of Sahara, the omurambas of the Kalahari desert and the creeks in Australia remain dry for several years, and huge flash floods suddenly occur after heavy rains. If large capacity reservoirs are built to impound such flows, they could serve the same purpose as groundwater.

Some amount of groundwater may reach the sea directly by moving below the level of the river basins. Besides, groundwater in the mountain regions near the coast may reach the sea directly, particularly if the region concerned is karstic (as in Italy). Kalinin and Zektser (quoted by L'vovich 1979) estimated that the flow of groundwater directly into the oceans, bypassing the rivers, is approximately 2200 km$^3$ yr$^{-1}$, which is about 5% of the stream flow (including the runoff originating in polar glaciers).

Geological structure (e.g. existence of karst) and geographic characteristics (e.g. climate, soil cover, relief and vegetation) together have profound influence on the areal distribution and the quantum of replenishment of groundwater. In the karstic regions which are characterized by high permeability, water seeps down fast and is retained in the upper layers of rock. This has the consequence of reducing evaporation (due to non-exposure to atmosphere) and increasing the runoff and baseflow. For instance, rivers Letimbro and Sansobia in northern Italy have almost identical areas of the basin (about 40 km$^2$), altitude (about 500 m), and precipitation (about 1370 mm). They differ from each other only in the extent of karstification. Letimbro river which is slightly karstic (15%) has less runoff (744 mm) and more evaporation (616 mm). In contrast, Sansobia river which is almost entirely karstic has greater runoff (950 mm) and lesser evaporation (426 mm).

Volcanic tuffs are also highly permeable, and consequently they have the same effect as karst. For instance, in the case of Armenian highlands which have volcanic tuffs, melted snow water quickly seeps downward. There is hardly any runoff, but there are plenty of groundwater springs. If the surficial layers are composed of, say, loose Quaternary sediments or less permeable soils, the percolated water would get stored as soil moisture. If dry weather prevails in the area, the soil moisture would get evaporated fast. Under these conditions, the groundwater would receive only a small part of the percolated water.

Generally, the impact of geological factors on water cycle is more readily recognizable for limited areas drained by small rivers. Such a relationship cannot be readily traced in the case of large rivers, as their catchment area may be composed of several geological formations with different characteristics.

### 1.3.4  *The soil link*

Soil is involved in some form or other with all the phenomena connected with water balance. The soil link with the water cycle is manifested in the form of soil moisture. The soil cover is the interface through which a number of processes operate – infiltration which leads to the formation of soil moisture, evaporation through which soil moisture is expended, and feeding of the groundwater by the soil moisture.

Two aspects of the soil moisture need special mention. Firstly, soil moisture plays a critically important role in the biological processes. As is well known, the fertility and productivity of a soil are largely dependent of the soil moisture and humus content. Secondly, soil moisture is more directly influenced by weather, compared with groundwater. When it rains or when the snow melts, the percolation builds up the soil moisture, but the soil moisture gets rapidly evaporated under conditions of dry weather.

Evaporation from land includes not only unproductive evaporation from the surficial cover but also productive transpiration from the plants arising from their vital activity. Where the soil moisture is inadequate, irrigation is resorted to.

Soil moisture feeds the groundwater. This happens particularly in the case of forests, areas occupied by lakes and reservoirs, and under river bends. In the forest areas, the soil cover gets broken up by the root systems of plants, and consequently possesses high infiltration capacity and transmissibility. The infiltration capacity of the forest soils is 2-3 times greater than in the open fields. Experimental data quoted by L'vovich (1979, p. 37) show that while the surface runoff in the case of forest areas does not exceed 3% of the annual precipitation, it is as high 38% in the case of the open fields. On the other hand, the interflow and groundwater runoff constitute 43% of the precipitation in the case of forests, and only 18% in the case of open fields. This explains as to why the baseflow of rivers with forested catchments, is much higher than in cases where the forests do not exist or have been cut down. Also, it is important to ensure that timbering operations (e.g. use of heavy skidders) do not adversely affect the hydrophysical properties of the soils. Though it may be possible to regenerate the forest cover in 8-12 years, it may take decades to restore the original infiltration capacity of the forest soils. This explains as to why it is important to preserve the forests not only to prevent erosion, but also to protect the groundwater sources. In the arid areas, the amount of soil moisture is low, and consequently it is unable to feed the groundwater. Renewable resources of groundwater in arid areas is therefore low.

The hydrophysical properties of the soil, namely, infiltration capacity and water retention capacity, have a bearing on the variability of the elements of the water balance. They can operate independently, but yet jointly. Two situations can be considered, both of which do occur in nature (Fig. 1.2; source: L'vovich 1979, p. 34).

1. *Case 1* – where both the infiltration capacity and water retention capacity change in the same direction (Fig. 1.2a):

   – When the two parameters are of low magnitude, bulk of the precipitation would end up as surface runoff (*S*). Total stream flow (*R*) would be high, mainly in the form of surface flood water. Soil moisture would be negligible or absent, and consequently its contribution to evaporation (*E*) and groundwater would be very low. During the interregnum between the floods, the rivers would dry up, as there is no groundwater to feed them.

   – When the two parameters are of high magnitude, surface runoff (*S*) decreases, but soil moisture and evaporation (*E*) increases. The base flow (*U*) reaches the maximum, when the infiltration capacity and water retention capacity attain their optimum mean values.

2. *Case 2* – where the infiltration capacity and water retention capacity have inverse relationship (Fig. 1.2b).

   – When the two parameters are of low magnitude, surface runoff (*S*) drops sharply, while the groundwater runoff increases sharply. Evaporation (*E*) is low initially. It reaches the maximum and the total stream flow (*R*) reaches its minimum when the infiltration capacity and water retention capacity attain their optimum values.

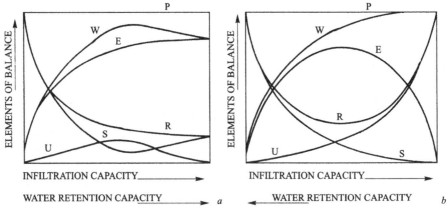

Figure 1.2. Hydrophysical properties of the soil and the water balance (source: L'vovich 1979, p. 34). P = Precipitation, R = Total stream flow, U = Base flow, S = Surface runoff, W = Total infiltration of the area, and E = Evaporation.

Despite the known perturbances caused by El Niño and La Niña, climate and atmospheric precipitation can be considered to be reasonably stable in contrast with the changes in the hydrophysical properties of the soil cover (caused by farming, growing of forests, draining of bogs, construction of civil engineering structures, etc.). When an area is built up, and the roads are paved, surface runoff increases, and it may also carry anthropogenic pollutants.

### 1.3.5 *The river link*

The river link is the most studied among all the links in the water balance, for the simple reason that human civilizations, permanent human settlements, growing of crops, use of water for irrigation, etc. started on the banks of rivers, such as, the Nile, Euphrates-Tigris, Indus and Ganges.

Rivers return to the oceans that part of the water which has been evaporated from the oceans, and precipitated on land. Rivers are fed from two sources: surface and underground. The surface runoff may arise from melting of the snow cover in the mountains (for instance, River Ganga originates from Gangotri in the snowclad Himalayas) or from falling rain. The spring high water from snow melt lasts three to four months. This is because all the snow does not melt at the same time – the melting starts at low reaches, and gradually extends to the higher parts of the catchment. An usual kind of river system forms during the short polar summers along the edges of the icecaps in Greenland and Antarctica. They flow through the middle of the ice field in channels under the ice. The channels are permanent but the flow is temporary.

In this respect, they resemble the rivers in the arid lands with permanent channels but transient flows.

Some rivers in the equatorial regions of evergreen forests (e.g. the Amazon, and the Congo) have floods most of the year. As a consequence of high precipitation, the succeeding flood comes in before the earlier one ends. In the case of the wadis of the Sahara, omurambas of the Kalahari desert in southern Africa, and the creeks in Australia, the stony or sandy river beds are dry most of the time. However, intense rainfall of short duration (a few days and sometimes even for a few hours), may give rise to immense flash floods. Flash floods are highly destructive in that they wash away the fertile top soil.

Rivers not only transport water to the sea or lakes, but also carry immense quantities of suspended and dissolved loads. The average total annual depth of denudation of land area has been estimated to be 0.08 mm, which is considerably more in the areas of high relief (caused by, say, tectonic upliftment).

Baumgartner & Reichel (1975) gave the following estimates of the continental runoff to the oceans and the corresponding drainage areas (Table 1.5).

The total flux of dissolved and particulate matter carried by the rivers is estimated to be of the order of $20 \times 10^{15}$ g yr$^{-1}$, composed of $15.5 \times 10^{15}$ g yr$^{-1}$ of solid load, and $4.0 \times 10^{15}$ g yr$^{-1}$ of dissolved load (Meybeck & Ragu 1997).

Table 1.5. Estimate of continental runoff to the oceans.

| Continent | Sum of drainage area (Mkm$^2$) | Sum of water discharge (km$^3$ yr$^{-1}$) |
|---|---|---|
| Africa | 17.56 | 3409 |
| N. America | 21.60 | 5540 |
| S. America | 16.40 | 11,039 |
| Asia | 31.45 | 12,467 |
| Europe | 8.27 | 2564 |
| Australasia | 4.70 | 2394 |
| Global total | 99.98 | 37,400 |

### 1.3.6   *The lake link*

The lake link in the water cycle is closely connected to the river link, because all lakes have stream inlets, and sometimes stream outlets as well. Lakes regulate the flow of stream outlets, making their flow more uniform with time. A classic example is that of the Great Lakes regulating the flow of the St. Lawrence river.

Evaporation always occurs on the surface of the lakes, even though there may not be any significant evaporation from the surrounding land area because of the aridity. For instance, in the case of the Caspian Sea, the annual depth of evaporation is almost 1 m, whereas the region surrounding it receives very low rainfall of 200-300 mm. Evaporation from lakes contributes 500 to 600 km$^3$ yr$^{-1}$ of water to the atmosphere, which constitutes less than 3% of the total water evaporated from the land area.

Some lakes and reservoirs represent more or less closed ecosystems. Where the lakes have a high rate of flow, the physical, chemical and biological processes approximate to those of rivers. But large lakes with relatively low flow (e.g. Baykal, Nyasa, Victoria, Superior, etc.), have unique ecosystems which are highly sensitive to anthropogenic impact.

### 1.3.7   *The biological link*

As is well known, life exists on earth because of the presence of liquid water. Animals and plants consists mostly of water. Human beings are composed of three-fourth water (about 40 l). The daily physiological consumption of drinking water varies from 1-4 l per capita per day, depending upon the climate (high in the arid regions), the kind of work a person does (a manual worker working in the open sun would need to drink more water, than a person working in an air-conditioned office) and social habits. The Tuaregs who live in the 130°F heat of the Sahara, sweat about 2 gallons (about 8 l) of water a day – 4 gallons (about 16 l) if walking! Unless the liquids lost are replenished, a person would suffer dehydration, leading to coma and death (source: *National Geographic*, Mar. 99).

At an average rate of 1000 l (1 m$^3$) per capita per year, the present population of the world (6 billion) would need about 6 km$^3$ yr$^{-1}$ of drinking water.

Animals need larger amounts of drinking water: 40 l d$^{-1}$ for cattle, horses and camels, 15 l d$^{-1}$ for hogs, and 10 l d$^{-1}$ for sheep. As should be expected, the largest land animal, the elephant, consumes huge quantities of food (about 300 kg of plant matter per day) and water (about 90 l d$^{-1}$, the same as six giraffe and 66 impala). The estimated consumption of drinking water by domestic animals in the world is of the order of 30 km$^3$ yr$^{-1}$. Animals can consume semi-potable water without any harm. For instance, the limit for Total Dissolved Solids (TDS) for potable water for human consumption is 500 mg l$^{-1}$. Water with TDS of, say, 700 mg l$^{-1}$ can be safely given to a camel.

The gross requirement of drinking water by humans and animals (domestic and wild) can be taken to be about 50 km$^3$ yr$^{-1}$ (L'vovich 1979, p. 43). This is just a minuscule fraction of the total renewable water resources of the world. The problem is not one of quantity, but of *quality* (i.e. potability).

The biological link of the water cycle involves aquatic animals and plants whose habitat is water. The catch of fish and marine animals and its quality depend upon the preservation of the habitat.

The role of the plants in the water cycle arises from transpiration which is a vital activity of the plant. The influx of the water into the plant not only brings in nutrients from the soil, but also has a role in regulating the temperature of plant. If there is no transpiration, plants would die of overheating. In hot weather, the stomata of the plants open wider, thereby facilitating greater transpiration and cooling of the plant.

The amount of water that a plant expends in transpiration depends, among other things, on three factors: the water needs of the plant itself, the weather conditions and the soil moisture. The transpiration ratio, which is defined as the amount of water required by a crop plant to form one unit weight of the dry matter, is a good indicator of the ability of a plant to transpire. For instance, 300-500 m$^3$ of water are spent in forming one tonne of grain and straw of wheat.

The depth of penetration of roots depends on the depth of the moist layer of soil or subsoil. For instance, the length of roots of wheat is about 40 cm in irrigated fields, and 2.5 m in the case of unirrigated land. The roots of some trees (e.g. oak) penetrate to a depth of 20 m to tap the groundwater.

The productive transpiration from plants and unproductive evaporation from the soil go hand in hand. In the case of forests, transpiration accounts almost wholly for the moisture released – there is hardly any evaporation from soil as the solar radiation is not able to penetrate through the forest canopy. As a crop (say, wheat) grows on farmland, productive transpiration increases from 25% to 60%, whereas unproductive evaporation from the soil decreases from 75% to 40%, because the soil is shaded by stalks.

Transpiration accounts for at least half of the evapotranspiration on land, i.e. about 30,000 to 35,000 km$^3$ yr$^{-1}$. This constitutes about 7% of the total evapotranspiration from the surface of the earth, including the oceans (L'vovich 1979, p. 49).

### 1.3.8   *The human link*

Water used for economic purposes is said to return back to the water cycle. We drink water and we sweat. We give water to a plant, and it transpires. This statement is valid only if we take into account the global water cycle. In actuality, the situation is far more complex. Water vapor produced in one place on earth does not precipitate in exactly the same place as rain. Evaporation from irrigated land in (say) central India may end up as rain (say) in the NE part of India which already experiences a very heavy rainfall!

There is another consideration. Suppose we use 4 km$^3$ of water for irrigation in a watershed. Let us say that about 1 km$^3$ of this evaporates and precipitates as rain in the same watershed. The actual runoff will be no more than 10-40% of this precipitation, i.e. 0.1-0.4 km$^3$.

The human intervention in the water cycle will be elaborated in the subsequent chapters.

## 1.4   WORLD WATER BALANCE

The World Water Balance is computed from the following equations (L'vovich 1979, p. 54):

For the land periphery, i.e. land abutting on the oceans:

$$E_p = P_p - R_p$$

For regions without access to the sea (endorheic):

$$E_a = P_a$$

For the oceans:

$$E_m = P_m + R_p$$

For the entire globe:

$$E = E_t + E_m = P$$

where $E_p$ = evaporation from the peripheral part of the land area, $P_p$ = precipitation on the peripheral part of the land area, $R_p$ = runoff from the peripheral part of the land area, $E_a$ and $P_a$ = evaporation and precipitation respectively, in the regions that have no runoff to the oceans, $E_m$ and $P_m$ = evaporation and precipitation respectively, of the oceans, $E$ and $P$ = evaporation and precipitation over the entire globe, $E_t$ = evaporation from the entire land surface.

Though ten parameters are mentioned above, it is possible to solve the problems of the world water balance from four parameters. Considerable uncertainty exists, however, in the magnitude of the parameters due to lack of accurate measurements. Precipitation data on land is reasonably accurate. Precipitation data on the oceans leaves much to be desired, because the measurements are made on oceanic islands and on shipboard. Data on evaporation and transpiration are scanty and not well distributed. The Annual World Water Balance is given in Table 1.6 (source: L'vovich 1979, p. 56).

El Niño and La Niña phenomena triggered several international initiatives to get more accurate measurements. For instance, a large number of anchored buoys in the oceans record and transmit data which, among other things, are relevant to the water balance studies.

### 1.4.1 *Residence times*

The rate of movement of water in the various parts of the hydrosphere varies very widely. The movement of the material between the pools is called the flux. The residence time ($t$) of an element in a reservoir, is given by the equation:

$$t = m \ (\mathrm{d}t/\mathrm{d}m)$$

where $m$ = mass of the element in the reservoir or the pool, and $\mathrm{d}t/\mathrm{d}m$ = rate of input (or output) to the pool.

The term, rate of water exchange, used by L'vovich (1979, p. 58) refers to the time required for 'hypothetical replacement of the entire volume of a given part of the hydrosphere in the process of the water cycle'. He defines the rate of water

Table 1.6. The annual world water balance (source: L'vovich 1979, p.56).

| Elements of the water balance | Volume (km$^3$) | Depth (mm) |
|---|---|---|
| Peripheral land area (116.8 M km$^2$): | | |
| – Precipitation | 106,000 | 910 |
| – Runoff | 41,000 | 350 |
| – Evapotranspiration | 65,000 | 560 |
| Enclosed part of the land area (32.1 M km$^2$): | | |
| – Precipitation | 7,500* | 238 |
| – Evapotranspiration | 7,500 | 238 |
| Oceans (361.1 M km$^2$) | | |
| – Precipitation | 411,600 | 1140 |
| – Inflow of river water | 41,000** | 114 |
| – Evaporation | 452,600 | 1254 |
| The Globe (510 M km$^2$) | | |
| – Precipitation | 525,100 | 1030 |
| – Evapotranspiration | 525,100 | 1030 |

*Including 830 km$^3$, or 26 mm of runoff. **The figure of 37,400 km$^3$ yr$^{-1}$ given by Baumgartner & Reichel (1975) is more widely used.

Table 1.7. Rates of water exchange in the various parts of the hydrosphere (source: L'vovich 1979, p. 58).

| Parts of the hydrosphere | Volume ($10^3$ km$^3$) | Elements of balance ($10^3$ km$^3$) | Rate of water exchange |
|---|---|---|---|
| Oceans | 1,370,000 | 452 | 3000 yr |
| Groundwater | 60,000 | 12 | 5000 yr* |
| Groundwater in the zone of active water exchange | (4000) | 12 | 330 yr** |
| Ice sheets | 24,000 | 33 | 8000 yr |
| Surface water on land | 280 | 39 | 7 yr |
| Rivers | 1.2 | 39 | 11 d |
| Soil moisture | 80 | 80 | 1 yr |
| Atmospheric vapor | 14 | 525 | 10 d |
| Hydrosphere as a whole | 1,454,000 | 525 | 2800 yr |

*The figure will be 4200 years if we take into account groundwater runoff directly into the oceans.
**The figure will be 280 years if we take into account the groundwater runoff directly into the oceans.

exchange ($A$) as the ratio of the volume of a given part of the hydrosphere ($\gamma$) to the input or output elements of its balance in the process of the water cycle ( $w$). Thus,

$$A = w/\gamma$$

where $A$ would correspond to the number of years required for the complete renewal of the water supply. Hence, the rate of water exchange of L'vovich corresponds to the residence time. The faster the turn over, the shorter the residence time.

Table 1.7 (L'vovich 1979, p. 58) gives the rates of water exchange for different parts of the hydrosphere. The data given in Table 1.7 leads us to several very significant conclusions:

−  Since the volume of the oceans is about 94% of the hydrosphere, the rate of exchange of water in the hydrosphere (2800 yr) is comparable to that of the oceans (3000 yr).
−  The rate of exchange of groundwater in the zone of active exchange (about 300 yr) is sharply different from that of the deep groundwater (5000 yr) which is mineralized and moves very slowly.
−  Water moves fastest in rivers (11 d) and in the atmosphere (10 d). It is not an accident that the two residence times are comparable, as runoff in rivers follows precipitation from the atmosphere. The short residence time of river water and precipitation means that these waters are fresh.

### 1.4.2  *Water balance of an area*

The water balance of an area is computed from the following equations:

$$P = S + U + E \qquad S + U = R \qquad W = P - S = U + E$$
$$K_U = U/W \qquad K_E = E/W$$

where $P$ = precipitation, $S$ = surface (flood) stream flow, $U$ = underground flow into the rivers (the stable part of the stream flow), $E$ = evapotranspiration, $R$ = total runoff, $W$ = total wetting of an area, $K_U$ and $K_E$ = groundwater runoff and evaporation coefficients respectively, which show what parts of annual infiltration end up as groundwater runoff and evapotranspiration.

It is not always possible to round off the water balance with a single river basin, for the simple reason that the boundaries of the surface catchment of a river may not be co-terminous with the boundaries of the underground basin. For instance, artesian water received in the basin of the Don river crops out in the basin of the Dnieper river in Russia. The scale of the phenomena is, however, not large. Similar phenomena are also observed in the karst regions. A comparison of the water balance of the river basins could reveal the possible existence of a passage of water by underground routes from one river basin to another (L'vovich 1979, p. 64).

The water balance of a land area is schematically shown in Figure 1.3 (L'vovich 1979, p. 65).

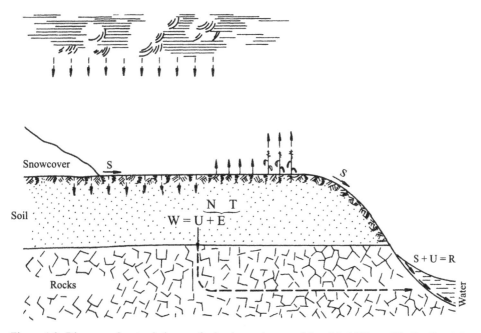

Figure 1.3. Diagram of water balance of a land area (source: L'vovich 1979, p. 65). P = Precipitation, R = Total runoff, U = Groundwater runoff, S = Surface runoff, W = Total wetting of the area (annual infiltration) including surface retention, N = Unproductive evaporation (evaporation proper), T = Transpiration of plants, and E = Evapotranspiration.

### 1.4.3   *Proportional water balance curves*

The following parameters are involved in the exercise:
$W$ = total wetting of the soil, $P$ = precipitation, $S$ = surface runoff; $U$ = groundwater runoff; and $E$ = evapotranspiration.

$$W = P - S$$
$$P = S + U + E$$

Combining the two equations, we get

$$W = (S + U + E) - S = U + E$$

The proportional balance curves developed by L'vovich (1979, p. 77-79) help us to understand the proportions of the soil moisture ($W$) that feeds the groundwater ($U$) and that get expended as evapotranspiration ($E$).

Figure 1.4 (source: L'vovich 1979, p. 79), which is a graph between $E$, $U$ (mm) versus $W$ (mm), has three curves.

1. $E = W$, this is limiting condition whereby the entire total wetting (soil moisture) is expended in evaporation, and there is no feeding of groundwater. This condition applies to deserts, where practically all the precipitation is absorbed by the soil ($W = P$) and then lost in evaporation ($E = P$).
2. $E = f(W)$, this is a more common situation. Depending upon the radiation balance, and the condition of the surface layer of the soil, the evapotranspiration increases with increasing soil moisture. The upper end of the proportional

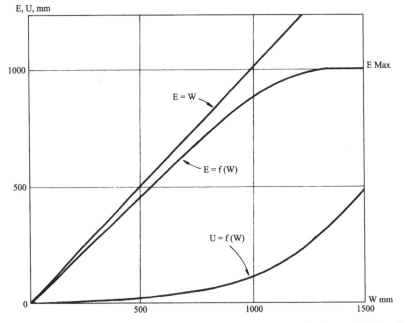

Figure 1.4. Proportional water balance curves (source: L'vovich 1979, p. 79). $W$ = Total wetting of the area, $E$ = Evapotranspiration, $U$ = Groundwater runoff, and $E_{max}$ = Potential evaporation.

curve reaches a plateau and asymptotically approaches the maximum possible evapotranspiration, $E_{max}$.

3. $U = f(W)$, there is a complementarity between $E = f(W)$ and $U = f(W)$, i.e. every position on the curve $U = f(W)$ has a corresponding point on the curve $E = f(W)$. The less the loss of soil moisture due to evapotranspiration, the greater would be the availability of soil moisture to feed the groundwater, and vice versa. The curve $U = f(W)$ thus rises steadily until it reaches a maximum corresponding to $E_{max}$.

Each geographical zone has its characteristic pair of $E = f(W)$ and $U = f(W)$ curves.

Figure 1.5 (source: L'vovich 1979, p. 101) shows how the linkages and interactions between the soil, vegetation, climate and relief, control the groundwater runoff into streams. Section 2.1 provides a case history of the application of these principles.

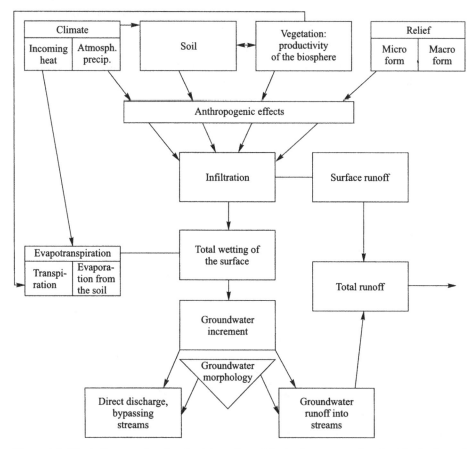

Figure 1.5. Block diagram showing the interaction of the basic processes involved in the formation of water balance (source: L'vovich 1979, p. 101).

### 1.4.4  *Water balance in various climatic zones*

L'vovich (1979, p. 196) classified the principal agro-climatic regions of the world into belts as follows (Tables 1.8 and 1.9).

An examination of the water balance data in regard to the various climatic zones in the world (Tables 1.8 and 1.9; Fig. 1.6; source: L'vovich 1979, p. 198) provides explanation for the hydrological phenomena that are observed in the various climatic zones.

In the tundra region, the potential evaporation is low. It should in the normal course of events, provide for greater groundwater runoff. But this does not actually happen as the ground is permanently frozen. The extreme cold does not allow vegetation growth, and consequently, the phytomass production is the lowest among all climatic zones (1-2 t ha$^{-1}$ yr$^{-1}$).

In the case of the desertic savannahs, the limited amount of precipitation (300 mm) is absorbed by soil (282 mm) and gets evaporated almost wholly (280 mm). Consequently there is virtually nil groundwater runoff (2 mm). The absence of soil moisture does not facilitate vegetation growth, and consequently the phytomass production is low (2-6 t ha$^{-1}$ yr$^{-1}$).

At the other end of the spectrum, the equatorial forests receive high rainfall (2000 mm), and the groundwater runoff (600 mm) is of the same order of magnitude as surface runoff (600 mm). This explains as to how the forests play a bene-

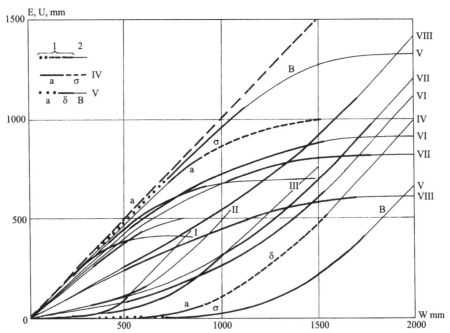

Figure 1.6. Patterns of water balance for various climatic zones (source: L'vovich 1979, p. 198). The legend is the same as given in Table 1.8.

Table 1.8. Principal climatic regions of the world (source: L'vovich 1979, p. 196).

| No* | Belt | Climatic zone |
|---|---|---|
| I | Subarctic | Tundra |
| II, III, IV-a, V-a | Temperate | Taiga, mixed forests, wooded steppes, prairies, steppes |
| IV-6, V-a, V-6, V-b, VI | Sub-tropical, tropical | Eastern, broad-leaved wet forests near the ocean, desertic savannah, dry savannah, wet savannah, wet monsoon forests near the ocean |
| VII | Equatorial | Perennially wet ever-green forests (hylean) |
| VIII | Mountain | Wet monsoon forests in mountains near the ocean |

*Notation used in Figure 1.6.

Table 1.9. Water balance in various climatic zones in the world (source: L'vovich 1979, p. 196).

| Climatic zone | 1 | 2 | 3 | 4 | 5 |
|---|---|---|---|---|---|
| Tundra | 370 | 110 | 40 | 70 | 300 |
| Taiga | 700 | 300 | 140 | 160 | 540 |
| Mixed forests | 750 | 250 | 100 | 150 | 600 |
| Wooded steppes, prairies | 650 | 120 | 30 | 90 | 560 |
| Steppes | 500 | 50 | 10 | 40 | 460 |
| Eastern, broad-leaved, wet forests near the ocean | 1300 | 420 | 120 | 300 | 1000 |
| Desertic savannah | 300 | 20 | 2 | 18 | 282 |
| Dry savannah | 1000 | 130 | 30 | 100 | 900 |
| Wet savannah | 1850 | 600 | 240 | 360 | 1500 |
| Wet monsoon forests near the ocean | 1600 | 820 | 320 | 500 | 1100 |
| Perennially wet, ever-green forests (hylean) | 2000 | 1200 | 600 | 600 | 1400 |
| Wet monsoon forests in the mountains near the ocean | 2200 | 1700 | 700 | 1000 | 1200 |

| Climatic zone | 6 | 7 | 8 | 9 | | 10 |
|---|---|---|---|---|---|---|
| Tundra | 260 | 0.13 | 400 | 300 | 575 | 1-2 |
| Taiga | 400 | 0.26 | 500 | 360 | 940 | 10-15 |
| Mixed forests | 500 | 0.17 | 700 | 400 | 1100 | 10-15 |
| Wooded steppes, prairies | 530 | 0.05 | 900 | 800 | 1480 | 8-12 |
| Steppes | 450 | 0.02 | 1300 | 900 | 1770 | 4-8 |
| Eastern, broad-leaved, wet forests near the ocean | 880 | 0.12 | 1000 | 600 | 1480 | 10-15 |
| Desertic savannah | 280 | 0.007 | 1300 | 900 | 1770 | 2-6 |
| Dry savannah | 870 | 0.03 | 1300 | 900 | 1770 | 6-12 |
| Wet savannah | 1200 | 0.16 | 1300 | 900 | 1770 | 10-20 |
| Wet monsoon forests near the ocean | 780 | 0.30 | 900 | 530 | 1680 | 15-30 |
| Perennially wet, ever-green forests (hylean) | 800 | 0.43 | 800 | 460 | 1620 | 30-50 |
| Wet monsoon forests in the mountains near the ocean | 500 | 0.58 | 600 | 750 | 1750 | 15-30 |

1 = Precipitation, 2 = Total runoff, 3 = Groundwater runoff, 4 = Surface runoff, 5 = Total surface wetting, 6 = Evapotranspiration, 7 = Groundwater runoff coefficient, 8 = Potential evapotranspiration, 9 = Parameters of the analytical expressions of relationships, 10 = Annual growth of the phytomass (t ha$^{-1}$).

ficial role in feeding groundwater and the streams. The high soil moisture and warm climate facilitate the luxurious growth of vegetation (with phytomass production of 30-50 t ha$^{-1}$ yr$^{-1}$).

## 1.5  HYDROGEOLOGY OF SURFACE WATER

### 1.5.1  *River runoff – surface water*

The river runoff has been estimated at 42,600 km$^3$. The runoff of the major rivers in the world (i.e. those whose average long-term runoff volume exceeds 100 km$^3$) is given by Shiklomanov (1998, p. 14). The five largest river systems in the world, namely, the Amazon, the Ganges with the Brahmaputra, the Congo, the Yangtze and the Orinoco, account for 27% of the world's water resources. The Amazon is the greatest river system in the world, accounting for 16% of the annual global river runoff. Surface water comprises water in the following situations:

- Flowing freshwater systems (called *lotic* water): rivers, streams, canals and ditches,
- Static freshwater systems (called *lentic* water): lakes, reservoirs and lagoons,
- Estuaries,
- Coastal waters,
- Runoff on land surface, and
- Water flowing through man-made drainage systems (such as, sewers).

Surface waters serve many purposes: they are sources of water for potable, agricultural and industrial uses; they serve as 'assimilative' sinks for effluent discharges; they have considerable recreational, amenity, aesthetic and landscape value; they support natural ecosystems, and they are used for transport and navigation. Changes in the runoff and discharges into the surface water, abstraction of water and drainage affect the quality, quantity and availability of water supplies at a given point in space and time (Petts & Eduljee 1994, p. 162-163).

Estuaries and coastal waters are of great ecological and economic significance. The dispersion and dilution of the effluents and contaminants in these environments are controlled by the nature of the circulation and the extent of mixing of freshwater and seawater. Most of the megacities in the world (e.g. New York, Tokyo, Bombay) are located on the coast. The estuaries are preferred locations for industries, such as oil refineries, due to the availability of water, convenience of transport, and discharge of effluents. Their ecological and environmental significance arises from the fact that, apart from the visual amenity, they act as routes for migratory fish. Mangrove forests in the tropical coastal areas serve as spawning grounds for fish and shrimp, besides protecting the freshwater resources on land.

The capacity of the surface waters to receive and assimilate the contaminants is determined by the physical, chemical and biological characteristics of the contaminants on one hand, and by the reaeration capacity of the waters on the other. When biodegradable pollutants are discharged into the waters, the breakdown of the contaminants depletes the oxygen content of water, thereby adversely affecting the oxygen-dependent flora and fauna. Since turbulence promotes greater interaction between water and air, flowing waters have a greater aeration capacity than static water. Hence eutrophication seldom occurs in the fast flowing rivers, whereas it is more common in lakes with their static waters.

Serious environmental problems are created when toxic substances (like heavy metals, and PCBs) enter into the surface waters. They are generally resistant to degradation, and they tend to accumulate in organisms, the magnitude of bioaccumulation being dependent upon the chemical characteristics of water. For instance, the less the hardness of water, the greater will be the toxicity to fish of heavy metals. Organic pesticides are both persistent and bioaccumulative. Suspended solids in water can reduce the light in water and thereby inhibit photosynthesis. The amenity value gets diminished, because it is not pleasant to look at dirty water!

### 1.5.2 *Catchment hydrology*

Hydrodynamic models are powerful predictive tools. Crawford & Linsley (1964) pioneered in the development of conceptual model of the hydrological cycle, which came to be known as the STANFORD Watershed Model (Fig. 1.7). This model has later been elaborated. Binley & Beven (1991) developed a variation of

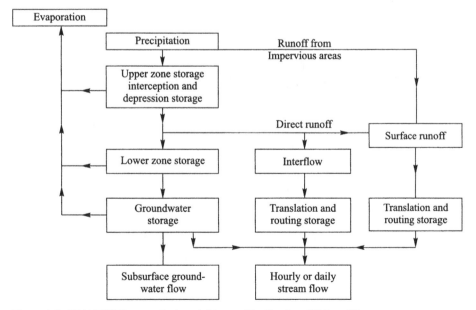

Figure 1.7. STANFORD watershed model (quoted by Sytchev 1988, p. 70).

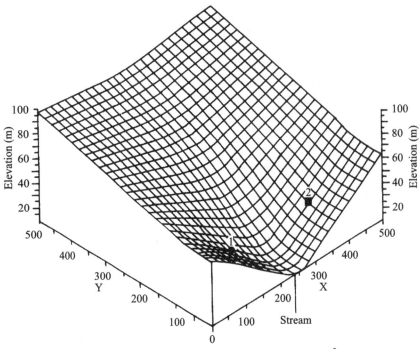

Figure 1.8. Topography of the headwater, with a plan area of 0.25 km$^2$, a small stream (60 m long) at the base of the slope, and an average depth of the soil over bedrock of 1 m (source: Binley & Beven 1991).

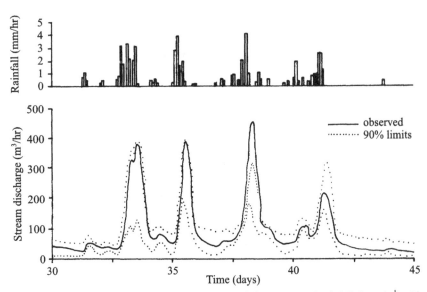

Figure 1.9. Headwater response in the context of the pattern of rainfall (mm h$^{-1}$). Observed discharge and 90% prediction lines are shown (source: Binley & Beven 1991).

the Monte Carlo simulation called GLUE (Generalized Likelihood Uncertainty Estimation) for modeling the catchment hydrology. On the basis of the elevation distribution in respect of the watershed (Fig. 1.8), and the rate of precipitation (mm h$^{-1}$), it is possible to predict the discharge (in m$^3$ h$^{-1}$) from the stream in a watershed (Fig. 1.9).

In the STANFORD model, precipitation is routed through various surface/subsurface reservoirs, where it undergoes evaporation, runoff and deep infiltration to different extents. This interactive model could indicate the extent of moisture availability to the roots, and the extent of evapotranspiration in a given watershed, etc.

Holtan & Lopez (1971) adapted the STANFORD model for agricultural watersheds. The following kinds of data are put into this model:
– Precipitation data taken from continuous weighted records for the watershed,
– Land capability in relation to the hydrological response, on the basis of the types and distribution of soils and pattern of land use,
– Evapotranspiration computed from pan experiments and adjusted to plant growth indices, and
– Infiltration capacity on the basis of the available storage capacity of the surface layer and a constant rate of infiltration.
Using this model, it is possible to calculate the proportion of infiltrated water to evaporation, the downward seepage and the lateral return flow in each flow regime. Where the hydrological conditions are particularly complicated, a combination of numerical and analog modeling can be used.


## 1.6 HYDROGEOLOGY OF GROUNDWATER


### 1.6.1 *Importance of groundwater*

Water managers should always bear in mind the important fact that *the quantity of fresh groundwater within drillable depth is about 67 times greater than all the waters in the rivers, lakes, reservoirs, etc. in the world put together* (Bouwer 1978).

Apart from quantity, the groundwater has the following three desirable characteristics:
1. *Quality* – The quality of groundwater is generally superior to surface water, because the soil column purifies the contaminants in water through processes such as anaerobic decomposition, filtration, ion exchange, absorption, etc.
2. *Evaporational losses* – The evaporational losses from the surface water reservoirs is so large (particularly, in the tropical regions) that they exceed the total municipal and industrial uses of water in the world (Shiklomanov 1998). Groundwater hardly suffers any evaporational losses. Besides, groundwater is less susceptible to pollution and less subject to seasonal fluctuations. Groundwater pumping may

be made gradually depending upon the water demand, whereas the construction of hydrotechnical structures requires a large, one-time expense.

3. *Use* – Groundwater is the main source of potable water for domestic purposes in most parts of the world (for instance, groundwater is the source of 75% of the municipal water supplies in USA). The use of groundwater for irrigation ranges from about 40% in the case of India, to about 70% in the case of Algeria and Libya.

It is not without significance that the International Hydrological Programme of UNESCO gives greater emphasis to groundwater vis-á-vis surface water (Zektser 1998).

Notwithstanding the above, the limitations of groundwater use should be clearly recognized. Firstly, groundwater alone cannot possibly cater to the large-volume users (such as large cities, which may need millions of m$^3$ per day). Secondly, overexploitation and pollution of groundwater may lead to irreversible changes in the quality and quantity of groundwater resources.

Artificial recharge of groundwater is an effective way to increase aquifer yield.

Groundwater is becoming an increasingly important natural resource. Abstraction of groundwater is increasing both in the Industrialized countries and the Developing countries.

USA withdraws 339 billion gallons (1314 M m$^3$) per day of groundwater. The importance of groundwater in the American economy could be judged from the fact that groundwater provides the domestic water requirements of about 50% of all households and 95% of the rural households, about half of the agricultural irrigation, and about one-third of the industrial water needs.

Across the European Community as a whole, groundwater provides 70% of the water supply.

In England and Wales, about 35% of the water demand is met from groundwater.

Groundwater is the principal source for private abstractors who cannot obtain water from public mains because of the location of their homesteads or who prefer to drink well water. It is some times used by food industries and breweries and farms requiring water for livestock and for crop irrigation.

In India, groundwater is extensively used for drinking and irrigation purposes, so much so the water-table is going down 1-3 m per year in most parts of India.

Apart from the reason of its providing water for potable, industrial and agricultural purposes, the protection of groundwater is of critical importance for the following reasons:
– If groundwater gets polluted, it is an extremely difficult and expensive task to clean it up; in some cases, remediation may be impossible,
– Aquifers serve as low-cost systems for storing potable water, which can quite often be used with little or no pre-treatment, and
– Groundwater seepages provide for stable runoff, and thus play a beneficial role.

### 1.6.2 *Underground hydrosphere*

The surface of the earth is the upper boundary of underground hydrosphere.

The Zone of Aeration is the topmost layer of the underground hydrosphere. Rainwater or snow meltwater percolates through this zone to reach the groundwater and replenish its resources. Thus, this zone constitutes the interface between the groundwater and the atmosphere. In this zone, the pore spaces are filled with air, water vapor, water which may be loosely or firmly held, and capillary water. The thickness of the Zone of Aeration may range from several centimeters to several tens of meters, depending upon the lithology of the rocks, the geomorphic setting and climatic conditions (precipitation, rate of infiltration). In arid zones, all groundwater at depths of less than 3 m, gets evaporated through the Zone of Aeration.

The following scenarios may be envisaged: Suppose it rained in an area of sandy sediments. Good part of the water will percolate through the zone of aeration to the groundwater table. On the other hand, if the surficial material is loamy, only a small portion of the water will percolate down, whereas most of the rainwater will flow to the nearest topographic low and accumulate there.

The Zone of Saturation (or Phreatic Zone) lies below the zone of aeration (or the frozen zone in cold countries). The cavities and pores in the rocks occurring in this zone are completely filled (saturated) with free and bound water. This part of the underground hydrosphere occurs at a depth where the critical point of water (374-450°C) would be reached, and below which the accumulation of liquid water becomes impossible. This boundary is generally to be found at depths of about 12-16 km. It may, however, occur at depths of 8-10 km in the areas of high thermal gradient, such as volcanic zones, or at depths of 30-35 km. in the Precambrian folded regions of low thermal gradients.

### 1.6.3 *Classification of underground waters*

The classification of underground waters on the basis of their mode of occurrence, head type, type of groundwater regime, water origin, and economic utilization is given in Table 1.10 (source: Klimentov 1983, p. 80-81). On the basis of the ionic species, waters are divided into three broad classes:

1. Hydrocarbonate (and carbonate) waters ($HCO_3^- + CO_3^{2-}$): comprising low-salinity waters of rivers and fresh water lakes, and many underground waters,
2. Chloride ($Cl^-$) waters, comprising saline waters of the seas, relict and continental lakes, groundwaters in the solonchak (saline soils) areas, deserts and semi-deserts,
3. Sulfate ($SO_4^{2-}$) waters, whose abundance and salinity are intermediate between the carbonate and chloride waters.

Each class is further subdivided, depending upon the predominance of $Ca^{2+}$, $Mg^{2+}$, and $Na^+ + K^+$.

Table 1.10. Classification of underground waters (source: Klimentov 1983, p. 80-81).

| Description | Water in the zone of aeration | Groundwater | Artesian water |
| --- | --- | --- | --- |
| Head type | Unpressured water | Usually unpressured water | Pressured water |
| Main water types | Swamp water, soil water, vadose water; Waters in solonchaks (saline soils) and solonetz (soils with high sodium saturation); Water in the active layer of the permafrost region | Water in Recent and Pleistocene fluvial deposits of river valleys Waters in glacial drift; Waters in steppes, deserts and semi-deserts; Waters in slope-wash, proluvial and other blanket deposits | Confined, pressured waters (including oilfield water, infrapermafrost water, mineral and thermal waters) |
| Ground-water regime | Usually temporary waters | Changes in water level depending upon the rate of infiltration of water from the surface, rate of underground evaporation, and locally, due to head transmission | Changes in water level due to head transmission |
| Origin | Mostly infiltrational; locally, condensational origin is also possible | Mostly infiltrational | Infiltrational, marine and juvenile |
| Economic utilization | Primarily used in agriculture, but some times used to provide seasonal water supplies for small industrial plants | Domestic water supplies, less frequently for irrigation and industries. | Fresh water is used for domestic supplies; high-salinity water for the production of salts and some elements; mineral water for medicinal purposes |

The combination of salts that indicate the fundamental properties of water, such as, alkalinity, hardness and salinity, are indicated in Figure 1.10 (source: Klimentov 1983, p. 101).

The geochemical characteristics of the waters are elaborated in the later chapters.

### 1.6.4   *Mobility of groundwater*

Physically bound water occurring in the upper 1.5 km. of the Zone of Saturation, is mobile. *Aquifers* are permeable, water-bearing rocks which allow the through-flow of water and which yield the water readily under the effect of gravity. An aquifer is defined as a relatively continuous (both laterally and vertically), isochronous or a dichronous sequence of rocks saturated with free gravitational water (Klimentov 1983, p. 74). Examples are: pebble gravel, gravel, poorly cement-

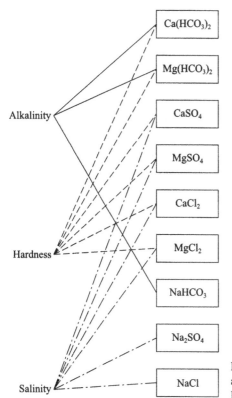

Figure 1.10. Chemical parameters that determine the alkalinity, hardness and salinity of waters (source: Klimentov 1983, p. 101).

ed conglomerates and sandstones, sands, siltstones, limestones, fractured rocks of igneous or metamorphic origin.

If the rock concerned is impermeable, it is called *aquiclude*. Examples are: clays, heavy loams, well-decayed compact peat, shales, mudstones, gypsum, non-fractured igneous and metamorphic rocks. Those which are poorly permeable are called *aquitards*. Thus the difference between aquicludes and aquitards is one of degree.

A *phreatic or unconfined aquifer* is an aquifer lying over an impermeable strata and whose uppermost part extends to the earth's surface. On the other hand, an aquifer which is confined by aquitards or aquicludes is called a *confined aquifer*. The term, *perched aquifer,* refers to small areas of water storage lying over impervious foundations (an example of this could be found in Figure 1.17 where a perched water-table exists in the sand overlying a sandy loam).

On the basis of the hydrologic regime, the aquifers may be classified into the following types: groundwater reservoirs, confined, unpressured aquifers, and artesian (confined, or pressured) aquifers. An *aquifer system* is a continuous (both vertically and laterally) water-bearing sequence of either isochronous (i.e. same age) or diachronous (i.e. of different ages) rocks of different composition. Such an

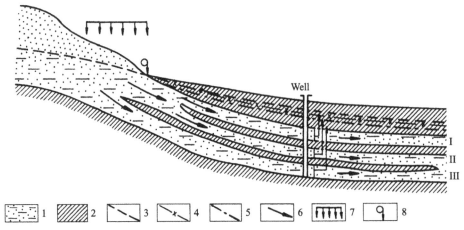

Figure 1.11. Schematic structure of an aquifer system (source: Klimentov 1983, p. 76). 1 = Permeable rocks (sands), 2 = Impermeable rocks, 3-5 = Isopotential lines for aquifers I, II and III respectively, 6 = Migration of groundwater, 7 = Recharge area of an aquifer system (rainwater infiltration), and 8 = Gravity spring.

aquifer system is bounded at the top and the bottom with the region ally continuous impermeable or nearly impermeable formations. The schematic structure of an aquifer system is shown in Figure 1.11 (source: Klimentov 1983, p. 76).

Groundwater levels from a number of wells are plotted on a geological or t opographic map, and then contoured. *Isohypses* or *water table contours* are lines connecting points with equal groundwater elevations or potentials. If an aquifer is isotropic, i.e. if it has the same properties in all directions, the direction of flow of the groundwater will be at right angles to the isohypses (vide Fig. 1.12; source: Turney & Goerlitz 1990).

### 1.6.5 *Porosity, permeability and transmissivity*

Porosity which arises from the presence of voids in rocks, is an important hydr ogeologic property of rocks. In the case of s andy rocks, water may be present in the pores, whereas in the case of hard rocks like quartzites and granites, water is co ntained in fractures or fissures (Fig. 1.13; source: Klimentov 1983, p. 37).

Total porosity ($n$) is the ratio between the volume space ($v_n$) of all the voids in a given sample (including voids which may not be connected with one another) and the total sample volume ($v$).

$$n = v_n/v \tag{1.2}$$

Porosity of a rock may be expressed as a fraction (say, 0.3) or as percent (o btained by multiplying the fraction by 100), say, 30%.

Permeability refers to the ability of the rocks to let the water through them u nder a pressure differential. Generally, the greater the porosity, the greater is the

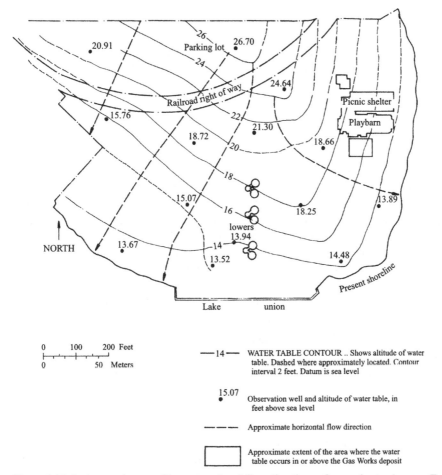

Figure 1.12. Isohypses (water table contours) and flow directions of groundwater (source: Turney & Goerlitz 1990).

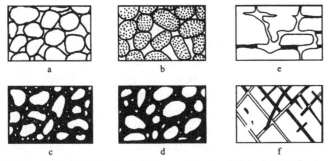

Figure 1.13. Various types of pores (openings) found in rocks (source: Klimentov 1983, p. 37). a and b = Pores in sands composed of well-rounded and well-sorted grains, c = Pores in inequigranular sands composed of poorly-rounded and poorly-sorted grains, d = Pores in sandstone, that are partly cement-filled, e = Pores in fissured (fractured) limestone, partly enlarged by leaching; and f = Joints in massive crystalline rocks.

Table 1.11. Permeability ($k$) and porosity ($\varepsilon$) of some sediments.

| Sediments | $k$ (m d$^{-1}$) | $\varepsilon$ (fraction) |
|-----------|------------------|--------------------------|
| Gravel | 200-2000 | 0.15-0.25 |
| Sand | 10-300 | 0.20-0.35 |
| Loam | 0.01-10 | 0.30-0.45 |
| Clay | $10^{-5}$-1 | 0.30-0.65 |
| Peat | $10^{-5}$-1 | 0.60-0.90 |

permeability. It may sometime happen, however, that a rock may be porous but not permeable, as the pores are not connected (as in pumice).

It may be noted that the porosity contrast between the more porous and less porous sediments is no more than three or four, the permeability contrast could be as high as a million times (Table 1.11; source Appelo & Postma 1996, p. 328).

The porosity of the various rocks is as follows: fine-grained granite (0.0006-0.007), coarse-grained granite (0.003-0.009), syenite (0.005-0.014), gabbro (0.006-0.007), basalt (0.006-0.013), basaltic lava (0.044-0.056), sandstone (0.032-0.152), loose sandstone (0.069-0.269), marble (0.001-0.002), limestone (0.006-0.169), chalk (0.144-0.439) (Klimentov 1983, p. 40).

Permeability may be defined as the ability of rocks/soils/sediments to let water through them under the pressure gradient. Evidently, the larger the pores and fissures in a rock, the more easily water moves. Depending upon the permeability, rocks are classified into three groups:
1. Permeable – pebble, gravel. sand, etc.,
2. Semi-permeable – argillaceous sand, sandy loam, light loam, loess, etc.,
3. Impermeable – clay, heavy loam, well-decayed tight peat, crystalline and sedimentary rocks without fissures (e.g. fine-grained granites, dolerites, massive quartzites).

Some hard rocks may have not only fissures but also large voids, through which water can penetrate and move, e.g. cemented, fractured limestones and marls.

Water yield is a measure of the ability of water-bearing rocks to produce water by way of the free gravity-driven draining. Specific yield is the volume of water that could be produced by 1 m$^3$ of rock (Klimentov 1983, p. 64). The greater the permeability of a rock, the greater would be its water yield (e.g. coarse-grained sand, gravel strata, etc.). Impervious materials (e.g. clay, peat, etc.) hardly yield any water. Water can be drawn from clay or peat only by suction.

The water yield (fraction) of different rock types is as follows: gravelly or coarse-grained sand (0.25-0.35), medium-grained sand (0.20-0.25), fine-grained sand (0.15-0.20), very fine sand or sandy loam (0.10-0.15), loam (<0.10), peat (0.05-0.15), sandstone cemented with argillaceous matter (0.02-0.03), jointed limestone (0.008-0.10).

It can happen that the actual porosity and permeability may differ from the parameters as measured. Hence the need to estimate effective porosity and permeability. *Effective porosity* is the porosity which contains water that partakes in the flow (as against total porosity which includes pores filled with stagnant water). Effective permeability is estimated on the basis of the quotient of discharge over hydraulic gradient (Appelo & Postma 1996, p. 328).

Transmissivity is the product of permeability and aquifer thickness.

### 1.6.6 *Darcy's law*

On the basis of a large number of experiments on the infiltration of water through sand filters, H. Darcy, the French hydraulics engineer, showed in 1856 that the rate of infiltration is directly proportional to the hydraulic head. Figure 1.14 (source: Klimentov 1983, p. 86) gives a schematic chart of the Darcy experiment.

$H_1$ and $H_2$ = water levels at the inlet and outlet of the sand-filled tube, $\Delta L$ = dimensions of the filter-length, $F$ = cross-sectional area, and $\Delta H = H_1 - H_2$ = difference in the piezometric heads = head loss over the seepage path.

The Darcy equation can be written as:

$$Q = K\,(\Delta H/\Delta L)\,F \tag{1.3}$$

where $Q$ = amount of water passing through the filter per unit time, $K$ = infiltrating index, which is a constant of proportionality factor depending upon the physical properties of the rock and the infiltrating liquid, and $\Delta H/\Delta L$ = hydraulic gradient = $I$.

Dividing both sides of the equation by $F$, we get

$$Q/F = K\,(\Delta H/\Delta L) = KI \tag{1.4}$$

This shows that $Q/F$ (the rate of infiltration) is directly proportional to the hydraulic gradient ($I$).

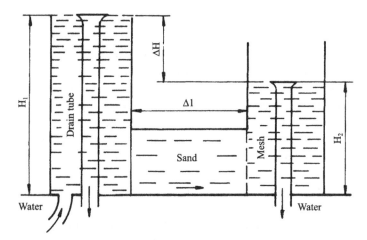

Figure 1.14. Schematic diagram of Darcy's experiment (source: Klimentov 1983, p. 86).

The Darcy law is used extensively in groundwater studies, such as, the estimation of the rate at which the water in a well will be replenished, the prediction of the fate of movement of the contaminants in groundwater, and so on.

### 1.6.7   *Recharge and discharge of groundwater*

Groundwater is water stored in the saturated zones in the subsurface. It gets recharged due to the infiltration of precipitated water (rainwater, meltwater, dew, etc.). The rate of infiltration is controlled by the nature and intensity of precipitation on one hand, and the permeability of rocks and soils in the zone of aeration on the other. Highest rate of rainwater infiltration and recharge of groundwater occurs when there are continuous rains of low intensity under conditions of high humidity. In the case of mountainous areas, not only rainwater and snow but also other types of precipitation, such as dew and hoarfrost, humidify the soil surface and thus contribute to the groundwater recharge. During their movement inland, moisture-bearing air masses from the sea rise when they encounter mountains. The air masses cool, and precipitate the moisture on the stony surface and soil. For instance, in the essentially rainless Namibia (SW Africa), both animals and men adopted ingenious methods of harvesting this precious moisture (insects face their body in such a manner that the moisture falling on the hind part slowly drips towards the mouth). In some desert areas, atmospheric water vapor condenses and precipitates on the cooled rock surface. Thus, in the Kara Kum desert in Russia moist sands can be observed even after three or four rainless months, and despite high evaporation.

Instances are known whereby artesian water from deeper aquifers contributed to the recharge of groundwater. If this is to happen, two conditions need to be satisfied: there should be no aquitard covering the artesian aquifer ('hydrologic window'), and the isopotential level of the artesian aquifer is higher than that of the groundwater table.

Figure 1.15 (source: Klimentov 1983, p.144) shows the vertical section of an aquifer (composed of sand) from the ground surface to the bottom aquitard (composed of clay). Rainwater and surface water infiltrate through the zone of aeration and accumulate in the zone of saturation (cd), which lies below the capillary fringe (bb). The surface of the groundwater is generally wavy, and follows the topographic surface. It dips towards the nearest depression (such as, a gully, ravine or river valley). The direction and rate of groundwater flow depends upon the slope of the water-table, and hydraulic conductivities of rocks. Thus, groundwater moves by gravity from the recharge areas to discharge areas (such as a gravity spring at c).

Groundwater basins tend to form only in the plains, where the groundwater plane is nearly horizontal (Fig. 1.16; source: Klimentov 1983, p. 144). In these situations, the uppermost aquifer (sand) occurs at depth, with margins (a, a) at approximately the same altitudes. The aquitard in this case is clay. Such basins seem

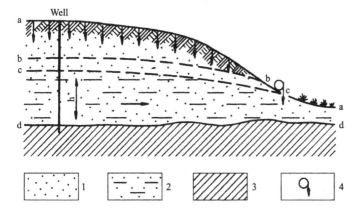

Figure 1.15. Zonal distribution of various water types in a vertical section (source: Klimentov 1983, p. 144). 1 = Sand, 2 = Water-bearing sand (aquifer), 3 = Clay (aquitard), 4 = Gravity spring (at c), aa = Ground surface, bb = Surface of the capillary fringe, cc = Groundwater table, dd = Surface of an aquitard, ab = Zone of aeration, bc = Zone of capillary saturation (capillary fringe), cd = Zone of saturation, and h = Height of the groundwater flow.

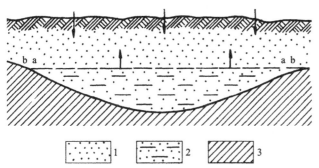

Figure 1.16. Schematic diagram of a groundwater basin (source: Klimentov 1983, p. 144). 1 = Sand, 2 = Water-bearing sand (aquifer), 3 = Clay (aquitard), aa = Groundwater table, bb = Surface of an aquitard.

to develop in areas where the rainwater infiltrating down cannot overfill the basin. This does not necessarily mean that the groundwater in such a basin always remains static. It can happen that there may be hydraulic communication between the groundwater basins, thus allowing the flow of groundwater.

In the river valleys, the groundwater movement tends to be in the same direction as the river. As should be expected, the groundwater level in the recharge area is at a greater altitude than that in the discharge area (such as, a ravine). The intersection of the groundwater flow by a topographic low, such as a river valley, leads to the discharge of the groundwater in the form of seepage faces, springs or small bogs. The flow of groundwater towards the discharge area, forms a curvilinear surface called the *depression surface*.

It should be self-evident that recharge of confined water is possible only in areas where there is no aquitard at the top. During years when there is excessive recharge, the usually unpressured confined water may get temporarily pressured in places.

Figure 1.17 (source: Klimentov 1983, p. 146) shows a combination of aquifers (sand), and aquitards (sandy loam, fractured limestone) across a portion of the river valley. Attention is drawn to the following features of the diagram:

– Recharge takes place through aquifer material (e.g. sand) outcropping at the surface, but no recharge takes place if an aquitard outcrops at the surface,
– Occurrence of *vadose* water: vadose water occurs as laterally limited body (of a sand aquifer), at a relatively shallow depth in the zone of aeration over a lenticular aquitard (in this case, sandy loam),
– Groundwater flow occurs below the aquitard,
– Confined, unpressured water occurs below the upper clay,
– Gravity springs form whenever the groundwater flow is intersected by a topographic low, and
– Pressured (artesian) water moves along the fractures in limestones and discharges into the fluvial deposits in the river valley. Near the surface, perched water-table exists in the sand overlying a sandy loam.

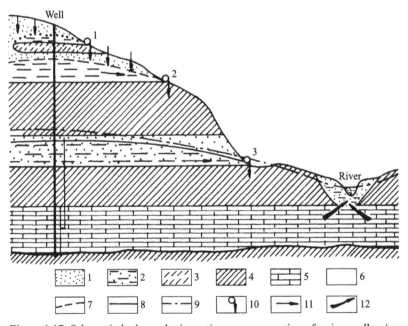

Figure 1.17. Schematic hydrogeologic section across a portion of a river valley (source: Klimentov 1983, p. 146). 1= Sand, 2 = Water-bearing sand, 3 = Sandy loam, 4 = Clay, 5 = Fractured limestone, 6 = Vadose water level, 7 = Groundwater level, 8 = Level of confined, unpressured water, 9 = Artesian water level, 10 = Gravity springs, 11 = Upressured groundwater flow, and 12 = Artesian water discharge into fluvial deposits.

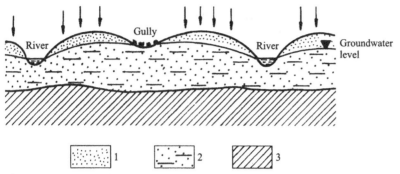

Figure 1.18. Correlation between surface topography and groundwater table (source: Klimentov 1983, p. 150). 1 = Sand, 2 = Water-bearing sand (aquifer), and 3 = Clay (aquitard).

Groundwater table generally tends to follow the surface topography, with few exceptions, such as, when the groundwater flow is drained by a river or when there is abrupt thickening of the aquifer. In the case of the river valleys, ravines and gullies, groundwater occurs in shallow depths corresponding to topographic lows. The depth to the groundwater table increases to several tens of meters below the hills and other topographic highs (Fig. 1.18; source: Klimentov 1983, p. 150). Thus, groundwater invariably flows from uplands to lowlands.

### 1.6.8 *Springs*

Springs are well-known manifestations of groundwater discharge. Most of the rivers in the world originate from springs. Springs are venerated in all cultures as sources of pure water. The formation of springs is controlled by the geological and structural setting and the drainage system. The existence of springs in an area gives an indication of the hydrological conditions in the area, the disposition of aquifers and aquitards and their stratigraphic and water contents, the quality of groundwater, etc. Springs outcropping at the surface are easily detectable during field studies on the basis of water oozing out or on the basis of the dampness of the soil, but when the springs are buried under loose Quaternary sediments, geo-physical studies or pitting or augering could reveal their presence. Some springs are active round the year, though with reduced discharge during the dry season. Some springs completely dry up during summer. Very high yields (of the order of several tens of $m^3 s^{-1}$) are characteristic of springs occurring the fractured or karstic terrains. The author has seen an incredibly copious spring in the Gaza province of Mozambique, which singly is irrigating about 400 ha of paddy fields.

Springs develop at places where erosional channels intersect an aquifer, or along traces of tectonic fractures. Gravity springs occur at the contact between an aquifer and an aquitard (formation of gravity springs at points 1, 2 and 3 could be seen in Fig. 1.17). Some springs occur in the gullies along the edge of a slope wash. Groundwater may outcrop when the river valley becomes narrower, thereby reducing the cross-sectional area of groundwater flow. Sometimes, facies change in the composition of alluvial deposit may create a situation favorable for water discharge. For instance, when an alluvial sand grades into sandy-clay, groundwater flow is retarded, the groundwater level rises, thus allowing the formation of springs at topographic lows.

Close to seepage sites, dampness to seepage is clearly visible along strips of valley slopes. Peat vegetation, and water-filled hollows may develop at these sites. In the case of arid lands, the seepages may be manifested in the form of very thin veneers of caliche-type deposits.

### 1.6.9 *Hydraulic connection of groundwater with surface water*

Almost invariably, groundwater has a hydraulic connection with surface water (rivers, lakes, reservoirs, ponds, etc.). River valleys may be filled with alluvial deposits or occasionally fluvio-glacial deposits, of sands and gravels. Such deposits often contain large quantities of high-quality groundwater. There is much diversity in the mode of communication between the river water and the groundwater. Figure 1.19 (source: Klimentov 1983, p. 151) depicts three kinds of situations:

1. When the groundwater is drained by a river: in humid and temperate climates, groundwater tables dip towards the river, and the river water is recharged with groundwater,
2. Groundwater is recharged by a river: in arid climates, groundwater levels usually rise in the direction of the river towards its banks, indicating that the groundwater is charged by the river water, and

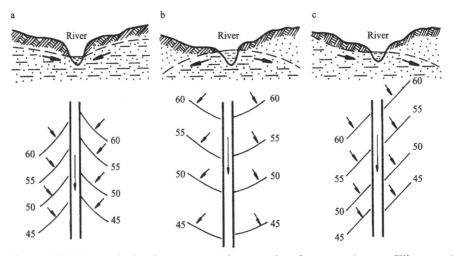

Figure 1.19. Communication between groundwater and surface water (source: Klimentov 1983, p. 151). a = Groundwater is drained by a river, b = Groundwater is recharged by a river, c = Groundwater is recharged by a river on its left bank, and is drained by a river on the right bank.

3. Groundwater is recharged by a river on its left bank, and is drained by a river on its right bank: this is a more complex situation than the previous two, and could occur some times.

An analysis of the hydrographs enables us to separate the groundwater and the surface components of the runoff. This is, however, a laborious task. The usual practice it to compute them from the figures for four years: two years of average rainfall, a dry year and a wet year. Hydrographs are plotted for these four years. This exercise is simple, and yields reasonably reliable figures.

Because of the hydraulic connection between the surface water and groundwater, what happens to the surface water affects the groundwater, and *vice versa*. During floods, the high water level in the river would raise the groundwater level. The afflux may be felt in the river banks for hundreds of meters, sometimes over several kilometers. When there is a drop in the water level in the river, there would be corresponding lowering of the groundwater level. When the groundwater in due course returns to the normal position, it will flow towards the river channel (in humid and temperate climates) (Fig. 1.20; source: Klimentov 1983, p. 152).

Generally, the groundwater runoff increases during periods of high water. It is low between rains (which cause surface runoff) or in the period of stable winter in the temperate zone. The discharge following high water tends to be greater than the discharge observed before the high water. The influx of groundwater into the river depends on the morphology of the river valley, its hydrogeological structure, the lithology of the rocks making up the bedrock of the banks and the nature of the alluvial deposits at the bottom of the river. If the conditions are favorable for the seepage of channel water into groundwater, the influx of groundwater into the river may decrease (L'vovich 1979, p. 67).

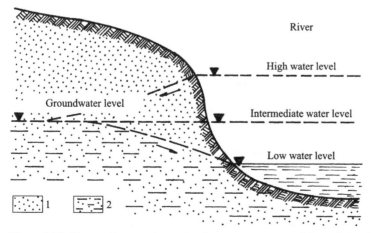

Figure 1.20. Changes in groundwater level as a consequence of the variation in the water level in the river (source: Klimentov 1983, p. 152). 1 = Sand, and 2 = Water-bearing sand.

With the drop in the level of flood water, the feeding of the groundwater from the channel flow decreases, and ultimately ceases. Now, the groundwater will start returning to the river, the water it has received during the flood time. All the water received cannot be returned because some water has been used up to saturate the bedrock of the banks and the alluvial deposits in the river valley to the minimum (field) moisture capacity.

As the floods are usually of short duration, and the movement of groundwater is slow, the seepage of channel water to feed the groundwater is generally not substantial. However, if rapid seepage of channel water could occur because of the high percolation coefficient of the valley floor material, correspondingly rapid return of the groundwater to the channel could occur.

The processes described above are schematically shown in the discharge versus time graph which enables us to determine the groundwater component of the runoff during the period of flood (Fig. 1.21; source: L'vovich 1979, p. 68). The surface runoff increases sharply due to floods, and after some time-lag, the surface runoff decreases again. The diagram clearly shows that the groundwater runoff after the flood is clearly higher than the groundwater runoff before the flood. Using this graph, it is possible to compute the magnitude of influx into the river of groundwater hydraulically connected to the channel water during floods (area below the straight line connecting points 1 and 1). It is of course possible a river may be fed by groundwater that is not hydraulically connected to the channel water. Line 4 in the graph reflects the boundary of the total influx of all types of groundwater runoff.

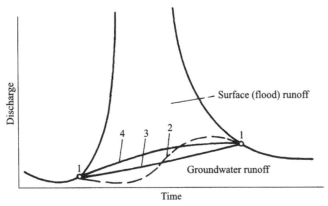

Figure 1.21. Diagram for determining the groundwater component of the runoff in a period of high water (source: L'vovich 1979, p. 68). 1 = The points correspond to groundwater runoff before and after the flood, 2 = The curve representing the natural course of the influx into the rivers of groundwater hydraulically connected to the river water channel – the first part representing the reduction, and the second part the increase of flow of the groundwater into the rivers, 3 = Line which approximately indicates the average influx of groundwater hydraulically connected to channel water, and 4 = Line adopted as the boundary between the groundwater and surface runoff and taking into account the feeding of the rivers with groundwater not connected hydraulically to channel water.

When large water reservoirs are constructed by damming the rivers, the groundwater afflux may take place within a much wider strip of the river. If the afflux is shallow, swamps may develop. It may be necessary to pump out the water to lower the level of groundwater and dry up the swamp.

Groundwater may be in hydraulic communication with bog water, which may itself be connected to vadose water. The bogs may be oligotrophic (upland) occurring in interfluvial plateaus, or eutrophic (lowland) bogs occurring in topographic lows, or mesotrophic (transitional) occurring in topographic lows. The bogs may be charged by different combinations of groundwater and rain water.

### 1.6.10  *Main types of groundwater occurrences*

Some important environments of groundwater occurrence are described as follows:

*River valleys* – Often, river valleys have alluvial sequences composed of two lithologic units: the lower coarse sandy and gravelly sediments, and the upper fine-grained sands, silty loams and clays. Generally, groundwater occurs in the lower unit. However, there can be great variability in the lithologies, both vertically and along the strike. Consequently, the aquifers may be discontinuous, and may contain both unpressured, and occasionally pressured, water. The existence of hydraulic communication between stream water and groundwater, results in the direction of the groundwater flow coinciding with the direction of movement of surface water. The recharge of water in the alluvium takes place through infiltration of rainwater, absorption of surface water that is flowing down the slopes and inflow from adjacent aquifers with higher hydrostatic heads. Alluvial waters are generally fresh, and are of the calcic hydrocarbonate type.

*Glacial drift* – By its very nature, the glacial drift material is highly uneven. On one hand, it may contain aquifer-forming material such as fluvioglacial sands, and on the other, aquitard material, such as unsorted boulder clay and loam. Fluvioglacial sands get recharged either from percolating rainwater or water flowing from adjacent highs in boulder clay and sandy loam. When thick fluvioglacial sands occur over large territories in humid climates, good quality (i.e. low-salinity) groundwater could accumulate in large enough quantities, so as to be able to supply large cities and industries.

*Steppes, deserts and semideserts* – These areas are characterized by low rainfall (annual average of 150-250 mm), high evaporation (about 2500 mm per year), and flash floods. The rivers are ephemeral, since they receive no influx along their course. Dry sands, and loess-like argillaceous rocks cover large areas of steppes, deserts and semideserts. Any rain falling on them during the warm season gets quickly evaporated. The infiltration is minimal, and hence these areas do not allow accumulation of groundwater. The occasional flash floods also do not recharge groundwater to any significant extent, because the precipitation occurs so fast that most of it goes as surface runoff. The groundwater in the arid and semi-arid regions is scanty and salty.

There is evidence to show that Sahara desert, presently the largest desert in the world, once enjoyed a temperate, humid climate. SLAR (Side-Looking Airborne Radar) images show the existence fossil river channels under the sand cover. Fresh, fossil groundwater has been found to occur in the alluvial valleys of pa- laeostreams, and is being tapped (as in Libya).

*Mountain regions* – Groundwater in the mountain regions occurs in the waste mantles (loose, Quaternary, arenaceous and coarse clastic deposits) and in frac- tures cutting the pre-Quaternary rocks. The water may occur in unpressured or ar- tesian conditions. As a consequence of the highly dissected relief, groundwater moves fast in the mountain regions. Springs are ubiquitous – they are generally of low yield, and tend to occur at topographic lows. Groundwater occurs in large quantities in the alluvial deposits of mountain rivers. In the case of endorheic (without access to the sea), intermontane depressions, groundwater salinities tend to increase from the margins of the depressions to the central area, and the depth to the groundwater level decreases in the same direction.

*Sandy seashores* – Fairly fresh groundwater often occurs in the dunes along the sandy seashores. The groundwater table generally follows the dune topography (Fig. 1.22; source: Klimentov 1983, p. 167). Rainwater and in some cases, water of condensation, percolates from the surface, and accumulates in the porous mate- rial overlying the denser saline seawater (density of seawater is $1.024\,\mathrm{g\,cm^{-3}}$). The intermixing of the fresh water with seawater through diffusion is a slow proc- ess. Consequently, the water could remain fresh for long periods of time. The thickness of a fresh groundwater body in the central part of a sand island can be estimated by a simple calculation. The difference in altitudes between the fresh- water level in the central part of the island and the sea level ($h$) is 0.024 times that of $H$ which is the depth to the bottom of a fresh water body from the sea level.

### 1.6.11  *Artesian water*

The word, Artesian, is derived from the province of Artois (Artesium, in Latin) in northern France which is the site of the first flowing well dug in Europe in 1126. Artesian water has been defined as the 'underground water present in the aquifers

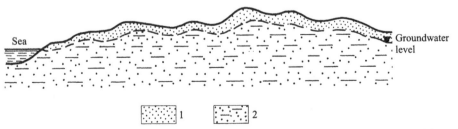

Figure 1.22. Groundwater table in dunes (source: Klimentov 1983, p. 167). 1 = Sand, and 2 = Wa- ter-bearing sand.

(aquifer systems) overlain and underlain by aquitards or aquicludes, and having a hydrostatic head which leads to the rise of water level to the top of such aquifers when this water is struck by boreholes or other excavations' (Klimentov 1983, p. 169). Artesian waters are highly prized because if the structural and hydrogeologic conditions are favorable, water will flow out by itself when a borehole is made.

Artesian water occurs in a variety of structures, such as, syneclises, depressions, foredeeps, monoclines, intermontane troughs, grabens and tectonic fracture zones. The host geological formations are generally of pre-Quaternary age, though occasionally, artesian water may be struck in Quaternary formations. An artesian basin comprises a set of aquifers occurring in a synclinal depression.

The three components of an artesian basin, namely, (1) Recharge area, (2) Discharge area, and (3) Transitional area, are briefly described as follows:

1. *Recharge area* – This is the area where the hydrological basement outcrops at maximal elevations. Precipitated water and surface runoff contribute to the recharge. The recharge area may have an outer segment consisting of folded regions or uplifts adjacent to the artesian basin, and an inner segment where the artesian aquifers are either exposed at the surface, or covered with loose younger sediments. There may be hydraulic communication between groundwater and artesian water. Depending upon the relative altitudes of the piezometric surfaces of the artesian water vis-á-vis groundwater, artesian water could replenish the groundwater, or vice versa (Fig. 1.23; source: Klimentov 1983, p. 172).

2. *Discharge area* –Where the aquifers outcrop at lower elevations relative to the recharge area. The discharge may be either open or hidden. Open (natural) dis-

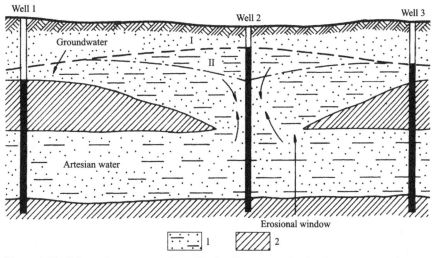

Figure 1.23. Schematic representation of hydraulic communication between groundwater and artesian water (source: Klimentov 1983, p. 172). 1 = Water-bearing rocks, 2 = Impervious rocks, I= Groundwater level over an erosional window: artesian water replenishes groundwater, and II= Groundwater level over an erosional window: groundwater replenishes artesian water.

charge sites may be of the erosional type (e.g. river valleys, or endorheic depressions), or barrier type (where a barrier obstructs the movement of the artesian water flow) or structural type (e.g. fracture zones, anticlines in folded mountains). Hidden discharges may occur under fluvial deposits or on the seafloor.

3. *Transitional area* (also known as *pressure head area*) – This constitutes the principal segment of the artesian basin. The aquifers present in the area are characterized by *isopotential* or *potentiometric* levels, which always stand above the top of the aquifer. Since all artesian basins are *ipso facto* synclinal structures, the lower aquifers could be generally expected to have greater heads and outcrop at higher elevations. Consequently, flowing wells are characteristic of artesian basins where the surface relief is not inverted.

### 1.6.12   *Groundwater in fractured and karstified rocks*

Jointing in igneous and metamorphic rocks may be caused by tectonism, weathering, leaching, dissolution and other processes. The joints may be open, or filled with sandy-clayey minerals or vein minerals. Leach joints and solution joints are typical of carbonate rocks. The character, size and origin of the joints in rocks control the distribution and migration of groundwater. The water in the joints may be unpressured or pressured. Intersecting joints occurring in recharge area at higher latitudes facilitate the infiltration of rainwater and surface water, and thus contribute to the building of hydrostatic head in the groundwater. On the other hand, artesian springs are related to joints at greater depths.

Where underlain by breccias or other porous material, volcanic rocks (e.g. basalts and andesites) yield much groundwater through the cracks. In the Deccan Traps basaltic terrains in western and central India, a given flow may be separated from the flow overlying it, by a weathered zone if it suffered subaerial exposure, or by sediments if it went under water (inter-trappean bed). In the Deccan basalt country of Sagar, MP, India, the tuffaceous and weathered top of a particular flow is such an unfailingly prolific producer of water, that the standard procedure to locate groundwater in Sagar area is to look for the particular flow. It never failed! High yielding springs occur where this horizon outcrops in topographic lows.

The aggregate yield of the springs in Hawaii is an incredible 110-140 $m^3 s^{-1}$ ($3.5$-$4.4 \times 10^9$ $m^3$ $yr^{-1}$) which is adequate to supply drinking water to a large city. The yield of a single well, 152-185 m deep, in Hawaii could be as much as 7.5 $m^3 s^{-1}$. No wonder, the water supply for the city of Honolulu comes from basalt aquifers.

In the case of granites and other crystalline rocks, limited amounts of groundwater occurs in weathering joints to a depth of 30-50 m, very rarely 100 m. Groundwater can be expected to occur in the weathered mantle of the granite in the humid regions.

Carbonate rocks, which are generally karstified, are the largest sources of groundwater. The term, karst, is derived from the Cretaceous limestone of the

Karst Plateau in the Dinaric Alps near the Adriatic coast between Italy and Yugo-slavia. Karstified landscape is characteristic of leachable or soluble rocks, such as, limestone, dolomite, marble, gypsum, anhydrite, rock salt, etc. Carbonate rocks are not by themselves highly leachable, but if the water contains carbon dioxide, the leaching process would get greatly accelerated. Sinkholes in limestones may be 5-10 to 30-50 m in diameter, and 10-25 m deep. Some karstified areas may have 100-200 sinkholes per km².

Four kinds of groundwater movement in karstified lands are illustrated in Figure 1.24 (source: Klimentov 1983, p. 198):

1. If the karstified rocks are underlain by non-karstified permeable rocks (such as gravels and conglomerates), groundwater flows downward along vertical joints in the karstified rocks, and could emerge as springs issuing from the permeable rocks,
2. Inclined lateral flow in gently dipping, thin karstified rocks interbedded with non-karstified formations,

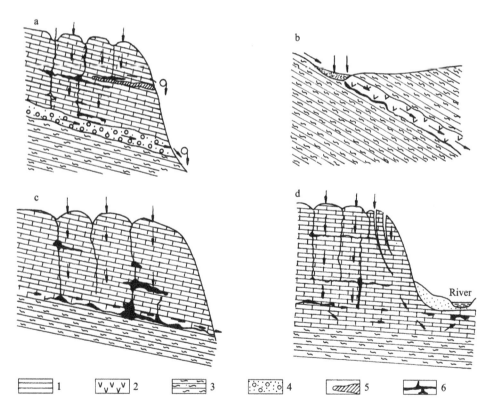

Figure 1.24. Groundwater movement in various karst situations (source: Klimentov 1983, p.198). 1 = Carbonate rocks, 2 = Sulfate rocks (gypsum and anhydrite), 3 = Clay, 4 = Gravel and conglom-erate, 5 = Local confining bed, and 6 = Karst caverns.

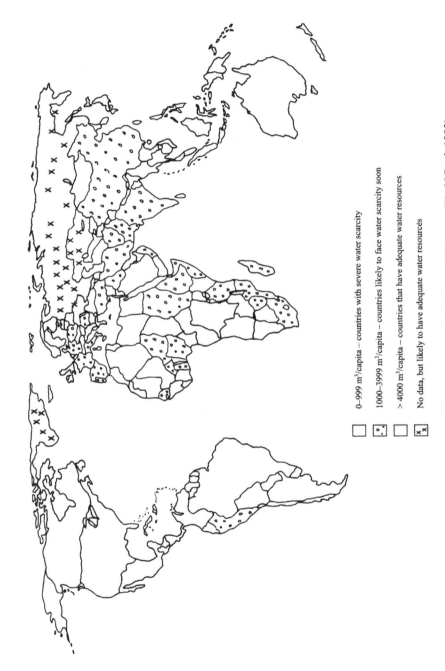

Figure 1.25. Annual renewable water resources of the various countries in the world (source: World Bank 1992).

3. Both vertical and lateral flow of water: such flows are characteristic of thin sequences of carbonate rocks, overlying impermeable strata, and
4. Vertical (downward), lateral and siphonal flows. The siphonal flows may discharge into the limestone caverns occurring beneath river beds, and occasionally, they may manifest themselves as submarine springs.

The Vaucluse spring in France is the most famous spring occurring in karst terrain (Neocomian limestone). Its annual average yield is $17 \text{ m}^3 \text{ s}^{-1}$, with peak discharge of $152 \text{ m}^3 \text{ s}^{-1}$ during the spring season.

## 1.7 DIMENSIONS OF WATER RESOURCES MANAGEMENT

The World Bank projects that between 1990 and 2030, the world's population will grow by 3.7 billion, demand for food will almost double, and industrial output and energy use will probably triple worldwide, and increase six-fold in developing countries (source: World Development Report 1992).

The United Nations Report entitled, 'Comprehensive Assessment of the Freshwater Resources of the World' (1997), draws a sombre picture of how water shortages in the 21st century could constrain economic and social development, and become sources of conflict between countries. Water use has been increasing twice as fast as population during this century. Pollution has made the matters worse. Total water use in the world rose from $1000 \text{ km}^3 \text{ yr}^{-1}$ in 1940 to $4130 \text{ km}^3 \text{ yr}^{-1}$ in 1990. It is expected to rise to $5000 \text{ km}^3 \text{ yr}^{-1}$ in 2000. At least one-fifth of the world's people lack access to safe drinking water. About 80 countries in the world, making up 40% of the world's population, are already suffering from serious water shortage, which had become a limiting factor in their economic and social development. The Report forecasts that by 2025, as much as two-thirds of the world's population will be affected by moderate to severe water scarcity, unless appropriate mitigation measures are taken.

According to Gustafsson (1982), the annual per capita need of water for habitation and *subsistence* crop production is $1250 \text{ m}^3$. Figure 1.25 (source: World Development Report 1992, p. 217, of the World Bank) depicts the annual renewable water resources of the various countries in terms of $\text{m}^3$ per capita (the average amount of water available per person per year is calculated by dividing the country's annual internal water resources by its population). Four categories are shown: $< 1000 \text{ m}^3$ per capita, $1000\text{-}3999 \text{ m}^3$, $4000\text{-}29,999 \text{ m}^3$, and $> 30,000 \text{ m}^3$.

The figure shows that a fairly large number of countries (e.g. Botswana, Kenya, Rwanda-Burundi, North African Saharan countries, Middle East, etc.) suffer from severe water scarcity right now. The problem is going to get worse not only for these countries, but also for several other countries as well, since the per capita availability of water in a country goes down with increasing population.

Presently, agricultural irrigation accounts for 70-80% of all water use. Increasing population would need more food and hence more water would be needed for irrigation. The Report argues that water must be perceived as a marketable commodity, with its use being subject to the market laws of supply and demand. It has been estimated that in USA, one acre-foot of water would yield an income of USD 400 when used in agriculture, and USD 400,000 when used in manufacturing. So in the American context, when water is scarce, it would be used for drinking and manufacturing purposes. Food is to be grown in areas where water is plentiful.

A complicating factor is the global trend towards urbanization – urban population as a percentage of total population, varies from 100% in city-states like Singapore, 80-90% in several countries of Western Europe (e.g. UK: 89%) and South America (e.g. Argentina: 86%), 20-30% in most African countries (e.g. Mozambique: 30%), with Bhutan having the lowest percentage (5%). Globally, urban populations are increasing at a faster rate than the general population growth, and this is particularly evident in the developing countries where this differential can be as high as three to four. This has serious implications for water resources planning. Cities cannot be shifted to places where resources like water and fertile soil are available. On the other hand, services like drinking water, habitation, sanitation, roads, etc. have to be organized keeping in mind the existing location of towns.

### 1.7.1   *World water requirements*

Shiklomanov (1998) estimated the future requirements of water in the world. Water withdrawal is expected to grow at the rate of 10-12% per decade, with a 1.38-fold increase by 2025. The water consumption is expected to grow at a slower pace, with a 1.26-fold increase by 2025 (see Table 1.12).

Presently, Asia with its extensive irrigation systems, accounts for about 57% of the total water withdrawal and 75% of the total water consumption. During the next few decades, water withdrawal in Africa and South America is expected to go up by 1.5-1.6 times. The smallest increase in water withdrawal is expected to occur in Europe and North America (1.2 times).

In 1995, the agriculture sector accounted for 67% of the total water withdrawal and 86% of its consumption. Whereas the demand for agriculture is expected to increase by 1.3 times by 2025, much higher increases are forecast for industry (1.5 times) and public water supply (1.8 times).

Table 1.12. Projected requirements of water in the world.

| Year | Total global water withdrawal $(km^3\,yr^{-1})$ | Total actual consumption $(km^3\,yr^{-1})$ | Consumption of withdrawal (%) |
|------|------|------|------|
| 1995 | $3750\ km^3\,yr^{-1}$ | $2270\ km^3\,yr^{-1}$ | 61% |
| 2025 | $5100\ km^3\,yr^{-1}$ | $2860\ km^3\,yr^{-1}$ | 56% |

An important statistic to be borne in mind is that globally the quantity of water lost by evaporation from the reservoirs is more than the total industrial and potable use. Water losses due to evaporation are the highest in the African continent (33-35% of the total water consumption). This shows how much can be gained by taking steps to minimize evaporation.

In Europe in 1995, industry accounted for 44% of the total water withdrawal. This is expected to increase to 50% by 2025. In the case of North America, the current water demand by industry which is 40% of the total withdrawal, is expected to decrease to 37% by 2025, because of the intense growth of irrigation in Central America, which is a part of the North America. The percentage of demand in the industrial sector may not exceed 20% in South America, 13% in Asia and 6% in Africa.

### 1.7.2 *World Commission on Water for the 21st century*

The UN General Assembly constituted a World Commission on Water for the 21st century, to provide an assessment of current and future global freshwater resources. The Commission is preparing a document entitled, 'World Water Vision 2025' (Strzepek 1999) (summarized in Fig. 1.26).

The extent of withdrawal (%) and the consumptive use in regard to various sectors are summarized in Table 1.13.

The water demand of a family/community/industry is not absolute – it is partly controlled by price. For instance, the requirement of water for drinking would be inelastic and independent of the price of water, whereas the requirement for other purposes (say, gardening) would be elastic and would depend upon the price.

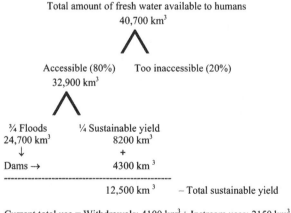

Total amount of fresh water available to humans
40,700 km³

Accessible (80%)        Too inaccessible (20%)
32,900 km³

¾ Floods                ¼ Sustainable yield
24,700 km³                  8200 km³
    ↓                          +
Dams →                     4300 km³
-------------------------------------------------
                12,500 km³      – Total sustainable yield

Current total use = Withdrawals: 4100 km³ + Instream uses: 2150 km³ = 6250 km³

Figure 1.26. Sustainable yield of water resources in the world.

Table 1.13. Sector-wise withdrawal and consumptive use of water.

| Sector | Withdrawals (%) | Consumptive use (%) |
|---|---|---|
| Agriculture | 65 | 87 |
| Reservoirs | 4 | 7 |
| Municipality | 7 | 2 |
| Industry | 24 | 4 |

Global forecast of water demand in 2000 = 5000-7000 $km^3$.

### 1.7.3  *Continental aquatic systems*

The following account of the impacts on continental aquatic systems, and the issues arising from them, is largely drawn from Meybeck (1998) (Table 1.14).

The continental aquatic systems – rivers, lakes, wetlands, estuaries, groundwater and coastal zones – constitute a critically important part of the biogeochemical functioning of the earth system. They are studied in the context of the changing fluxes, levels and transfer of water, sediments, carbon and nutrients (such as, N, P and Si). The impacts on the continental aquatic systems manifest themselves in the following ways:

*Human health* – Diseases may be water-related (such as malaria) or waterborne (such as cholera). Diseases such as typhoid, cholera and diarrhea are carried in infected drinking water; others are spread when people wash themselves in contaminated water. Unsafe water is implicated in the deaths of 3 million people, mostly children, and causes about 900 million episodes of illnesses each year. At any one time, more than 900 million are afflicted with roundworm infections and 200 million people with schistosomiasis. Besides, frequent attacks of diarrhea can leave a child more vulnerable to illness and death (source: World Development Report 1992, of the World Bank, p. 48-49). Human health may be impaired by toxic chemicals (such as, Cd, Pb, Hg, organo-metallic complexes) in drinking water, and toxic algae development resulting from eutrophication.

*Water availability* – May be affected because of enhanced evaporation, enhanced use of water for agriculture, and fragmentation of river networks.

*Global carbon balance* – Carbon species may get trapped either in organic and inorganic dissolved form or in the form of particulate. On the other hand, organic carbon may get released through permafrost melts. Weathering of silicate minerals by atmospheric $CO_2$ affects the global carbon fluxes.

*Fluvial morphology* – May be affected as a consequence of shift of river courses or due to erosion and sedimentation in the river beds.

*Aquatic biodiversity* – Several rivers and streams have lost their pristine headwaters. The segmentation of river networks and degradation in the quality of water have an adverse effect on aquatic biodiversity.

*Coastal zone impacts* – The rates of discharge of water, sediment and nutrients by rivers into the coastal zone are getting reduced, whereas the levels of pollutants and microbial pathogens are increasing.

Table 1.14. Major threats to coastal aquatic systems, and the issues arising therefrom (source: Meybeck 1998).

| Environmental state changes | Major impacts | Issues arising therefrom |
|---|---|---|
| 1. Climate change | Development of non-perennial rivers | B C D E F G |
| | Segmentation of river networks | E F |
| | Development of extreme flow events | B E F G |
| | Changes in wetland distribution | A B C D F G |
| | Changes in chemical weathering | D G |
| | Changes in soil erosion | D E G |
| | Salt water intrusion in coastal groundwater | B |
| | Salinization through evaporation | B C F |
| 2. River damming and channelization | Nutrient and carbon retention | D G |
| | Retention of particulates | D E G |
| | Loss of longitudinal & lateral connectivity | F |
| 3. Land-use change | Wetland filling and draining | C D F |
| | Change in the sediment transport | D E G |
| | Alteration of first order streams | E F |
| | Nitrate and phosphate increase | A C D G |
| | Pesticide increase | A C G |
| 4. Irrigation and water transfer | Partial to complete decrease of river fluxes | E F G |
| | Salinization through evaporation | B C |
| 5. Release of industrial and mining wastes | Heavy metals increase | A C |
| | Acidification of surface waters | C F |
| | Salinization | A C F |
| 6. Release of urban and domestic wastes | Eutrophication | A C D F G |
| | Development of water-borne diseases | A |
| | Organic pollution | A C F |
| | Persistent organic pollutants | A C G |

A = Human health, B = Water availability, C = Water quality, D = Carbon balance, E = Fluvial morphology, F = Aquatic biodiversity, and G = Coastal zone impacts.

It has been projected that in the next 25 years, the average country water demand will increase by a factor of 1.4 to 2.9 for developed countries and 3-10 for developing countries. Irrigation will continue to be the most important user of water (50 ha of irrigated land is needed to feed 1000 persons in the arid regions). The World Bank estimates that there are now 160 Mha (million hectares) of irrigated land in the developing countries. Though this represents only 20% of all land harvested, it makes use of 60% of all fertilizer and accounts for 40% of all crop output. Global irrigated land is projected to increase from 254 Mha now to 330 Mha in 2025. Global expenditure in irrigation initiatives is currently running at USD 15 billion, and that is growing at the rate of 2% per year.

The most spectacular example of anthropogenic impact on continental aquatic systems is the diversion of the waters of Amu Darya river and the consequent disastrous degradation of the Aral Sea. Several of the world's major rivers (e.g. the Nile, Colorado, Zambeze, Indus, etc.) are already impacted by the construction of large dams and water diversion activities (the present day global area of reservoirs is estimated at 500,000 $km^2$, with a volume of 6000 $km^3$). Global sediment and particulate organic trapping is estimated at 20%, and the global aging of river water transfer exceeds one year in most cases. Such a situation has an adverse impact on biodiversity, particularly of the fish species, and manifests itself in severely reduced discharge of water, nutrients or sediments to the coastal zone.

As a consequence of the pollution of the river water from various sources, the dissolved inorganic nitrogen and phosphorus inputs to the coastal zone increased globally by a factor of 2.5 and 2.0 respectively, whereas dissolved silica may have decreased as a result of riverine eutrophication. Similar increasing trends could be noted in regard to the inputs to the coastal zone of toxic elements (e.g. Cd, Hg, Pb, Zn, etc.), and organic pollutants (e.g. PCBs, PAHs, etc.)

About 61% of the withdrawn water (3750 $km^3$ $yr^{-1}$) is consumed, i.e. evaporated. This happens to be more than 5% of the global river input into the ocean. This percentage is higher for some regional seas such as the Mediterranean (source: Shiklomanov 1998).

## 1.8  GLOBAL DATABASES FOR WATER RESOURCES

WMO (World Meteorological Organization, Geneva) realized quite early that global sets of meteorological, climatological and hydrological data have to be secured and integrated not only to run global circulation models, and assess the potential impacts of climate change, but also to address the problems of the scarcity of freshwater resources. In the context of several governments of the Industrialized countries commercializing or privatizing the services provided by national meteorological and hydrological agencies, WMO has to tread a careful path in regard to its traditional policy of free (i.e. no cost) and open transfer of hydrological data and products (Askew 1999).

The principal data sources for global water resources and their management, are described below.

All the industrialized countries and some of the developing countries put out annual reports about river discharges, and quality of surface and groundwater, etc. Several developing countries (e.g. Burma, Mozambique) are, however, not in a position to issue annual reports regularly due to lack of infrastructure and trained personnel.

A number of international organizations, UNESCO, UNEP, WHO, WMO, ICSU, etc., established global programmes and are publishing reports on various aspects of water resources and their management: UNESCO River Discharges (1971), UNEP/ UNESCO Regional Seas Reports, River discharges by the Global

Runoff Data Centre (WMO, Koblenz), World Register of River Inputs (WORRI) of UNEP/UNESCO, etc.

During 1998, UNESCO brought out two seminal documents on the issue of water resources: A volume entitled, *Water: A looming crisis?* (edited by H. Zebidi 1998) and a lucid summary by Shiklomanov (1998) entitled, *World Water Resources: A new appraisal and assessment for the 21st century.*

In 1978, UNEP launched the Global Environmental Monitoring Program for continental waters (GEMS-water) in cooperation with WHO, UNESCO and WMO. This Program involves the collection of data, training of technicians, creation of software and publication of handbooks and assessments in four major languages, etc. (WHO/UNEP 1991; Fraser et al. 1995). The focus of the program has been on surface waters, streams, rivers and lakes.

In 1986, the International Council of Scientific Unions (ICSU) established the International Geosphere – Biosphere Program (IGBP), 'to describe and understand the interactive physical, chemical and biological processes that regulate the total Earth system, the unique environment that it provides for life, the changes that are occurring in this system, and the manner in which they are influenced by human actions'. In 1987, the IGBP Secretariat was established at the Royal Swedish Academy of Sciences, S-104 05, Stockholm (e-mail: <sec@igbp.kva.se>; website: http://www.igbp.kva.se/). IGBP has been highly successful in mobilizing and coordinating a variety of international activities which have relevance to water resources (see the lucid and elegant summary by Berrien Moore, in IGBP Newsletter, Dec. 1999). The Land-Ocean Interactions in the Coastal Zone (LOICZ) of IGBP, in cooperation with GEMS-Water, established the GEMS-GLORI (Global Register of River Inputs) (Meybeck & Ragu (1997). GEMS-GLORI, circulated by UNEP for review in 1996, contains information about 555 major rivers of the world, discharging to the oceans: $>10 \text{ km}^3 \text{ yr}^{-1}$, or drainage area: $>10,000 \text{ km}^2$, or sediment discharge $>5 \text{ Mt yr}^{-1}$, or basin population: $>5$ M people.

LOICZ is located at Netherlands Institute for Sea Research, P.O. Box 59, 1790 AB Den Burg, Texel, Netherlands (e-mail: loicz@nioz.nl; website: http://www. nioz/nl/loicz/. The Global Data bank of GEMS-Water is managed at the WHO Collaborating Center, Burlington, Ont., L76 4A6, Canada (e-mail: gems@cciw. ca; website: http://www.cciw.ca/gems).

In 1998, IGBP has set up a Water group, to 'understand the role of the continental aquatic systems – rivers, lakes, wetlands, estuaries, groundwater and coastal zones – in the biogeochemical functioning of the Earth system' (Meybeck 1998). This Group works in close cooperation not only with technical programmes dealing with coastal zones (LOICZ), atmosphere (IGAC), soil (CGTE), river impoundments (BAHC), etc. but also with socio-economic programs dealing with human dimensions(IAHD).

According to Michel Meybeck (e-mail: meybeck@biogeodis.jussieu.fr) and Charles Vörösmarty (e-mail: charles.vorosmarty @unh.edu) (1999), IGBP Water Group is attempting an analysis of the 'past, contemporary and future state of terrestrial water systems together with the biogeochemical constituents they store, transform and transport to the coastal ocean'. Making use of the existing repositories, such as WMO/ GRDC (Global Runoff Data Centre, c/o Federal Institute of Hydrology, Koblenz, Germany, e-mail: grabs@bafg.de) and WHO/GEMS – Water, and UNESCO / FRIEND (Flow Regimes from International Experimental and Network Data, e-mail: ihp@unesco.org), IGBP Water Group is developing an electronic River Archive Series, wherein the water discharge, chemical constituents, biophysical and socio-economic data are integrated for each drainage basin.

The Global Hydrological Archive and Analysis System (GHAAS) contains a wide array of hydrometeorological data including river network topology, runoff data, etc.. All GHAAS data is available free of charge from http://www.csrc.sr. unh.edu/hydro/welcome.html.

Table 1.15 gives the key biogeophysical datasets that are used to build typology models (source: Vorosmarty et al. 1999).

Table 1.15. Key datasets for use in typology models.

| Geophysical parameter | Space/time scale | Source/archive |
|---|---|---|
| Surface hydrology attributes | | |
| – River networks and Basin | 30 minute | Vorosmarty et al. |
| – Boundaries (STN30 v. 5.12) | | |
| – Discharge | 30 minute | Fekete & Vorosmarty |
| | Station data | RivDis, R-HydroNET, |
| | | R-ArcticNET |
| | | Vorosmarty et al. |
| – Runoff | 30 minute | Fekete & Vorosmarty |
| – Digital topography | 1 km | Hydro 1K EDC |
| Surface attributes | | |
| – Land cover | 1 km | EDC (IGBP 1996) |
| – Soils | Vector | FAO (1996) |
| Geology/lithology | Vector | UNESCO (1999) |
| N deposition (NOy and NHx) | 10° | Dentener et al. |
| N equivalent fertilizer use/input | 30 minute | CSRC-UNH |
| Livestock | 1° | Lerner et al. |
| Population | 30 minute | CSRC-UNH |
| Climatological Data | | |
| – Air temperature | 30 minute | Wilmott & Matsuura |
| – Precipitation | 30 minute | Wilmott & Matsuura |
| Evapotranspiration | 30 minute | Fekete & Vorosmarty |
| Lake density and volume | Vector | ESRI |
| Reservoir induced aging | Basin scale | Vorosmarty et al. |
| Constituent data (various forms of N, P, C, | Station Data | |
| Si, susp. solids and discharge) | Mean Annual (530) | GEMS/GLORI |
| | Mean monthly (52) | GEMS/WATER |

# REFERENCES

## *Suggested reading

*Appelo, C.A.J. & D. Postma 1996. *Geochemistry, Groundwater and Pollution*. Rotterdam: A.A. Balkema.

Askew, A.J. 1999. National and international policies on access to hydrological data: Abstract no. HW1/29/A5, IUGG, Birmingham.

Baumgartner, A. & E. Reichel. 1975. *The World Water Balance*. Amsterdam: Elsevier.

Binley, A.M. & K.J. Beven 1991. Physically-based modelling of catchment hydrology: A likelihood approach to reducing predictive uncertainty. In D.G. Farmer & M.J. Rycroft (eds), *Computer modeling in environmental sciences*: 75-88. Oxford: Clarendon Press.

*Bouwer, H. 1978. *Groundwater Hydrology*. New York: McGraw-Hill.

Crawford, N.H. & R.K. Linsley 1964. *A conceptual model of the hydrological cycle*. IAHS Publ. no. 63.

Fraser, A.S., M. Meybeck & E.D. Ongley 1995. Water quality of the world river basins. *UNEP Environmental Library*, no. 4.

Gustafsson, Y. 1982. *Water demand and water supply*. Proc. UN Inter-regional seminar on Rural Water Supply. Uppsala, Sweden.

Holtan, H.N. & N.C. Lopez 1971. *USDAHL – 70 model of the watershed hydrology*. US Dept. of Agriculture.

*Klimentov, P.P. 1983. *General Hydrogeology*. Moscow: Mir Publishers

*L'vovich, M.L. 1979. *World Water Resources and their future*. Washington, DC: Am. Geophy. Uni.

Meybeck, M. & A. Ragu 1997. Presenting the GEMS-GLORI, a compendium of world discharge to the oceans. in *Freshwater Contamination*. IAHS Publ. no. 243: 3-14.

*Meybeck, M. 1998. The IGBP Water Group: a response to a growing global concern. *IGBP Newsletter*, p. 8-12.

Meybeck, M., and C. Vörösmarty 1999. Global Data Bases in support of the newly created IGBP Water Group. Abstract no. HW1/05/A4, IUGG, Birmingham.

*Petts, P. & G. Eduljee 1996. *Environmental impact assessment for waste treatment and disposal facilities*. Chichester: John Wiley.

Ridder, T.B. 1978. *On the chemistry of precipitation* (in Dutch). Royal Met. Inst. Rep., 78-4.

*Shiklomanov, I.A. 1998. *World Water Resources – A new appraisal and assessment for the 21st century*. Paris: UNESCO.

Strzepek, K. 1999. Global water resources, agricultural and socio-economic databases for water resource and environmental policy assessment: current issues and future needs. Abstact no. HW1/19/A4, IUGG Birmingham.

*Sytchev, K.I. 1988. *Water Management and Geoenvironment*. Paris-Nairobi: UNESCO-UNEP.

Turney, G.L. & D.F. Goerlitz 1990. Organic contamination of groundwater at Gas Works Park, Seattle, Washington. *Ground Water Monitoring Rev*. 10: 187-198.

UNDP1997. *Human Development Report*.

Vörösmarty, C., et al. 1999. The Global Hydrological Archive and Analysis System. (GHAAS). Abstract no. HW1/25/A 5, IUGG, Birmingham.

WHO/UNEP 1991. Water Quality. Progress in the implementation of Mar del Plata Action plan and a strategy for the 1990's. *GEMS Report*, WHO, Geneva.

World Bank1992. *World Development Report*.

*Zebidi, H. 1998. *Water: A looming crisis?*. Paris. UNESCO.

Zektser, I.S. 1998. Groundwater resources of the world and their use. In H. Zebidi (ed.), *Water: A looming crisis?*, p. 55-65. Paris: UNESCO.

# CHAPTER 2

# Dynamics of interactions involving water

## 2.1 DYNAMICS OF WATER ENVIRONMENTS

Water is a part of the *Geoenvironment*. It is linked to and is influenced by, atmosphere, climate, rocks and soils, and biota. Surface water and groundwater are by far the most dynamic components of the geoenvironment, and are most affected by the activities of man. Any strategy to protect water resources from depletion and pollution has to be based on an understanding of the linkages and dynamics of water environments.

Figure 2.1 (source: National Institute for Environmental Studies, Tsukuba, Japan) depicts the space – time magnitudes of and linkages between, various environmental phenomena. For instance, hydrological changes are linked with and affected by, human activities, and changes in land-use, and ecosystems, meteorological changes and ocean changes. Spatially, hydrological changes may affect catchments, and sometimes whole countries, for time periods ranging from days to decades.

Figure 2.2 (source: Land-surface model of NCEP-OSU-Air Force-Office of Hydrology, NOAH, USA) is a comprehensive depiction of the various phenomena involved in the movement of water:

- Atmospheric forcing near the surface, involving precipitation, temperature, humidity, surface pressure, and wind,
- Radiation forcing at surface involving downward solar and long wave radiations,
- How precipitation + condensation + snowmelt water get expended as surface runoff, infiltration, evaporation and transpiration,
- Moisture flux and moisture budget, and the parameters involved, and
- Heat flux and heat budget, and the parameters involved.

### 2.1.1 *Hydrological dynamics of heterogenously structured drainage basins*

Flügel (1995) developed the concept of Hydrological Response Units (HRUs), in order to model the hydrological dynamics of heterogeneously structured basins. A HRU is an ensemble having a common land use and pedo-topo-geological associations. The evapotranspiration, infiltration, surface runoff, interflow, groundwater

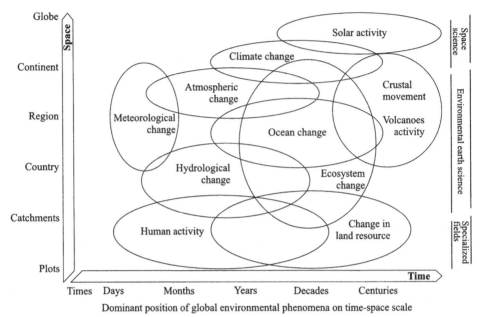

Figure 2.1. Space – time magnitudes and linkages in regard to hydrological changes (source: National Institute for Environmental Studies, Tsukuba, Japan).

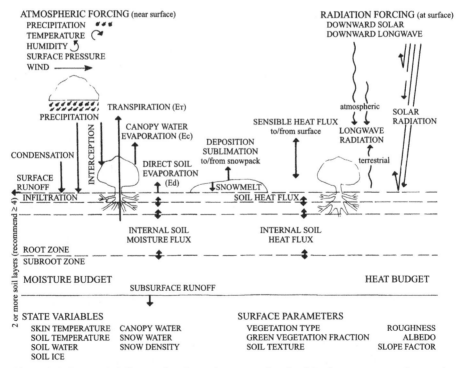

Figure 2.2. Forces and fluxes of various phenomena involved in the movement of water (source: NOAH, USA).

recharge and runoff generation of a river basin are dependent upon the kinds of soil catena, vegetation and atmospheric conditions in the basin. A given soil catena is a consequence of the operation of the processes of weathering, and erosion in the context of particular topography, soils, geology and land use. Thus, a Hydrological Response Unit arises from a particular pedo-topo-geological association and has characteristic hydrological dynamics. The soil catena generally determines the land use. Fertile, developed soils in the valley floor and gentle hills are most suited for agriculture. On the other hand, shallow, less productive soils on the slopes and on the plains can only be used for forestry, and rangeland. For instance, gleysols are formed in lowlands (wetlands) with level topography, with shallow groundwater. In tropical and subtropical regions, they are widely used to grow rice. They are rarely to be found on the slopes. Figure 2.3 (source: Flügel 1995, p. 80) is a block diagram of the basin subsystem storages and their interlinked water fluxes.

Flügel (1995) gave a case study of the application of HRU concept to the drainage basin of the River Bröl in Germany, with the following features: catchment area: 216 km$^2$; topography: middle mountain range of Rheinische Schiefergebirge; geology: impermeable Devonian shale, with negligible deep percolation; climate: oceanic climate, with rainfall varying between 800-1200 mm depending upon the elevation; soil catenas: hydromorphic brown soils on high plains, brown soils on adjacent slopes, and gleysols on the valley floors; precipitation: convective storms during summer, and advective rainfall during the winter; hydrometeorological data: rainfall, temperature, radiation, and discharge.

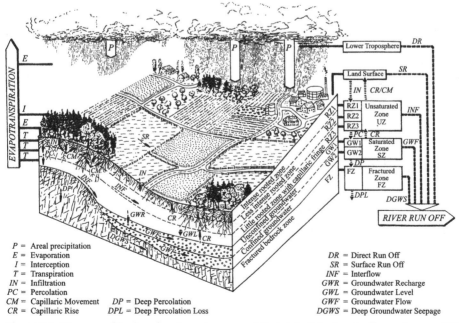

$P$ = Areal precipitation
$E$ = Evaporation
$I$ = Interception
$T$ = Transpiration
$IN$ = Infiltration
$PC$ = Percolation
$CM$ = Capillaric Movement  $DP$ = Deep Percolation
$CR$ = Capillaric Rise  $DPL$ = Deep Percolation Loss

$DR$ = Direct Run Off
$SR$ = Surface Run Off
$INF$ = Interflow
$GWR$ = Groundwater Recharge
$GWL$ = Groundwater Level
$GWF$ = Groundwater Flow
$DGWS$ = Deep Groundwater Seepage

Figure 2.3. Diagram of basin subsystem storages and their interlinked water fluxes (source: Flügel 1995, p. 80).

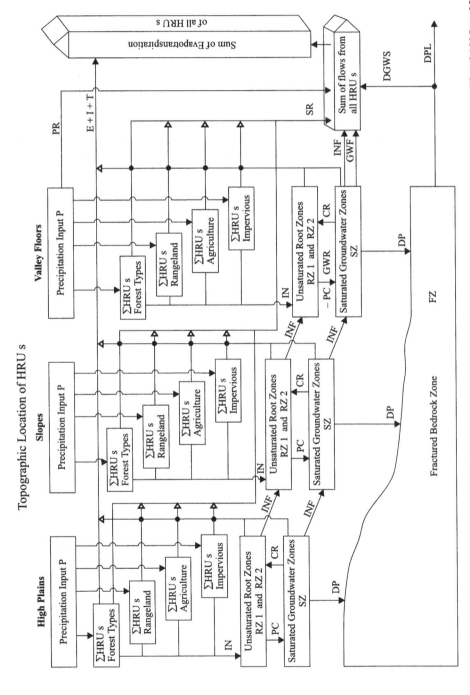

Figure 2.4. Interflows between the various Hydrological Response Units (HRUs) for different topographic locations (source: Flügel 1995, p. 85).

A GIS database was built by digitizing map data regarding soils, land use, etc. Topography, soils and geology were grouped into four different pedo-topo-geological associations:
1. Plains (0-10% slope and hydromorphic brown soils),
2. Slope 1 (10-20% and brown soils),
3. Slope 2 (> 20% and brown soils), and
4. Valley floor (> 0% and gley soil).
Twenty-three HRU's were delineated with their particular physiographic properties (area, av. elevation, topography, slope, soil, type, land use, and aspect).

Figure 2.4 (source: Flügel 1995, p. 85) shows the interflows between the various HRUs for different topographic locations, namely, high plains, slopes and valley floors.

The 23 HRUs of the Bröl basin were tested using the hydrological basin model PRMS/MMS originally developed by the US Geological Survey (Leavesley et al. 1983). The fit between the simulated and observed discharges has a high correlation coefficient, $r = 0.91$. As should be expected, the model has been found to be particularly sensitive to vegetation cover and the land use.

### 2.1.2 *Hydrological dynamics of hill slopes*

Bronstert et al. (1998) developed a model to simulate the hydrological dynamics of hill slopes, and applied it to the Weiherbach catchment in southwest Germany, which is a hilly, agricultural area of deep loess soil. Figure 2.5 (source: Bronstert et al. 1998) is a schematic representation of the parameters involved in a two-dimensional, hill slope model. The hydrological flow processes considered are infiltration (including the effects of macropores), (unsaturated) subsurface storm flow and surface runoff. The water dynamics in the unsaturated soil matrix of the hill slope have been computed through appropriate equations.

## 2.2 TEMPORAL AND SPATIAL CHANGES IN WATER COMPOSITION

Paces (1976) gave a good account of the changes that the chemical composition of water may undergo in time and space. Figure 2.6 (source: Paces 1976, p. 92) is a schematic sequence of the temporal changes that water composition may undergo. The interaction of water with rock, air and biota, and the units involved are shown in Figure 2.7 (source: Paces 1976). Some changes in water environment may persist over geological time, or may change rapidly enough to be observed directly. The temporal change may be periodic, abrupt or systematic. An environment may have virtually unchanging physico-chemical properties, as is the case with many sandstone and limestone aquifers. The environment may change periodically, such as due to monsoons, or it may change abruptly, such as when acid mine discharge is let into an oxidized aquifer. In some situations, the environment

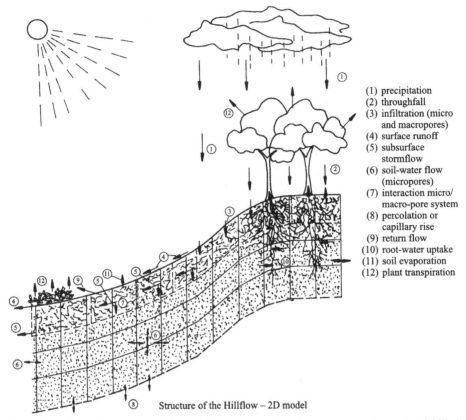

(1) precipitation
(2) throughfall
(3) infiltration (micro and macropores)
(4) surface runoff
(5) subsurface stormflow
(6) soil-water flow (micropores)
(7) interaction micro/ macro-pore system
(8) percolation or capillary rise
(9) return flow
(10) root-water uptake
(11) soil evaporation
(12) plant transpiration

Structure of the Hillflow – 2D model

Figure 2.5. Schematic representation of the parameters involved in a two-dimensional hill-slope model (source: Bronstert et al. 1998).

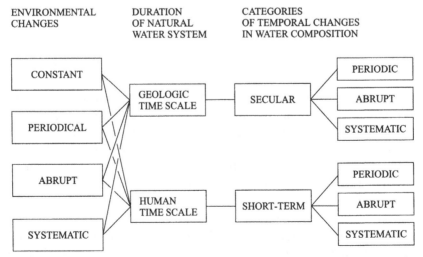

Figure 2.6. Interaction of water with rock, air and biota, and the parameters involved (source: Paces 1976).

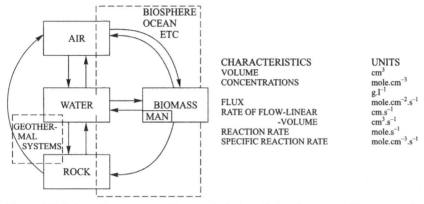

Figure 2.7. Schematic sequence of temporal changes that water composition may undergo (source: Paces 1976, p. 92).

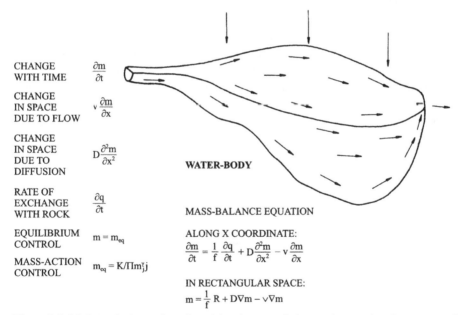

Figure 2.8. Mathematical equation of spatial and temporal changes in regard to the concentration of the dissolved component in the flowing water (source: Paces 1976, p. 98).

may change systematically in one direction – e.g. the temperature increase with depth of a thick or vertical aquifer due to geothermal gradient.

The changes that groundwater undergoes while flowing through a rock or soil is a good example of the continuous, non-periodic changes that water undergoes from an initial state to a final state. Mathematical modeling of such a process is useful not only to validate a real situation, but also to quantify the parameters. Figure 2.8 (source: Paces 1976, p. 98) gives a schematic diagram and mathematical equation of the temporal changes in the concentration of the component dissolved in flowing water within space characterized by coordinates, $X$, $Y$ and $Z$.

The following parameters are involved: $m$ = molarity of the component in water; $t$ = time; $v$ = linear flow velocity of water; $D$ = diffusion coefficient of the component in water, $q$ = moles of the component exchanged between water and unit volume of rock; $m_{eq}$ = equilibrium molarity of the component in water; $K$ = equilibrium constant for the overall reaction affecting the exchange of the component between rock and water: $m_j$ = molarities of species participating in reversible overall reaction affecting $m$; $v_j$ = stoichiometric coefficients of participating species in the reversible reaction; $f$ = porosity of the rock environment.

The following sub-units are calculated from the above data:
– Change with time: $(\partial m/\partial t)$, change in space due to flow (v. $\partial m/\partial x$), and
– Change in space due to diffusion ($D$. $\partial^2 m/\partial x^2$); rate of exchange with rock ($\partial q/\partial t$); equilibrium control ($m - m_{eq}$); mass action control ($m_{eq} = K/\Pi m_j v_j$).

These sub-units are used in the mass balance equations along the x-axis, and in rectangular space (vide Fig. 2.8).

There is a change in the dissolved components in the runoff as water moves towards the bottom of a watershed. The concentration of a component at a given point in time and space reflects the mixtures of precipitation, discharge from the soil, aerated and saturated zones, and the frequency distribution of the flux of water through a rock environment (reacted to the lengths of residence in each environment). Figure 2.9 (source: Paces 1976, p. 106-107) gives a schematic model and mathematical equations for the purpose.

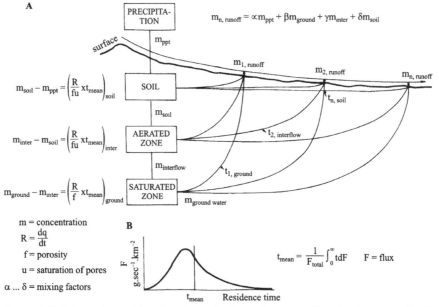

Figure 2.9. Schematic model and mathematical equations regarding the frequency distribution of flux of water through a rock environment (source: Paces 1976, p. 106-107).

## 2.3 EVOLUTION OF WATER CHEMISTRY IN THE WATER CYCLE

Rainwater is a source of surface water and the principal source of groundwater. Starting with rainwater, the chemical composition of water in the hydrological cycle evolves as a consequence of operation of a variety of processes, such as, evaporation, transpiration, selective uptake by vegetation, oxidation / reduction, cation exchange, dissolution of minerals, precipitation of secondary minerals, mixing of water, leaching of fertilizers and manure, pollution, and lake / sea biological processes. Appelo and Postma (1996, p. 35) gave a schematic overview of these processes. Thus the chemical composition of water at a given point in time and space would depend upon the nature and extent of the process(es) to which the water was exposed.

An examination of the distribution curves for different chemical species in terrestrial waters in USA (vide Fig. 2.10; source: Davies & De Weist 1966) helps us to trace the evolution of terrestrial waters from rainwater in the following ways:

– In the case of Total Dissolved Solids, 10% of the samples contain less than 100 ppm of TDS, while 95% of the samples contain less than 1000 ppm. Evidently, these high concentrations of TDS have been added on to the rainwater through interaction with soils and rocks.
– The steep gradients of the frequency curves of some species, such as $Ca^{2+}$, $HCO_3^-$, $SiO_2$, $K^+$ and $F^-$, indicate that the solubility of the mineral places an upper limit on the maximum concentrations of the species in natural waters.
– The concentrations of the species with lower gradients tend to follow a normal distribution. This may be a consequence of the low availability of the concerned minerals in rocks or a slow rate of dissolution of the minerals or the biological processes being at work.

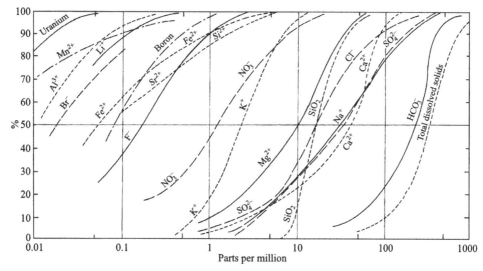

Figure 2.10. Frequency plot of concentrations of different solutions in terrestrial waters in USA (source: Davies & De Wiest 1966).

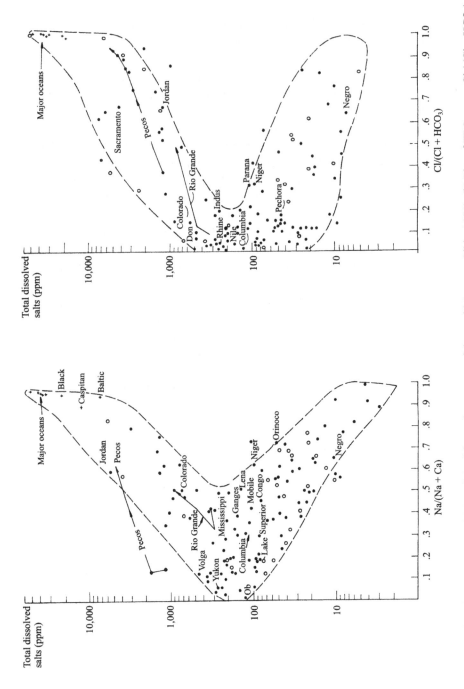

Figure 2.11. (a) Plot of TDS versus Na / (Na + Ca) of the surface waters of the world (source: Gibbs 1970). (b) Plot of TDS versus Cl / (Cl + HCO₃) of the world surface waters (source: Gibbs 1970).

In his classic paper, Gibbs (1970) developed an elegant method to identify the mechanisms that control the solute composition of the freshwaters. His system is based on three kinds of waters:

1. Seawater whose chemistry is dominated by $Na^+$ and $Cl^-$, and which has a high content of total dissolved solids (TDS),
2. Rainwater which is essentially dilute seawater but with very low TDS, and
3. Terrestrial waters which have acquired $Ca^{2+}$ and $HCO_3^-$ as a consequence of interaction of rainwater with rocks and soils.

When TDS (ppm) is plotted against Na/(Na + Ca) (Fig. 2.11a; source: Gibbs 1970), seawater plots at the terminal part in the upper right-hand side, and the rainwater plots at the end of lower branch. The waters of the various rivers plot somewhere in between depending upon their TDS and Na and Ca contents. When TDS (ppm) is plotted against Cl/(Cl + $HCO_3$) (Fig. 2.11b; source: Gibbs 1970), seawater and rainwater plot at the same places as in the previous diagram, but the waters of the rivers like the Jordan, Colorado and Rio Grande, which are subject to severe evaporation plot at different places in the two diagrams. That is so because when evaporation takes place, the least soluble minerals precipitate. Thus the precipitation of $CaCO_3$ would lead to the decrease of $Ca^{2+}$ and $HCO_3^-$ and this leads to a shift in the position of the concerned river in the two plots and within the same plot.

## 2.4 ROLE OF WATER IN THE SECONDARY PROCESSES

### 2.4.1 *Redox processes*

Redox reactions involve electron transfer from one atom to another. The inorganic reactions involved are generally slow. They tend to get accelerated when mediated by bacterial catalysis. For instance, the presence of *Desulfovibrio* catalyses the rapid reduction of sulfate by organic matter, which otherwise would proceed very slowly.

Though water is the most ubiquitous compound on earth, it is by no means an ordinary substance. Its structure allows it to be the universal solvent that it is. Two parameters (pH and Eh/pE) arise from its two components (hydrogen and oxygen).

The redox reaction in the case of water may be written as

$$H_2O \leftrightarrow 2e^- + 2H^+ + \tfrac{1}{2} O_{2(g)} \tag{2.1}$$

Hydrogen ion concentration (pH) is $-\log [H^+]$. The range of pH is 0 to 14. Pure water has a pH of 7. Acid pH is below 7, and alkali pH is above 7.

Redox potential is expressed as Eh or pE. On the analogy of pH, pE is defined as $pE = -[\log e^-]$, where $[e^-]$ is the 'activity' of electrons. $pE = Eh/0.059$ at 25°C. Just as in the case of Eh, high positive values of pE indicate oxidizing conditions and low negative values reducing conditions.

Generations of students learnt the basics of low-temperature geochemistry through the pH – Eh diagrams of Garrels & Christ (1965). This was followed by the specialized text of Brookins (1988) on the subject.

The environments which involve interactions with water are constrained by the stability fields of water in terms of pH and pE (Fig. 2.12; source: Garrels & Christ 1965). $H_2O$ exists as liquid water above pE of 0 and below pE of 20.60 – below pE of 0, water gets reduced to $H_2$, and above pE of 20.60, water gets oxidized to $O_2$.

pE – pH diagrams have been found to be immensely useful not only in the fundamental understanding of the processes of nature, but also in designing ameliorative measures for various environmental problems (nitrate contamination of water caused by the excessive use of fertilizer, removal of iron from mine water to make it potable, etc.).

How pE – pH diagrams could be made use of to ameliorate an environmental problem could be illustrated with the example of the mitigation measures for arseniasis. Drinking water with high arsenic content induces skin lesions, bronchitis, noncirrhotic portal fibrosis, and polyneuropathy. Arsenic is found in natural waters in the form of As (V) and As (III). The undissociated forms for As (V) and As (III) are $H_3AsO_4$ and $H_3AsO_3$ respectively. Appelo & Postma (1996, p. 249) gave the partial pE – pH stability diagram for dissolved arsenic species (see Fig. 10.3).

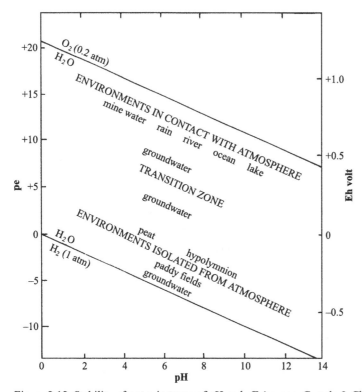

Figure 2.12. Stability of water in terms of pH and pE (source: Garrels & Christ 1965).

Toxicity of arsenic is strongly speciation-specific; As (III) is far more toxic than As (V). As (III) is the most mobile of all As species in water. *It is also more toxic.* Thus, a tubewell which taps an aquifer with high Eh characteristics, would contain lesser quantities of the more toxic As (III), and therefore should yield safer water. The arsenic content of water could be brought down to 30 μg As $l^{-1}$, by treating the water with an oxidant (say, $Cl_2$) and a coagulant (say, $FeCl_3$).

### 2.4.2 *Redox zoning*

In their monumental work, 'Aquatic Chemistry', Stumm & Morgan (1981) gave the sequence of important redox processes at pH 7 that take place under highly oxidizing to highly reducing conditions (Fig. 2.13; source: Stumm & Morgan 1981). The inset of the figure gives the combination of reduction and oxidation processes and their energy yield. Thus, the figure indicates that $O_2$ reduction takes place before nitrification, and that reduction of Mn-oxides takes place before the reduction of Fe-oxides, and so on. The diagram can be used to reconstruct the pE of palaeo-environment. For instance, if in a palaeo-environment, Mn exists in Mn (II) form, the palaeo-environment had a pE of +10.

The sequence of redox reactions takes place in the order of decreasing energy. When oxygenated water (i.e. exposed to the atmosphere) comes into contact with an aquifer rich in organic matter (i.e. which is under reducing conditions), the oxygenated water will lose its oxygen content first, and then its nitrate content. Then sulfate will be reduced and finally methane will appear. On the other hand, if the organic-rich leachate from a landfill enters an oxic aquifer or a body of freshwater exposed to the atmosphere, Mn-oxides in the leachate would first be reduced, followed by Fe-oxides. Then sulfate is reduced and methane forms at the end of the cycle.

Sulfate reduction involves both chemical and microbial reactions. Considerations of decreasing energy for the sequence of processes hold good for microbial mediated reactions as well. For instance, *Desulphovibrio*, which is a sulfate-reducing anaerobic bacteria, will become active only after the anoxic conditions are reached.

pE changes with depth, and so do the ionic species that are stable at a particular depth or pE. Redox zoning manifests itself in the form of changes in the chemical composition of groundwater with depth (Berner 1981). In the case of $O_2$, $NO_3^-$ and $SO_4^{2-}$, the redox zoning shows up in the form of the disappearance of a reactant. In other cases (such as, $Mn^{2+}$, $Fe^{2+}$, $H_2S$ and $CH_4$), it manifests itself in the form of appearance of a reaction product.

Berner (1981) used the amount of dissolved $O_2$ as a criterion to distinguish between *oxic* and *anoxic* environments – oxic if the dissolved $O_2$ is more than $10^{-6} M$, and anoxic, if it is less than $10^{-6} M$. The anoxic environments are further subdivided into *post-oxic* (corresponding to the reduction of nitrate, Mn-oxide and Fe-oxide), *sulfidic* (corresponding to sulfate reduction) and *methanic* (corresponding to the formation of methane). The characteristic mineral phases pertaining to the different redox zones are given in Figure 2.13.

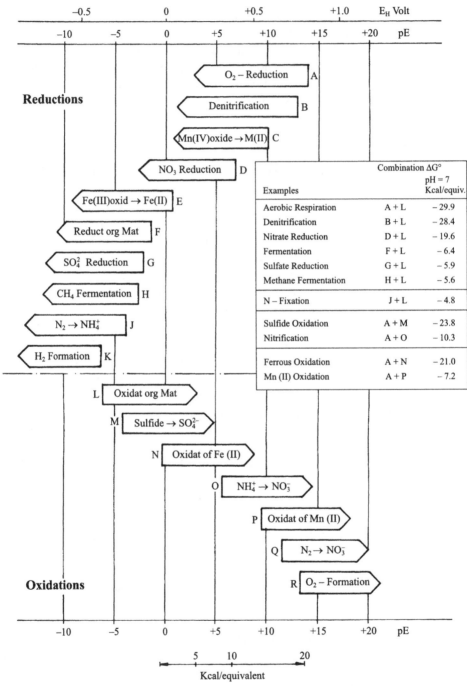

Figure 2.13. Sequence of important redox processes at pH 7 in natural systems (source: Stumm & Morgan 1981).

In the case of groundwater environments, a further division of the post-oxic zone is useful (Appelo & Postma 1996, p. 259):

*nitric* zone $\quad(m_{NO_3}^-$ $(10^{-6}\,M;\ m_{Fe^{2+}} \leq 10^{-6}\,M)$

*ferrous* zone $(m_{NO_3}^-$ $(10^{-6}\,M;\ m_{Fe^{2+}} \geq 10^{-6}\,M)$

The groundwater hardly ever passes the post-oxic state. Generally, all the redox sequences may not be present. The sulfidic zone may follow the oxic zone directly, with the post-oxic zone being present marginally.

### 2.4.3 *Ion exchange and sorption*

During the process of movement of water, soils and aquifers sorb chemicals present in the water. Sorption includes:
– Adsorption, whereby a chemical adheres to the surface of the solid,
– Absorption, whereby a chemical gets incorporated into the solid, and
– Ion exchange, which involves the replacement of one chemical by another one at the solid surface.

The different sorption processes are shown pictorially in Figure 2.14 (source: Appelo & Postma 1996, p. 143).

Sorption and exchange reactions are not open-ended – they are limited by the sorption and exchange capacity of the solid for a particular ion. For instance, clay minerals, organic matter, and oxides/hydroxides have all a specified exchange capacities. The transport of pollutant chemicals in aquifers and soils is controlled by the operation of the sorption and ion exchange processes, and hence their great importance in environmental studies.

The relevance of cation exchange is illustrated with an application, namely monitoring of saltwater intrusion in coastal aquifers.

As is well known, $Ca^{2+}$ and $HCO_3^-$ dominate the coastal waters, whereas seawater is dominated by $Na^+$ and $Cl^-$. When seawater intrudes into a coastal aquifer, the following exchange of cations takes place:

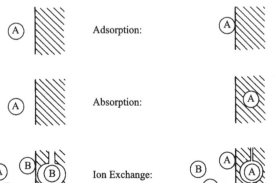

Figure 2.14. Pictorial representation of different sorption processes (source: Appelo & Postma 1996, p. 143).

$$Na^+ + \tfrac{1}{2} Ca\text{-}X_2 \rightarrow Na\text{-}X + \tfrac{1}{2} Ca^{2+} \qquad (2.2)$$

where $X$ indicates the soil changer.

Thus, sodium is taken up by the exchanger, and $Ca^{2+}$ is released. The water quality changes from NaCl type to $CaCl_2$ type.

On the other hand, when a salinized aquifer is ameliorated by flushing it with freshwater, the following reaction is involved:

$$\tfrac{1}{2} Ca^{2+} + Na\text{-}X \rightarrow \tfrac{1}{2} Ca\text{-}X_2 + Na^+ \qquad (2.3)$$

In this reaction, $Ca^{2+}$ is taken up from water in exchange for $Na^+$, thus resulting in $NAHCO_3$ type water.

Netherlands provides a case study of the successful implementation of the second process. A few decades ago, large quantities of freshwater was pumped out from the coastal dunes for purposes of drinking. This led to incursion of salt water which was threatening to salinize the freshwater. To reverse the situation, large quantities of freshwater from the Rhine river was infiltrated through numerous canals. The saltwater was flushed backwards, thus freshening the groundwater and restoring it to $NAHCO_3$ type. Figure 2.15 (source: Stuyfzand 1985) gives a section through the Dutch coastal dunes near Castricum, showing how brackish water was 'freshened'.

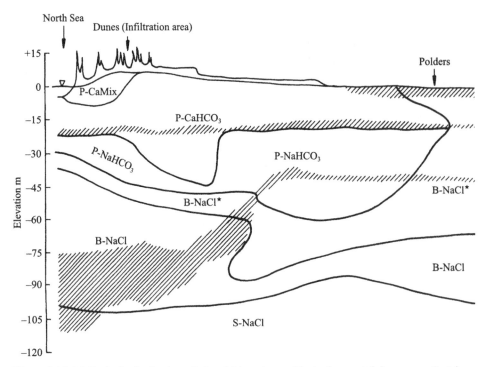

Figure 2.15. Method of refreshening of a brackish water aquifer in the coastal dunes near Castricum, Holland (source: Stuyfzand 1985). The prefixes P, S and B stand for fresh, salt and brackish water.

Use of the Stiff diagrams to understand the hydrogeochemical processes is dealt with in Section 3.2.4.

## 2.5 CHEMICAL AND ISOTOPIC TRACING

Josephus (AD 37-100) recorded an experiment made by Philip, tetrach of Trachonitis, about 2000 years ago to trace the source of River Jordan (of late, Banias Spring has been in the news in the context of the water resources of the Golan Heights that Israel is due to return to Syria). Philip was convinced Ram crater lake was the feeder of the Banias spring which was believed to be one of the sources of the Jordan River. He threw chaff into the crater lake, and it reportedly came to the surface at Banias. Modern studies involving Cl, $\delta D$ and $\delta^{18}O$ conclusively showed that Ram lake water which has a high Cl content of 19 mg $l^{-1}$ and is isotopically heavier (due to evaporation), cannot possibly be the source of the Banias and other springs, which have an average Cl content of 9.5 mg $l^{-1}$ and are of isotopically light composition (attributable to their origin from the precipitation on Mount Hermon) (Fig. 2.16; source: Mazor 1976).

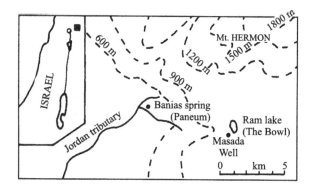

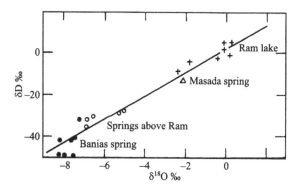

Figure 2.16. Location, and $\delta D$ and $\delta^{18}O$ of Ram lake and Banias Spring in Israel (source: Mazor 1976).

King Philip could not have realized that chaff cannot move along with ground-water through pores in subsurface rocks. Though the experiment itself is a failure, it is still laudable as it is the forerunner of the modern day studies involving dyes, salts, and radioactive tracers (such as, [32]P).

Mazor (1976) gave a comprehensive account of the various tracers used in hydrological surveys: dissolved ions, stable isotopes of hydrogen and oxygen, tritium and carbon-14 measurements, radium-226 and radon-222, radiogenic helium-4 and argon-40, and the atmospheric noble gases. A given technique may not be able to provide an unequivocal answer, but if a combination of techniques lead to the same answer, greater reliance can be placed on such an answer.

Eriksson (1976) found that chloride in the groundwater in the Delhi region in India was derived from airborne sources (dust and rainfall). He estimated the recharge rates of groundwater in the region on the basis of the harmonic mean of the chloride concentrations in the groundwater, and rainfall rates and rate of chloride deposition from the atmosphere. He came up with a groundwater recharge rate of 3-4 cm $yr^{-1}$, which is less than 10% of the annual precipitation.

On the analogy of *critical loads* concept in regard to acid rain, Leibundgut (1998) sought to quantify vulnerability of the headwaters in the mountainous regions in terms of *hydroecological capacity* (vide Fig. 2.17; source: Leibundgt 1998). Artificial tracers are useful in quantifying parameters which have a bearing on the protection and management of water resources in the mountain environments. To trace the fast flow path in springs, naphtionat (a fluorescent dye) was injected into the unsaturated zone of active talus. The data was plotted in a graph, concentration of the tracer (mg $l^{-1}$) versus time since injection (in hours). A very quick response was measured 88 m down slope, resulting in a short residence time of 10 hrs. Springs originating from debris covered slopes are characterized by direct runoff components and high fluctuations in discharge, whereas the yield was generally constant in the case of springs originating from deep fissured aquifers.

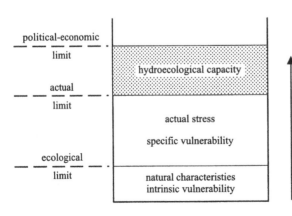

Figure 2.17. Hydroecological capacity of headwaters in the mountainous regions (source: Leibundgt 1998).

### 2.5.1 *Stable isotopes*

The isotopes most commonly used in the hydrological studies are stable isotopes. Light elements with mass numbers of less than 40 (such as H, C, O and S) undergo fractionation due to physical, chemical and biological factors.

The extent of fractionation in stable isotopes is quite small, and it is customary to express the enrichment/depletion of the heavier isotopic pair in relation to the standard pair, in parts per mille, i.e. parts per thousand ‰ (Table 2.1).

For instance, $\delta^{13}C$ (‰) is calculated as follows:

$$\delta^{13}C\ (‰) = \frac{\left(^{13}C/^{12}C\right)_{sample}}{\left(^{13}C/^{12}C\right)_{standard}} \times 1000 \tag{2.4}$$

Oxygen and hydrogen isotope studies have been extensively used for a variety of hydrological investigations:
- Characterization of water masses, and the origin of surface and subsurface water, e.g. salinity of groundwater in some regions,
- Interconnection of surface and subsurface water bodies, e.g. source of groundwater from possible aquifers,
- Estimation of leakages from reservoirs, and the contribution of reservoir water to groundwater recharge,
- Tracing of pollutants in an aqueous system, and
- Water balance studies in lakes, etc.

Industrial pollutants and their biological modifications can be traced using the S, C, O and H isotopes (Aswathanarayana 1986, p. 325).

### *Hydrogen and oxygen isotopes*

In natural waters, there are approximately 2000 molecules of $^{1}H_2{}^{18}O$ and 320 molecules of $^{2}H^{1}H^{16}O$ among $10^6$ molecules of $^{1}H_2{}^{16}O$. Isotopic exchange of water with rock brings about isotopic fractionation.

The concentration of hydrogen and oxygen isotopes in natural waters are given in Table 2.2 (source: Moser & Stichler 1980).

When the precipitation seeps into the ground, the $^{2}H$ and $^{18}O$ contents of the waters remain largely unchanged (with minor changes in the case of thermal wa-

Table 2.1. Fractionation in stable isotope pairs.

| Element | Isotope pair | Fractionation | Standard |
|---|---|---|---|
| H | $^{1}H$, $^{2}H$ (D) | $\delta^{2}H$ (D) | SMOW |
| C | $^{12}C$, $^{13}C$ | $\delta^{13}C$ | PDB |
| O | $^{16}O$, $^{18}O$ | $\delta^{18}O$ | SMOW, PDB |
| S | $^{32}S$, $^{34}S$ | $\delta^{34}S$ | Troilite |

SMOW = Standard Mean Oceanic Water; PDB = Pee Dee Formation, Belemnite; Troilite: FeS phase of the Canyon Diablo meteorite

ters). Thus investigations employing these isotopes could be made use of to trace the paths of the groundwater. Figure 2.18 (source: Moser & Stichler 1980) shows how the $^2$H and $^{18}$O contents of the waters get modified due to evaporation, condensation and exchange processes. Since SMOW is the standard, $\delta^2$H (‰) and $\delta^{18}$O (‰) of seawater is zero. The 'meteoric waters' line is the locus for meteoric waters throughout the world. The line corresponds to the equation:

$$\delta^2H\,(D) = 8\,\delta^{18}O + d \qquad\qquad (2.5)$$

where $d$ is usually +10 in most cases worldwide, but it could vary to a limited extent.

The deviation of terrestrial waters from the meteoric water, and the extent of exchange between oxygen in water and oxygen in the rock are studied using the $\delta$D versus $\delta^{18}$O plot. Figure 2.19 (source: Clayton et al. 1966) depicts the isotopic composition of the brines in the sedimentary basins of North America. That the brines in all the basins must have been derived from a common single source, namely rainwater, is evident from the fact that the data lines for each basin can be extrapolated to the meteoric water line. Hydrogen is not a common constituent of rock forming minerals (barring some clay minerals and hydroxides), and therefore

Table 2.2. Hydrogen and oxygen isotopes.

| Isotope | Water molecule | Isotope ratio | Parts per million | Range of concentration |
|---------|---------------|---------------|-------------------|------------------------|
| $^2$H | $^1H^2H^{16}O$ | $^2H/^1H$ | 90- 170 | –430 to +75 (‰) |
| $^{18}$O | $^1H_2^{18}O$ | $^{18}O/^{16}O$ | 1880-2010 | – 55 to +10 (‰) |

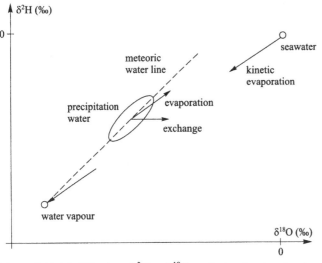

Figure 2.18. Modification of $^2$H and $^{18}$O contents of waters due to various processes (Moser & Stichler 1980).

the shift in δD as a consequence of interaction of rainwater with rocks is either nil or minimal. On the other hand, oxygen is a common constituent of rock-forming minerals. Therefore a significant shift in $\delta^{18}O$ can be expected as a consequence of interaction between the rainwater and the silicates. The chloride in the brines must have originated by the dissolution of salts contained in the sediments or rocks. The Gulf Coast samples are indistinguishable from connate seawater which did not undergo much evaporation. On the other hand, the brines in the igneous rocks in the Canadian Shield seem to have acquired their $Ca^{2+}$ content as a consequence of interaction with anorthite (Ca- feldspar) in the rocks.

Fritz et al. (1976) found that leachwater leaving a solid waste disposal is enriched in deuterium and oxygen-18 relative to the surrounding groundwater ($\delta^{18}O$ in unpolluted groundwater and leachwater samples collected from wells are −8.5 ± 0.2‰ and −6.4‰ respectively). They outlined the pollution plume on the basis of the $\delta^{18}O$ data and used the information to develop control measures (Fig. 2.20; source: Fritz et al 1976).

*Sulfur isotopes*
Sulfur isotopes can be made use of to trace the process of sulfate reduction. Fresh waters are generally low in sulfate. Sulfate may get added on to fresh waters

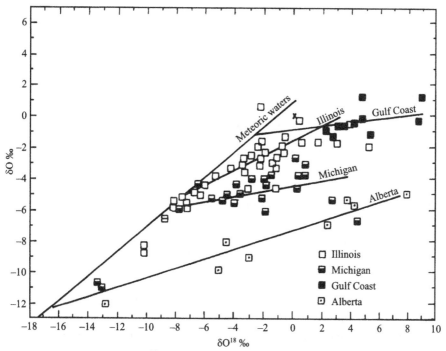

Figure 2.19. Plot of δ D and $\delta^{18}O$ of saline formation waters in relation to meteoric water (source: Clayton et al. 1966).

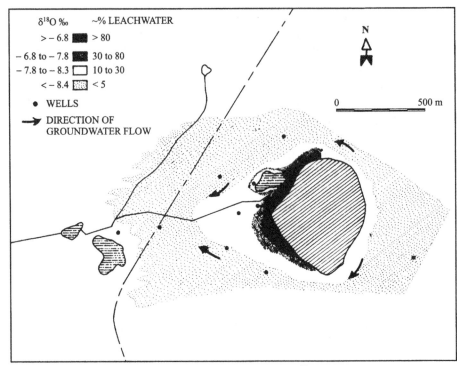

Figure 2.20. Outlining of the pollution plume on the basis of $\delta^{18}O$ data (source: Fritz et al. 1976).

through the dissolution of gypsum, oxidation of pyrite, mixing of freshwater with seawater, or due to acid rain and fertilizers. Organic matter (symbolized by $CH_2O$) can bring about the reduction of sulfate to sulfide, through the bacterial mediated catalysis by the bacteria, *Desulfovibrio*.

$$2\,CH_2O + SO_4^{2-} \rightarrow 2\,HCO_3^- + H_2S \tag{2.6}$$

$H_2S$ in drinking water is undesirable not only because of its foul smell of rotten eggs but also because it is toxic.

Sulfur isotopes can be made use of to trace the process of sulfate reduction. During the process of sulfate reduction, $^{32}S$ is preferentially consumed compared with $^{34}S$.

$$\delta^{34}S\,(\permil) = \frac{^{34}S/^{32}S_{sample} - {}^{34}S/^{32}S_{standard}}{^{34}S/^{32}S_{standard}} \times 1000 \tag{2.7}$$

It may be noted that sulfate reduction can only take place in the anoxic zone.

Atmospheric $^{34}S$ values are less than $+10\permil$. Increase in $\delta^{34}S$ values and along with decreasing sulfate concentrations are indicators of sulfate reduction.

Robertson et al. (1989) provide a good case study of sulfate reduction in a sandy aquifer, by plotting the variation with depth and age (tritium dates) of

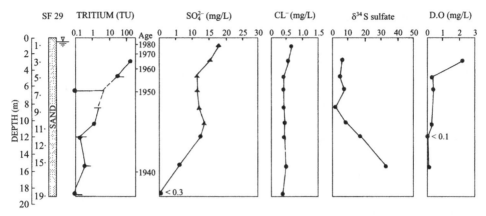

Figure 2.21. Temporal variation of $\delta^{34}S$ with depth (source: Robertson et al. 1989).

$SO_4^{2-}$ (mg l$^{-1}$), Cl (mg l$^{-1}$), $\delta^{34}S$ (‰), and Dissolved Oxygen (mg l$^{-1}$) (Fig. 2.21). The sulfate reduction is manifested by increase in $\delta^{34}S$ and decrease in $SO_4^{2-}$ at lower depths. There is hardly any change in Cl. The availability of tritium dates (1980 to 1940) allows the quantification of the rate of sulfate reduction.

### 2.5.2 *Radiogenic isotopes*

Radiogenic isotopes have been made use of to estimate the rate of some processes of hydrogeological relevance. For instance, the magnitude of $^{87}Sr/^{86}Sr$ ratio is made use of to trace the interaction between the atmospheric deposition and the bedrock. The method is based on the principle that the contribution due to weathering can be estimated from the difference between $^{87}Sr/^{86}Sr$ in deposition ad run-off, assuming isotopic equilibration between the soil solution, exchange complex and runoff (Åberg et al. 1989, Jacks et al. 1989).

The weathering rate is calculated from the following equation:

$$R_w = P_{bc} \frac{q-a}{s-q} \qquad (2.8)$$

where $R_w$ = the base cation weathering rate, $P_{bc}$ = the base cation deposition rate, $a$ = the $^{87}Sr/^{86}Sr$ ratio in the deposition, $q$ = the $^{87}Sr/^{86}Sr$ ratio in the runoff, and $s$ = the $^{87}Sr/^{86}Sr$ ratio in the mineral matrix.

### 2.5.3 *Cosmic-ray produced isotopes*

#### *Tritium ($^3H$) tracing*
Tritium ($^3H$) is produced naturally by cosmic-rays or artificially through nuclear weapons testing and nuclear industry. It has a half-life of 12.3 years. It gets incorporated in rainwater as HTO. It can be used to determine the age of groundwaters.

Tritium profiles can be used to model the flow conditions and trace the lateral heterogeneities.

The natural tritium content of pre-bomb rainwater was 5-10 TU. During the period of bomb testing, 1954-1963, tritium content of rain went upto 800 TU in the northern hemisphere, and about 60 TU in the southern hemisphere. After the cessation of the bomb testing in the atmosphere, the TU content of rainwater came down to a few tens. The presence of natural and man-made tritium allows to distinguish between two categories:

1. Pre-bomb, low TU water older than 20 years, and
2. Post-bomb high TU water younger than 20 years and then determine the age accordingly.

Tritium content can also be used as a tracer. The contribution of a dam or a lake to an adjacent well can be evaluated if there are high $T$ values in the nearby wells and low values in the wells farther away.

*Carbon-14*

Carbon-14 is a cosmic-ray produced isotope of carbon. It enters precipitation as $^{14}CO_2$. Carbon-14 is a beta-emitter with a half-life of 5730 years. Groundwater dating by radiocarbon is based on the dissolution of fossil soil carbonate by contemporary biogenic carbon dioxide during the process of groundwater generation.

The usual practice is to combine Tritium and carbon-14 measurements in respect of three categories of ages:

1. Ages younger than 30 years – indicating local recharge or karstic movement,
2. Older than 30 years, but younger than 30,000 years – this includes all hydrologically active waters, and
3. Older than 30,000 years – indicating trapped static water.

Carbon-14 could also be used as a regular tracer. Recharge from nearly open water bodies is manifested in post-bomb high $^{14}C$ contents (upto 120 pmc – percent modern carbon).

## 2.6 THERMODYNAMICS OF HYDROGEOCHEMICAL PROCESSES

Thermodynamics deals with the energetics of systems. Energy considerations determine whether a given reaction is possible and will proceed at what rate. There are excellent texts on basic thermodynamics (Lewis & Randall 1961) and their application to aqueous systems (Nordstrom & Munoz 1986, Appelo & Postma 1996).

A general reversible reaction can be written as follows:

$$aA + bB \leftrightarrow cC + Dd \tag{2.9}$$

For this equation, we may write:

$$\Delta G_r = \Delta G_r^0 + RT \ln \frac{(C)^c (D)^d}{(A)^a (B)^b} \tag{2.10}$$

where $\Delta G_r$ = the change in Gibbs free energy (kJ mol$^{-1}$) of the reaction. The standard Gibbs free energy of the reaction, $G_r^0$, becomes equal to $\Delta G_r$ when each product or reactant is present at unit activity and at a standard state (25°C, and 1 atm. pressure). [$i$] denotes the activity of $i$, $R$ is the gas constant (8.314 × 10$^{-3}$ kJ mol$^{-1}$ deg$^{-1}$) and $T$ is the absolute temperature in Kelvin (°C + 273.15).

The change in Gibbs Free Energy, $\Delta G_r$, would determine how the reaction will proceed.

If $\Delta G_r > 0$, the reaction proceeds to the left
If $\Delta G_r = 0$, the reaction is at equilibrium
If $\Delta G_r < 0$, the reaction proceeds to the right

$G_f^0$ is the free energy of formation i.e. energy needed to produce one mole of a substance from pure elements in the most stable form. $G_f^0$ values are available in the literature for various mineral water systems (e.g. Cox et al. 1989). $G_r^0$ values are calculated from the $G_f^0$ values in the following way.

$$G_r^0 = \Sigma \Delta G_f^0 \text{ products} - \Sigma \Delta G_f^0 \text{ reactants} \qquad (2.11)$$

## 2.7 ROLE OF MICROORGANISMS IN HYDROCHEMICAL INTERACTIONS

The role of microorganisms in the secondary processes have been dealt with by Krumbein (1983) and Zehnder (1988).

The pE – pH diagram (Sposito 1989, p. 107) shows the domain in which the soil microorganisms function, and the electron and proton activity levels commonly observed in the soils. At pH 7, oxic soils have pE > +7, suboxic soils have pE in the range of +2 to +7, and anoxic soils have pE of less than +2. pE and pH have an inverse relationship. If pE increases, pH must decrease. This inverse relationship between pE and pH is manifested by the slanting lines separating the domains of oxic, suboxic and anoxic soils. Thus, a higher pE is needed for oxic conditions in acid soils than in alkaline soils.

C, N, O, S, Mn, and Fe are the elements most affected by soil redox reactions. When a soil is contaminated by anthropogenic activities, other elements, such as As, Se, Cr, Hg, and Pb, enter the picture. There are well-defined pE limits for the functioning of microorganisms in the soil. For instance, aerobic microorganisms do not function below pE 5. Denitrifying bacteria function in the pE range of +10 to 0. Sulfate-reducing bacteria cannot function at pE values above +2, and so on. Thus, pE – pH diagrams help us to understand the stability conditions of chemical and microbial species in the soil.

Soil pH could cause direct toxicity for plant roots (for instance, the Al, Fe, and Mn toxicity in acid soils causes the loss of root membrane integrity), and to microorganisms. It indirectly affects bioavailability as protonation could enhance the production of free metal cations in the soil solution.

pE has both *direct* and *indirect* effects on the bioavailability of plant nutrient elements.

The mineralization of soil organic matter and the precipitation of hydroxy solids constitute the most important direct results of increasing soil pE. This process enhances the concentration of free ion species of N, P and S in the soil solution, and increases their bioavailability. At the same time, the process leads to the immobilization of Fe and Mn, thus decreasing their bioavailability. In the pE range of 3-9 (which corresponds to the transition interval for oxygen-nitrogen respiration by soil microorganisms), the bioavailability of nitrogen is profoundly influenced by the microbially catalyzed processes of nitrification, nitrogen fixation, and ammonia volatilization.

Adsorption phenomena constitute the most significant indirect effects of decreasing soil pE (Sposito 1989, p. 258). Under suboxic to anoxic conditions (pE < 3), the Fe and Mn hydrous oxide adsorbents get destabilized, thus facilitating the increase in the free-ion concentrations of anionic nutrients. In general, low pE enhances the bioavailability of anionic nutrient elements (particularly, B, inorganic C, P, and Mo). It does not seem to affect the micronutrient metals.

Nitrogen plays an important role in biogeochemical cycles. Nitrogen is found in the atmosphere, in the organic matter and as dissolved species in water. It is rarely found as minerals in nature, as nitrates are highly soluble. Bacteria are involved in nitrification and denitrification. The process of nitrification involves the bacterial oxidation of amines from organic matter to nitrite and nitrate. Denitrification involves the reduction of nitrate to $N_2$. Bacteria bring about the denitrification through a complicated pathway involving intermediates like nitrite. It should, however, be pointed out that bacteria cannot oxidize $N_2$ to $NO_3^-$. The reduction of nitrate to $NH_4^+$ also occurs in nature.

Bacterial-mediated catalysis has a profound influence in the rates of some reactions. For instance, the iron oxidizing bacteria, *Thiobacillus ferrooxidans*, increases the $Fe^{2+}$ oxidation rate by up to five orders of magnitude. The sulfide in pyrite is oxidized by *Thiobacillus denitrificans*, while $Fe^{2+}$ is subsequently oxidized by nitrate with *Gallionella ferruginea*.

Bacteria of the genus, *Desulfovibrio*, catalyses the reduction of sulfate.

### 2.7.1 *Val Roseg Project – A case study of an alpine stream ecosystem*

Alpine stream ecosystems are known to be sensitive indicators of anthropogenic impacts and climate change. Ward et al. (1998) made a holistic ecological study of Val Roseg which is situated in the Bernina Massif of the Swiss Alps (Fig. 2.22; source: Ward et al. 1998, p. 426). The flood plain is dominated by glacial meltwater during summer and groundwater-controlled system during winter. Several parameters (turbidity, temperature, nutrients, organic matter, surface water – groundwater interactions, etc.) were measured at close intervals of about 200 m, and interpreted in terms of habitat heterogeneity.

The temporal and spatial variability of the benthic algal biomass is directly related to the seasonal changes in the environmental conditions. During the summer high flow (6-13 $m^3 s^{-1}$), biomass increased from 0.3 mg chl. *a* $m^{-2}$ in the unstable channels below the glacier to 9.3 mg chl. *a* $m^{-2}$ in the flood plain. The trend got reversed at the end of September when the biomass decreased downstream from 138 to 11.3 mg chl. *a* $m^{-2}$. Figure 2.23 (source: Uehlinger et al. 1998, p. 424) shows the relative abundance of phytobenthic taxa in different types of channels during different parts of the year. For instance, *Hydrurus foetidus* is virtually the only species to be found in the main channel in the proglacial area, whereas diatoms and cyanobacteria dominate the algal community in the groundwater channels, though their relative abundances vary depending upon the season.

Springs are ecotones which link groundwater with surface water. Spring biota of pristine mountain areas are well suited for environmental monitoring (Fig. 2.24;

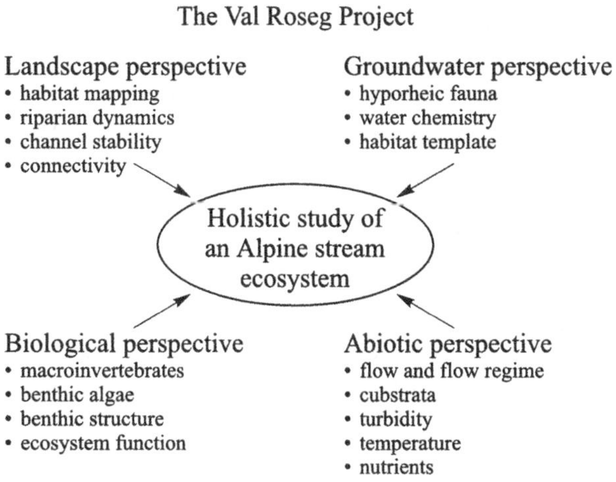

The Val Roseg Project
Landscape perspective
• habitat mapping
• riparian dynamics
• channel stability
• connectivity
Groundwater perspective
• hyporheic fauna
• water chemistry
• habitat template
Holistic study of an Alpine stream ecosystem
Biological perspective
• macroinvertebrates
• benthic algae
• benthic structure
• ecosystem function
Abiotic perspective
• flow and flow regime
• cubstrata
• turbidity
• temperature
• nutrients


Figure 2.22. A holistic ecological study of Val Roseg, Swiss Alps (source: Ward et al. 1998, p. 426).

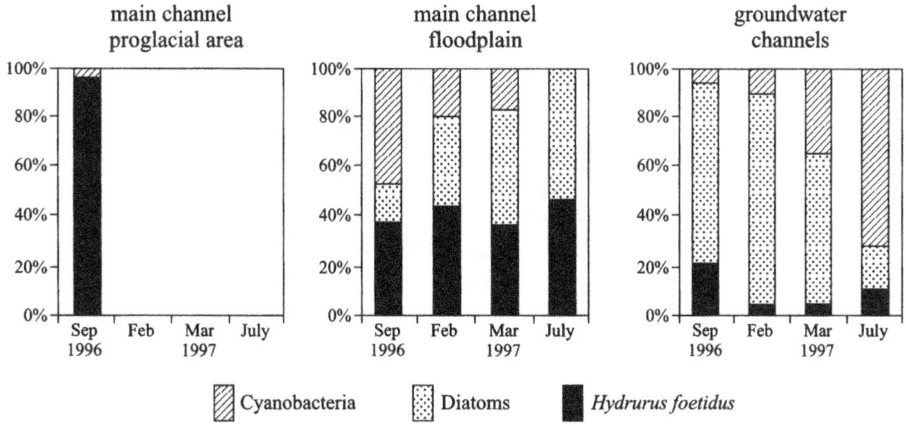

main channel proglacial area
main channel floodplain
groundwater channels
100%
80%
60%
40%
20%
0%
Sep 1996
Feb
Mar 1997
July
Cyanobacteria
Diatoms
Hydrurus foetidus


Figure 2.23. Relative abundance of phytobenthic taxa in different types of channels in different seasons (Uehlinger et al. 1998, p. 424).

source: Cantonati & Ortler 1998, p. 382). The authors recommend the use of water mites for checking the persistence of stable conditions, and diatoms for monitoring acidification and eutrophication.

## 2.8 HYDROCHEMISTRY AND FRESHWATER ECOSYSTEMS

### 2.8.1  *Uptake of ions by vegetation*

Likens et al. (1977) made a comparative study of the yearly fluxes (in kg ha$^{-1}$ yr$^{-1}$) of different elements in rain, streamwater output, biomass, etc. in the Hubbard Brook ecosystem in USA (Table 2.3; source: Likens et al. 1977).

The data in the above table clearly shows that barring Cl which has no role in the biological processes, the flux of the elements within the biomass far exceeds those that enter the soil by precipitation or leave the soil through the stream water output.

The ratio of the flux of vegetation uptake to the flux of the stream water output is as follows: Ca: 62/13.7 = 4.5; Mg: 9/3.1 = 2.9; Na: 35/7.2 = 4.9; K: 64/1.9 = 33.7; N: 80/3.9 = 20.5; S: 25/17.6 = 1.4; P: 9/0.01 = 900. Thus, the more closely an element is involved in the biological processes (e.g. P, K, N), the more heavily

Table 2.3. Yearly fluxes in the Hubbard Brook ecosystem (USA).

|  | Ca | Mg | Na | K | N | S | P | Cl |
|---|---|---|---|---|---|---|---|---|
| Bulk precipitation output | 2.2 | 0.6 | 1.6 | 0.9 | 6.5 | 12.7 | 0.04 | 6.2 |
| Stream water output | 13.7 | 3.1 | 7.2 | 1.9 | 3.9 | 17.6 | 0.01 | 4.6 |
| Vegetation uptake | 62 | 9 | 35 | 64 | 80 | 25 | 9 | small |
| Root exudate | 4 | 0.2 | 34 | 8 | 1 | 2 | 0.2 | 1.8 |
| Weathering release | 21 | 4 | 6 | 7 | 0 | 1 | ? | small |

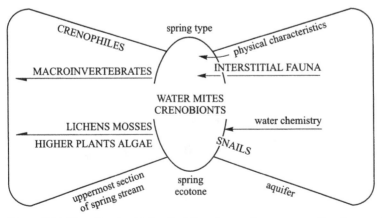

Figure 2.24. Use of spring biota of pristine streams for purposes of environmental monitoring (Cantonate & Ortler 1998, p. 382).

it is recycled in the biomass. Elements, such as Mg and S, which are less important in the biological cycle, are recycled to a lesser extent.

Vegetation is thus very effective in storing essential elements such as P, K and N. This also implies that small disturbances in the flux of the essential elements in the biomass would have profound impact on the smaller flux of the dissolved constituents leaving the biosphere.

As is well known, vegetation absorbs $CO_2$ from air in the process of photosynthesis. When vegetation absorbs gases such as $SO_2$, $NH_3$ and $NO_2$, part of the gases may be incorporated in the plant, and the remaining flushed away by rain.

### 2.8.2 *Decay of organic matter*

While the formation of biomass involves the uptake of ions, the decay of biomass involves the release of ions in the process of the oxidation of organic matter.

Though the composition of organic matter is highly complex, a good approximation of it is carbohydrate ($CH_2O$). Thus the oxidation of the organic matter could be represented by the following equation:

$$CH_2O + O_2 \rightarrow H_2O + CO_2 \tag{2.12}$$

The oxidation process may release other constituents, such as P, K, N, and S, which the organic matter may be containing. The decomposition of the organic matter in an aquifer could lead to the production of iron oxides, sulfates and nitrates and the formation of methane. The production of $CO_2$ will have a profound effect on reactions involving carbonate minerals.

### 2.8.3 *Groundwater chemistry below agricultural land*

The large-scale application of nitrogenous fertilizers to promote crop yields, has resulted in the severe nitrate contamination of soils and waters. Appelo (1985) gave a description of the changes that the chemistry of groundwater undergoes under an agricultural land (Fig. 2.25; source: Appelo 1985). Attention is drawn to the high $NO_3^-$ concentrations in the upper part of the borehole. These are obviously attributable to the heavy application of manures and fertilizers to the fields which are located upstream. Thus, in a way the fertilizer contamination may be used as a tracer for the movement of groundwater. Research is going on to find out whether the nitrate behaves as a conservative component or whether it undergoes any degradation during its movement with groundwater.

## 2.9 GEOCHEMISTRY OF WATER – ROCK INTERACTION

The chemical composition of precipitation gets modified when it comes into contact with rocks and soils. Some minerals such as carbonates and evaporates, dis-

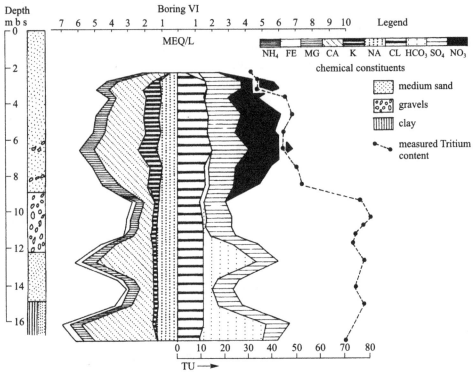

Figure 2.25. Groundwater chemistry below agricultural land (source: Appelo 1985). Concentrations are plotted cumulatively.

solve quickly, and significantly affect the geochemistry of water. The role of the carbonate reactions in determining the composition of groundwater is manifested by the significant presence of dissolved calcium and carbonate in groundwater. The relationship between total dissolved solids and concentrations of individual ions in average river water composition from different continents is given in Figure 2.26 (source: Garrels & Mackenzie 1971). Evidently, high TDS values are a reflection of high contents of $Ca^{2+}$ and $HCO_3^-$ derived from carbonate dissolution. On the other hand, silicate minerals dissolve slowly, and consequently their impact on the chemical composition of water is not that much evident. Nevertheless, it has been estimated that weathering of silicate minerals contribute about 45% of the dissolved load of river waters (Stumm & Wollast 1990). Thus, in areas where there are no limestones and dolomites, buffering of the acidification of the soil- and groundwater has to depend upon the products of silicate weathering.

A given ion, say $Na^+$, in unpolluted water may be derived not only from the minerals in the rocks of the catchment, but also from other sources (such as, atmosphere). Table 2.4 gives the normal range of concentrations of various ions and the sources from which they might have been derived (source: Appelo & Postma 1996, p. 36):

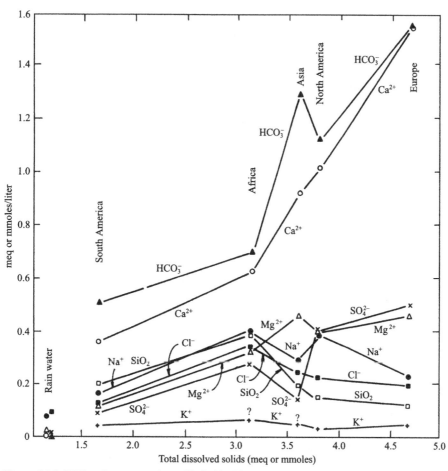

Figure 2.26. TDS and concentration of individual ions in average river water composition in different continents (source: Garrels & Mackenzie 1972).

Table 2.4. Concentrations and sources of various ions.

| Element | Concentrations, mmol $l^{-1}$ | Source |
|---|---|---|
| $Na^+$ | 0.1-2 | Feldspar, rock-salt, zeolite, Atmosphere |
| $K^+$ | 0.01-0.2 | Feldspar, mica |
| $Mg^{2+}$ | 0.05-2 | Dolomite, serpentine, pyroxene, amphibole, olivine, mica |
| $Ca^{2+}$ | 0.05-2 | Carbonate, gypsum, feldspar, pyroxene, amphibole |
| $Cl^-$ | 0.05-2 | Rock-salt, Atmosphere |
| $HCO_3^-$ | 0-5 | Carbonates, Organic matter |
| $SO_4^{2-}$ | 0.01-5 | Atmosphere, Gypsum, sulfides |
| $NO_3^-$ | 0.001-0.2 | Atmosphere, Organic matter |
| $SiO_2$ | 0.02-1 | Silicates |
| $Fe^{2+}$ | 0-0.5 | Silicates, siderite, hydroxides, sulfides |
| $PO_4^-$ Total | 0-0.02 | Organic matter, Phosphates |

The processes by which different ions are generated, and their concentrations are constrained are summarized in Table 2.5 (source: Appelo & Postma 1996, p. 41).

As should be expected, the chemical characteristics of soil water and groundwater depend upon the kind of rock type with which water has come into contact (Table 2.6).

The relationship between the rock types and the geochemistry of groundwater can be delineated using a Piper diagram. The diagram has three components:
1. A trilinear plot of cations ($Ca^{2+}$, $Mg^{2+}$, $Na^+ + K^+$),
2. A trilinear plot of anions ($HCO_3^-$, $SO_4^{2-}$, $Cl^-$), and

Table 2.5. Processes of generation and concentration limits of ions.

| Element | Process | Concentration limits determined by |
|---|---|---|
| $Na^+$ | Dissolution, Cation exchange in coastal aquifers | Kinetics of silicate weathering |
| $K^+$ | Dissolution, absorption, Decomposition | Solubility of clay minerals; Uptake by vegetation |
| $Mg^{2+}$ | Dissolution | Solubility of clay minerals |
| $Ca^{2+}$ | Dissolution | Solubility of calcite |
| $Cl^-$ | Evapotranspiration | None |
| $HCO_3^-$ | Soil $CO_2$ pressure, Weathering | Organic matter decomposition |
| $SO_4^{2-}$ | Dissolution, oxidation | Removal by reduction |
| $NO_3^-$ | Oxidation | Uptake, removal by reduction |
| Si | Dissolution, adsorption | Chert, chalcedony solubility |
| Fe | Reduction | Redox potential, $Fe^{3+}$ solubility, siderite, sulfide |
| $PO_4$ | Dissolution | Solubility of apatite, Fe, Al phosphates; Biological uptake |

Table 2.6. Rock type and the chemical characteristics of water.

| Rock type | Dominant ions | Total ion concentration | pH range | $SiO_2$ content |
|---|---|---|---|---|
| 1. Granite, rhyolite | $Na^+$, $HCO_3^-$ | Low | 6.3-7.9 | Moderate to high |
| 2. Gabbro, basalt | $Ca^{2+}$, $Mg^{2+}$ $HCO_3^-$ | Moderate | 6.7-8.5 | High |
| 3. Sandstone, arkose, graywacke | $Ca^{2+}$, $Mg^{2+}$, $Na^+$, $HCO_3^-$ | High | 5.6-9.2 | Low to moderate |
| 4. Siltstone, clay, shale | $Na^+$, $Ca^{2+}$, $Mg^{2+}$ $HCO_3^-$, $SO_4^{2-}$, $Cl^-$ | High | 4.0-8.6 | Low to moderate |
| 5. Limestone, dolomite | $Ca^{2+}$, $Mg^{2+}$, $HCO_3^-$ | High | 7.0-8.2 | Low |
| 6. Slate, schist, gneiss | $HCO_3^-$, $Ca^{2+}$, $Na^+$ | Low to moderate | 5.2-8.1 | Low |

3. A diamond diagram which combines the cation and anion data points, by drawing lines parallel to the outer boundary units until they intersect in the diamond.

Appelo & Postma (1996, p. 37) show how a Piper diagram of the European bottled waters can be made use of to identify the rock type with which the water concerned has been in contact (Fig. 2.27).

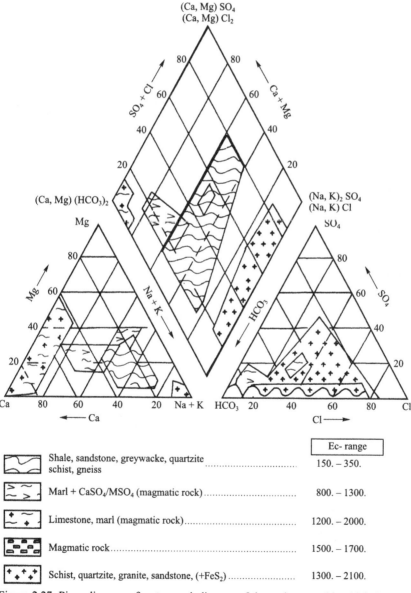

| | | Ec- range |
|---|---|---|
| | Shale, sandstone, greywacke, quartzite schist, gneiss | 150. – 350. |
| | Marl + CaSO₄/MSO₄ (magmatic rock) | 800. – 1300. |
| | Limestone, marl (magmatic rock) | 1200. – 2000. |
| | Magmatic rock | 1500. – 1700. |
| | Schist, quartzite, granite, sandstone, (+FeS₂) | 1300. – 2100. |

Figure 2.27. Piper diagram of waters as indicators of the rock type with which the water concerned has been in contact (source: Appelo & Postma 1996, p. 37).

The ways in which the precipitation interacts with rocks and soils can be examined in different ways. Walling (1980) examined the effect of the bedrock geology in a watershed on the composition of the annual runoff, by plotting the maximum Total Dissolved Solids (TDS) concentration (mg l$^{-1}$) against the mean annual runoff (mm) (Fig. 2.28). It was found that TDS does vary depending upon the rock type. In the case of limestone, volcanics, sand and gravel, TDS is virtually independent of the amount of runoff, whereas in the case of granite, shale and sandstone, schist and gneiss, the TDS sharply decreases with the runoff. These differences are attributed to the rate of dissolution of the minerals present in the parent rock.

The insoluble quartz pebbles in the gravel are highly resistant to dissolution, and nothing gets added on to water. In the case of volcanics and limestone, two kinds of factors are at work. These rocks do contain soluble minerals, and hence TDS concentrations are high, but their dissolution times are so fast (of the order of days) relative to the residence time of water in the drainage area that TDS concentrations show little change. Granites and sandstones consist principally of quartz, feldspar and mica, which dissolve very slowly. Thus the dissolution times in the case of granites and sandstones are of the same order as the residence time of water in the drainage area.

### 2.9.1   *Weathering of silicate minerals versus composition of groundwater*

The weathering of silicate minerals in igneous and metamorphic rocks is manifested in the composition of groundwater in these rocks (Fig. 2.29; source: Hem 1985;

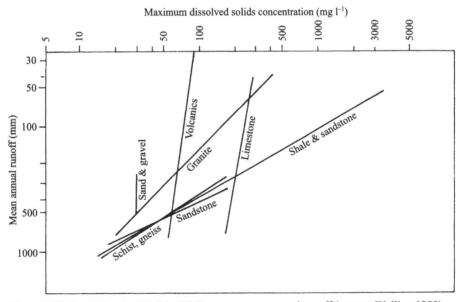

Figure 2.28. Total Dissolved Solids (TDS) versus mean annual runoff (source: Walling 1980).

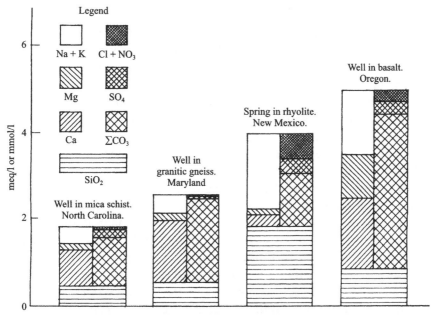

Figure 2.29. Groundwater compositions in igneous and metamorphic rocks (source: Hem 1985).

Appelo & Postma 1996, p. 204). Groundwaters invariably contain high quantities of silica, which is an indication of active degradation of silicate minerals. Highest concentrations of silica are to be found in the groundwaters associated with siliceous volcanic rocks, such as rhyolites. Sodium (with minor contribution from potassium) is present in all waters. It is derived from the weathering of soda feldspar (albite) and plagioclases (Na – Ca feldspars). That the sodium content of the groundwaters is of terrestrial origin is indicated by the fact that it is not balanced by chloride, which would have been the case if sodium was of marine origin. $Ca^{2+}$ may be derived from plagioclase feldspars, pyroxenes and amphiboles.

The total dissolved concentrations in groundwaters in silicate rocks are generally low, because of the slow dissolution rate of the silicate minerals. Further, as the movement of groundwater in silicate rocks tends to be through fractures, this limits the interaction of water with weathered silicate minerals.

Laterites and bauxites are products of deep, tropical weathering. Their formation is facilitated by a combination of high rainfall, warm temperatures, intense leaching, strongly oxidizing environment, long duration of weathering, and chemically unstable rock. Lateritisation involves the depletion of alkalis, alkaline earths and silica, and enrichment of major elements such as, Fe, Al and Ti and trace elements, such as Ni, Ga, Cr, Nb, etc. Any model for the formation of bauxites and laterites have to account for the following:

– Removal of silica: unless silica is removed, neither laterite nor bauxite could form, and
– Separation of Al and Fe.

*Removal of silica*

In water, silica may occur in the form of colloidal silica or dissolved molecules of Si $(OH)_4$ or as ions. Experimental studies show that the solubility of Si $(OH)_4$ is strongly dependent upon temperature (for instance, the solubility increases four-fold from 0°C to 73°C), and pH (at 22°C, the solubility of Si $(OH)_4$ increases sharply after pH 9). Increased dilution at constant Al/Si ratio results in the poor adsorption of silica. Thus, alkalinity, warm temperatures, and dilute solutions favor the transport and migration of silica.

Thus the type of weathering product that could develop depends not only on the rate of mineral weathering but also on the hydrological conditions. This can be illustrated with the example of weathering of albite ($NaAlSi_3O_8$) to different kinds of clays, with the release of $Na^+$ and silicic acid ($H_4SiO_4$). Montmorillonite [$Na_{0.5}Al_{1.5}Mg_{0.5}Si_4O_{10}(OH)_2$] develops from albite through desilication by 11%. It takes place in relatively dry climates, where the flushing rate of the soil is low, and where the parent is a volcanic rock containing rapidly dissolving material. Water resides in the soil for longer periods, and hence acquires greater concentrations of dissolved ions. Kaolinite [$Al_2Si_2O_5(OH)_4$] develops from albite through desilication to the extent of 67%. The formation of gibbsite [$Al(OH)_3$)] involves total removal of silica. Gibbsite typically develops in tropical areas with high rainfall and under well drained conditions. As the residence time of water in the soil is short, the content of dissolved ions in water tends to be low. The sequence, montmorillonite – kaolinite – gibbsite, thus represents increasing intensities of leaching and desilication (Fig. 2.30; source: Berner 1971).

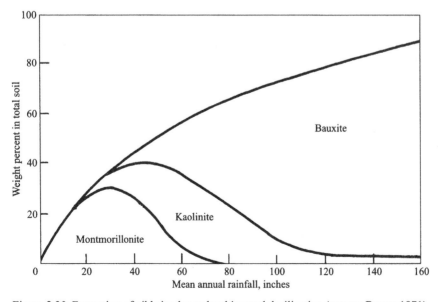

Figure 2.30. Formation of gibbsite due to leaching and desilication (source: Berner 1971).

*Separation of silica and alumina*

Under conditions of acid pH (less than four), alumina is highly soluble, whereas silica is sparingly so. This mechanism cannot, however, be invoked to bring out the separation of silica and alumina, as such low pH is uncommon in nature (except under conditions of acid rain). In the pH range of 5-9, which is more common in nature, alumina is virtually insoluble (and could remain behind), whereas the solubility of silica and hence its transport increases steadily. This is a likely mechanism to bring about the separation of $SiO_2$ and $Al_2O_3$.

*Separation of iron and aluminum*

The presence of organic acids increases the solubility of Al vis-á-vis Fe. For instance, Ong et al. (1970) showed that under the same level of organic acid content (say, 16 ppm C) and pH (7-9), the solubility of Al is markedly higher than that of Fe. This mechanism can bring about the separation of Fe and Al, and could result in the formation of Fe-enriched laterites and Al-enriched bauxites. Field observations support this view – in a given area, bauxite development is more evident in the vegetated locations than in locations devoid of vegetation.

Permeability of a porous rock (say, a sandstone) is profoundly influenced by the morphology of the clay that is formed during authigenesis. The layered booklets of kaolinite have only a moderate effect on the permeability, whereas the bushy aggregates of illite tend to clog the pores and reduce the permeability significantly.

When clay minerals are produced through silicate weathering, they tend to form coatings on mineral grains. Some times, the movement of water may wash down the clay minerals from the soil into lower horizons. The relative distribution of the illite, chlorite and kaolinite in the unsaturated zone of a Quaternary sand deposit is given in Figure 2.31 (source: Ohse et al. 1984). The percentage composition of kaolinite is found to decrease from the upper layers downwards. This is because kaolinite is a product of more intense leaching than chlorite and illite, and

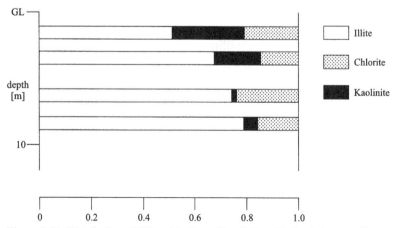

Figure 2.31. Distribution of illite, chlorite and kaolinite with depth (source: Ohse et al. 1984).

the surface layers are naturally subject to more intense leaching than the lower layers.

### 2.9.2   *Mass balance approach to silicate weathering*

The Mass Balance approach is a good way to study the impact of the dissolution and precipitation of minerals on the water chemistry through an examination of their budgets (Appelo & Postma 1996, p. 215). The reaction between mineral and water can be generalized as follows:

$$\text{Reactant phase} \rightarrow \text{weathering residue} + \text{dissolved ions} \qquad (2.13)$$

The reaction between a limestone (essentially, $CaCO_3$) and carbonic acid can be written as:

$$H_2CO_3 + CaCO_3 \rightarrow Ca^{2+} + 2\ HCO_3^- \qquad (2.14)$$

Thus, when water containing $CO_2$ percolates through a limy soil, two moles of $HCO_3^-$ are produced for each mole of $Ca^{2+}$.

It is useful to analyze the composition of stream or groundwater by reconstructing the sources of the water constituents. This exercise is easy if the watershed concerned is composed of monominerallic rocks. But this is rarely the case. Figure 2.32 gives the composition of water resulting from the alteration of different silicate minerals as a consequence of interaction with carbonic acid (source: Garrels 1976). This shows that reaction of carbonic acid with a particular silicate does not yield a unique water composition, but that $HCO_3^-$ is present in all waters, irrespective of the silicate involved.

Figure 2.33 (source: Garrels 1976) shows the 'reconstructed' waters from rhyolites and basalts shown in terms of constituent minerals. For instance, $Ca^{2+}$ in water in a basalt terrain is derived almost entirely from plagioclase feldspar and pyroxene. Let us look at the bicarbonate column in the reconstructed rhyolite. Most of the bicarbonate is derived from the weathering of the Na-feldspar, and only a trace from K-feldspar. Hornblende makes a significant contribution. The remaining bicarbonate is derived from the Ca-content of the plagioclase feldspar. The order of abundance of the minerals in rhyolite is

K-feldspar > Na-Ca feldspar >> hornblende

whereas the order of contribution of bicarbonate is

Na-Ca feldspar > hornblende >> K-feldspar

This discrepancy arises because the order of the rate of alteration is

hornblende > Na-Ca feldspar >> K-feldspar

Mass balance calculations essentially involve the solving of a set of linear equations. This can be done conveniently with the computer programs (BALANCE by Parkhurst et al. 1982) or NETPATH (Plummer et al. 1991).

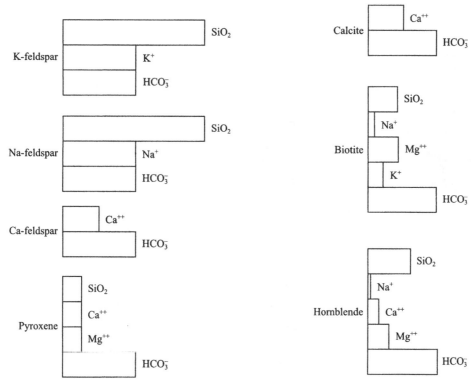

Figure 2.32. Composition of water resulting from the alteration of different silicate minerals (source: Garrels 1976).

Attention is drawn to the limitations of the mass balance calculations. Just because the two sides are chemically balanced does not necessarily mean that the reactions can actually take place as indicated by the equations, unless such reactions are demonstrably consistent thermodynamically and kinetically. Despite these limitations, the mass balance studies do indicate the possible products of reaction of water with various minerals, rocks and soils. They therefore constitute the first step in the elucidation of the hydrogeochemical processes.

### 2.9.3 *Kinetics of silicate weathering*

Way back in 1938, Goldich proposed a mineral stability series in the secondary environment, which happens to be the exact reverse of the Bowen's Reaction Series. Goldich found that olivine and calcic plagioclase (the earliest formed minerals in the discontinuous and continuous reaction series respectively) are the least stable in the weathering environment. On the other hand, quartz which is the last formed mineral, is most stable in the weathering environment. This explains why we hardly ever find olivine and anorthite in the soil.

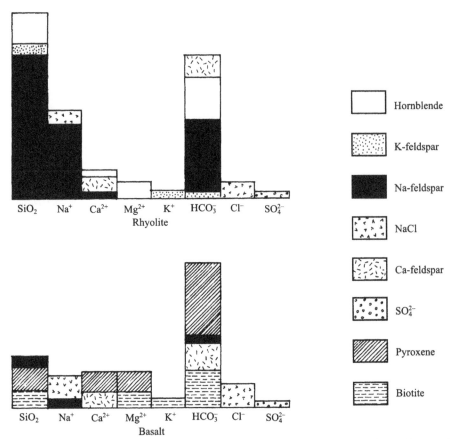

Figure 2.33. 'Reconstructed' waters from rhyolites and basalts in terms of constituent minerals (source: Garrels 1976).

Experimental studies in the dissolution kinetics of different minerals give a quantitative indication of the observations of Goldich. Lasaga (1984) calculated the mean lifetime in years of 1 mm crystal of various minerals at 25°C and pH 5 as follows:

| Quartz | = 34,000,000 | Enstatite | = 8800 |
|---|---|---|---|
| Muscovite | = 2,700,000 | Diopside | = 6800 |
| Forsterite | = 600,000 | Nepheline | = 211 |
| K-feldspar | = 520,000 | Anorthite | = 112 years |
| Albite | = 80,000 | | |

The range in lifetimes is enormous – for instance, anorthite dissolves about 300,000 times faster than quartz. Thus the laboratory dissolution experiments are in good agreement with the Goldich weathering sequence.

The dissolution rates of various minerals are strongly dependent upon pH. In most cases, the rate is minimum at neutral pH, with high rates in acid pH and relatively less so under alkali pH (Fig. 2.34; source: Sverdrup & Warfvinge 1988).

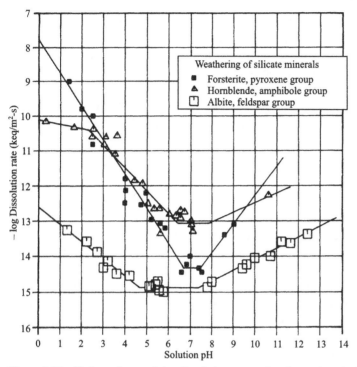

Figure 2.34. pH dependence of the dissolution rates of various minerals (source: Sverdrup & Warfvinge 1988).

Table 2.7. Hydrogeochemical zones in ground water.

| Zone | Processes at work | Geochemical characteristics |
|---|---|---|
| Upper | Active groundwater flushing through relatively well-leached rocks | Dominant anion is $HCO_3^-$; low content of Total Dissolved Solids (TDS) |
| Intermediate | Less active groundwater circulation | High TDS; Dominant anion is $SO_4^{2-}$ |
| Lower | Very sluggish groundwater flow; Characterized by the presence of highly soluble minerals | High TDS; High $Cl^-$ concentration |

## 2.10 EVOLUTION OF GROUNDWATER QUALITY IN A WET ZONE

The evolution of the groundwater quality in a wet zone is illustrated with a case history of Sri Lanka.

Domenico (1972) identified three hydrogeochemical zones in groundwater, depth-wise (see Table 2.7).

The geochemistry of groundwater in humid tropics is controlled by the following processes:

- High rainfall leading to high rates of circulation of infiltrating meteoritic water, rapid dissolution of minerals, and high concentrations of dissolved silica, and
- Very high dilution effect, and generally low levels of salts (e.g. NaCl and $CaSO_4$) in the soil by transpiration (Foster 1993).

On the basis of about 10,000 chemical analyses of well samples in Australia, Chebotarev (1955) came to the conclusion that groundwater tends to evolve chemically towards the composition of seawater. Freeze & Cherry (1979) depicted this evolution in terms of anions, as follows:

Travel along the flow path $\rightarrow$

$$HCO_3^- \rightarrow HCO_3^- + SO_4^{2-} \rightarrow SO_4^{2-} + HCO_3^- \rightarrow SO_4^{2-} + Cl^- \rightarrow SO_4^{2-} \rightarrow Cl^-$$

Increasing age $\rightarrow$

It may be noted that the above hydrogeochemical evolution of groundwater with age corresponds to the change, carbonate $\rightarrow$ sulfate $\rightarrow$ chloride with depth, as proposed by Domenico (1972).

The conceptual model of the geochemical evolution of the groundwater in wet zone in the Kandy – Puttalam area of Sri Lanka is given in Figure 2.35 (source: Song et al. 1999). The spatial and temporal distribution of groundwater quality have been found to be dependent upon climate (precipitation and evaporation), topography (slope gradient, which determines the residence time of groundwater), geology of the area (mineralogical and chemical composition of the rocks and soils with which groundwater is in contact), etc.

As the precipitation (about 2050 mm annually) exceeds potential evapotranspiration (about 1340 mm annually) for ten months in a year, the groundwater is recharged virtually all the year round. The water table generally follows the topography. As the topography is undulating, the groundwater moves fast. As the reaction time of the groundwater with rock is short, very little leaching is effected. Consequently, the groundwater has low electrical conductivity (100-450 $\mu S\ cm^{-1}$), and low concentration of major ions. Surface water stored in tanks is characterized by

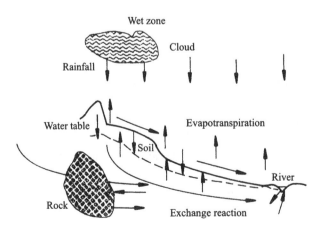

Figure 2.35. Geochemical evolution of the groundwater in a wet zone (source: Song et al. 1999).

very low electrical conductivity and low concentration of ions. Since groundwater is recharged by seepage from tanks, its major ion concentrations are similarly low. As should be expected, the groundwater is of the calcium bicarbonate type.

Rainwater and groundwater have been analyzed for $\delta D$ and $\delta^{18}O$ (Fig. 2.36; source: Song et al. 1999). The precipitation in the wet zone follows the Meteoric Water Line (MWL) with d of 10.5. The seasonal variations in $\delta^{18}O$ value are indicated in Table 2.8.

In both the basins, isotopically lighter water (with $\delta^{18}O$ values similar to that of the precipitation) occurs at the valley bottom in the rainy season. The slopes of the regression line between $\delta D$ and $\delta^{18}O$ and deuterium excess for groundwater

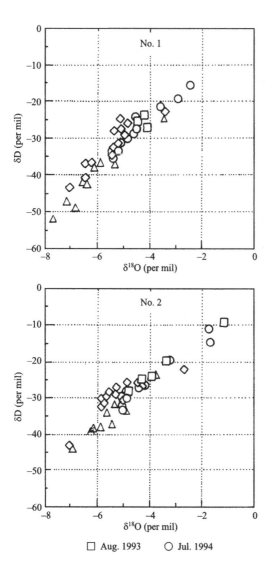

□ Aug. 1993    ○ Jul. 1994

Figure 2.36. Plot of $\delta D$ versus $\delta^{18}O$ of groundwaters in a wet zone (source: Song et al. 1999).

Table 2.8. Seasonal variation in $\delta^{18}O$ value.

| $\delta^{18}O$ value | Rainy season | Dry season |
|---|---|---|
| Basin no. 1 | –6.5 to –4‰ | –5.5 to –2.8‰ |
| Basin no. 2 | –6.5 to –3.5‰ | –5.0 to –1.0‰ |

are about 8 and 10 respectively. This clearly shows that groundwater movement is driven by precipitation during the rainy season.

## 2.11 BIOGEOCHEMCAL CYCLES

Elements are recycled within and between the various components of the geoenvironment (lithosphere, hydrosphere, atmosphere, and biosphere). The recycling processes are energized by solar radiation (ultra-violet, visible and infrared), mechanical (kinetic and potential), chemical, and thermal energy of the earth (part of which is derived from the decay of uranium, thorium and 40 – potassium). In most cases the elements released during the process of weathering, are transported and recombined in various ways. The transport mechanisms are described by Fergusson (1990, p. 147).

The toxic heavy elements (Pb, Cd, Hg, As, etc.) are introduced into the environment through the industrial processes. For each of these elements, there are natural emissions and anthropogenic emissions. The Mobilization Factor (MF) is given by: anthropogenic emissions/natural emissions.

Biological systems protect themselves from toxic elements by discriminating against them in the intake process. The Biopurification Factor (BPF) of a toxic element is estimated in comparison with a nutrient element, Ca.

$$\text{BPF of a toxic element} = \frac{\text{Toxic element/Ca}_{\text{in food}}}{\text{Toxic element/Ca}_{\text{in consumer}}} \quad (2.15)$$

The factor generally decreases further up the food chain, as the biological systems favor the nutrient element rather than the toxic element.

Fergusson (1990, p. 157) gave an account of the chemical and biochemical pathways of a toxic element (such as, mercury) in the hydrosphere.

## REFERENCES

*Suggested reading

Åberg, G., G. Jacks & P. J. Hamilton 1989. Weathering rates and $^{87}Sr$ / $^{86}Sr$ ratios: An isotopic approach. *J. Hydrol.*, 109, 65-78.

*Appelo, C.A.J. & D. Postma 1996. *Geochemistry, Groundwater and Pollution*. Rotterdam: A.A. Balkema.

Appelo, C.A.J. 1985. CAC, computer aided chemistry, or the evaluation o groundwater quality with a geochemical computer model (in Dutch). *H₂O*. 26, 557-562.

Aswathanarayana, U. 1986. *Principles of Nuclear Geology*. Rotterdam: A.A. Balkema.

Berner, R.A. 1971. *Principles of Chemical Sedimentology*. New York: McGraw-Hill.

Berner, R.A. 1981. Authigenic mineral formation resulting from organic matter decomposition in modern sediments. *Fortschr. Miner.*, 59, 1,117-135.

Brookins, D.G. 1988. *Eh – pH diagrams for geochemistry*. Berlin: Springer-Verlag.

Bronstert, A., B. Glüsing & E. Plate 1998. Physically-based hydrological modelling on the hillslope and micro-catchment scale: examples of capabilities and limitations. term monitoring. In K. Kovar (ed.) et al., *Hydrology, Water Resources and Ecology in Headwaters*. IAHS Publ. no. 248, p. 207-215.

Cantonati, M. & K. Ortler 1998. Using spring biota of pristine mountain areas for long-term monitoring. In K. Kovar (ed.) et al., *Hydrology, Water Resources and Ecology in Headwaters*. IAHS Publ. no. 248, p. 379-385.

Chebotarev, I.I. 1955. Metamorphism of natural waters in the crust of weathering. *Geochim. Cosmochim. Acta*, 8, 22-48, 137-170, 198-212.

Clayton, R.N. et al. 1966. The origin of the saline formation waters. 1. Isotopic composition. *J. Geophy. Res.*, 71, 3869-3882.

Cox, J.D., D.D. Wagman & V.A. Medvedev 1989. *CODATA Key values for thermodynamics*. Washington, DC: Hemisphere Publ. Corp.,

Davies, S.N. & R.C.M. De Weist 1966. *Hydrogeology*. New York: Wiley.

Domenico, P.A. 1972. *Concepts and models in groundwater hydrology*. New York: McGraw-Hill.

Erickson, E. 1976. The distribution of salinity in groundwaters of the Delhi region and recharge rates of groundwater. In *Interpretation of environmental isotope and hydrogeochemical data in groundwater hydrology*. p. 171-177. Vienna: IAEA

Fergusson, J.E. 1990. *The Heavy Elements: Chemistry, Environmental Impact and Health Effects*. Oxford: Pergamon Press.

Flügel, W.-A. 1995. Hydrological Response Units (HRUs) to preserve basin heterogeneity in hydrological modelling using PRMS / MMS – case study in the Bröl basin, Germany. In S.P. Simonovic (ed.) et al., *Modelling and Management of Sustainable basin-scale water resource systems*. IAHS Publ. no. 231, p. 79-87

Foster, S.S.D. 1993. Groundwater conditions and problems characteristic of the humid tropics. In *Hydrology of warm humid regions*. IAHS Publ. no. 216, 433-449.

Freeze, R.A. & J.A. Cherry 1979. *Groundwater*. Englewood Cliffs, N.J.: Prentice-Hall.

Fritz, P., G. Matthess & R.M. Brown 1976. Deuterium and oxygen-18 as indicators of leachwater movement from a sanitary landfill. In *Interpretation of environmental isotope and hydrogeochemical data in groundwater hydrology*. p. 131-142. Vienna: IAEA

Garrels, R.M. 1976. A survey of low-temperature, water – mineral relation. In *Interpretation of environmental isotope and hydrogeochemical data in groundwater hydrology*. p. 65-84. Vienna: IAEA

Garrels, R.M. & C.L. Christ 1965. *Solutions, Minerals and Equilibria*. New York: Harper and Row.

Garrels, R.M. & F.T. Mackenzie 1971. *Evolution of Sedimentary Rocks*. Norton

Gibbs, R. 1970. Mechanisms controlling world water chemistry. *Science* 170, 1088-1090.

Hem, J.D. 1985. Study and interpretation of the chemical characteristics of natural water. 3rd. ed. US Geol. Surv. Water Supply Paper 2254.

Jacks, G., G. Åberg & P.J. Hamilton 1989. Calcium budgets for catchments as interpreted by strontium isotopes. *Nordic Hydrol.*, 20, 85-96.

Krumbein, W.E. (ed.) 1983. *Microbial Geochemistry*. Oxford: Blackwells.

Lasaga, A.C. 1984. Chemical kinetics of water – rock interactions. *J. Geophy. Res.*, 89, 4009-4025.

Leavesley, G.H., B.M. Lichty, L.G.Troutman & L.G. Saindon 1983. Precipitation-runoff modelling system: user's manual. *USGS Water Resources Investigations Report 83-4328*, Denver, Colorado, USA.

Leibundgut, C. 1998. Tracer-based assessment of vulnerability in mountainous headwaters. In K. Kovar (ed.) et al., *Hydrology, Water Resources and Ecology in Headwaters*. IAHS Publ. no. 248, p. 317-325.

Lewis, G.N. & M. Randall 1961. *Thermodynamics*. 2nd. ed. New York: McGraw-Hill.

Likens, G.E. et al. 1977. *Biogeochemistry of a forested ecosystem*. Berlin: Springer-Verlag.

Mazor, E. 1976. The Ram Crater Lake – A note on the revival of a 2000 – year old groundwater tracing experiment. In *Interpretation of environmental isotope and hydrogeochemical data in groundwater hydrology*. p. 179-181. Vienna: IAEA

Moser, H. & W. Stichler 1980. Hydrological applications of analyses of stable isotope content. In *Proc. conf. on 'Application of isotope analyses in Archaeology, Hydrology and Geology'*, Zagreb, Oct. 1979. *Fizika*, 12, Supp. 2, 9-35.

Nordstrom, D.K. & J.L. Munoz 1986. *Geochemical thermodynamics*. Oxford: Blackwells.

Ohse, M., G. Matthess, A. Pekdeger & H.D. Schultz 1984. Interaction water-silicate minerals in the unsaturated zone controlled by thermodynamic equilibria. *Int. Assn. Hydrol. Sci.*, 150, 31-40.

Ong, H.L., V.E. Swanson & R.E. Bisque 1970. Natural organic acids as agents of chemical weathering. *US Geol. Surv. Prof. Paper*, 700C: C130-C137.

Paces, T. 1976. Kinetics of natural water systems. In *Interpretation of environmental isotope and hydrogeochemical data in groundwater hydrology*. p. 85-108. Vienna: IAEA

Parkhurst, D.L., L.N. Plummer & D.C. Thorstenson 1982. BALANCE – a computer program for calculating mass transfer for geochemical reactions in groundwater. *US Geol. Surv. Water Res. Inv.*, 82-14.

Plummer, L.N., E.C. Prestemon & D.L. Parkhurst 1991. An interactive code (NETPATH) for modeling net geochemical reactions along a flow path. *US Geol. Surv. Water Res. Inv.*, 91-4078.

Robertson,W.D., J.A. Cherry & S.L. Schiff 1989. Atmospheric sulfur deposition 1950-1985 inferred from sulfate in groundwater. *Water Resour. Res.*, 25, 1111-1123.

Song, X., I. Kayane, T. Tanaka & J. Shimada 1999. Conceptual model of the evolution of the groundwater quality at the wet zone in Sri Lanka. *Env. Geol.* 39 (2), 149-164.

Sposito, G. 1989. *The Chemistry of Soils*. New York: Oxford Univ. Press.

Stumm, W. & J.L. Morgan 1981. *Aquatic Chemistry*. 2nd. ed. New York: Wiley

Stumm, W. & R. Wollast 1990. Coordination chemistry of weathering. *Rev. Geophys.*, 28, 53-69.

Stuyfzand, P.J. 1985. *Hydrochemistry and hydrology of the dune area between Egmond and Wijk aan Zee* (in Dutch), KIWA, SWE. 85.012. Nieuwegein.

Sverdrup, H.U. & P. Warfvinge 1991. On the geochemistry of chemical weathering. In K. Rosen (ed.), *Chemical weathering under field conditions*. Rep. Forest Ec. Forest Soil 63, Swedish Univ. Agri. Sci., p. 79-119.

Uehlinger, U., R. Zah & H. Bürgi 1998. The Val Roseg project: temporal and spatial patterns of benthic algae in an alpine stream ecosystem influenced by glacier runoff. In K. Kovar (ed.) et al., *Hydrology, Water Resources and Ecology in Headwaters*. IAHS Publ. no. 248, p. 419-424.

Walling, D.E. 1980. Water in the catchment ecosystem. In A.M. Gower (ed.), *Water quality in catchment ecosystems*. New York: Wiley, p. 1-48.

Ward, J.V. et al. 1998. The Val Roseg Project: habitat heterogeniety and connectivity gradients in glacial flood-plain system. In K. Kovar (ed.) et al., *Hydrology, Water Resources and Ecology in Headwaters*. IAHS Publ. no. 248, p. 425-432.

Zehnder, A.J.B. (ed.) 1988. *Biology of anaerobic microorganisms*. New York: Wiley.

CHAPTER 3

# Water quality

## 3.1 HYDROLOGICAL CYCLE AND WATER QUALITY

Water quality gets modified in the course of movement of water through the hydrological cycle, through the operation of the following processes: evaporation, transpiration, selective uptake by vegetation, oxidation/reduction, cation exchange, dissolution of minerals, precipitation of secondary minerals, mixing of waters, leaching of fertilizers and manure, pollution, and lake/sea biological processes (Appelo & Postma 1996, p. 35) (Fig. 3.1). The 'natural' water cycle and the industrial and urban development influences on it are shown in Figures 3.2 and 3.3 (source: Petts & Eduljee 1994, p. 158).

## 3.2 DIMENSIONS OF WATER QUALITY

The quality of water must have been of concern to human beings and domestic animals from time immemorial. Forts were known to have been built to guard the sources of freshwater, and wars were fought for their possession. In the apocalypse, all the surface waters are supposed to be rendered bitter. It is to be hoped that human beings would have the vision to manage the water resources wisely and avoid the apocalyptic curse.

Water quality has no universally accepted definition, for the simple reason that it is use-specific and context-specific. For instance, semi-potable water which is unfit for human consumption, can be safely given to a camel. Water that is unsuitable for food industries can be used for industrial cooling, and so on. In earth sciences, water quality is expressed in terms of a set of concentrations, speciations and partitions of organic and inorganic substances (Meybeck & Helmer 1992). In ecology, apart from the chemical composition of water, consideration is given as to what extent the natural water composition has been modified at a given station, and the impact of water chemistry on the aquatic biota. For purposes of water resources management, the definition of water quality has been greatly extended. It takes into account water chemistry, particulate matter chemistry and health of the biological communities. It includes not only considerations of water composition for multiple human uses (such as drinking, and irrigation), but also in terms of its

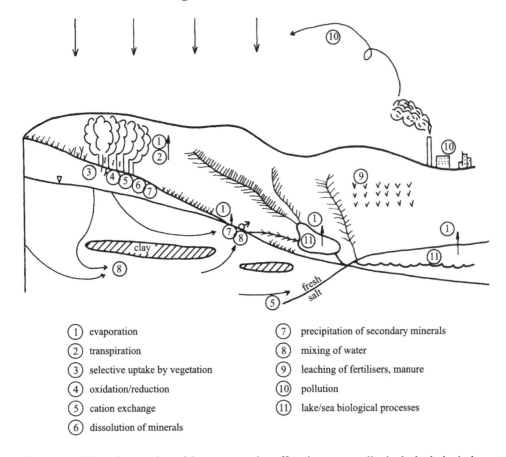

| ① | evaporation | ⑦ | precipitation of secondary minerals |
| ② | transpiration | ⑧ | mixing of water |
| ③ | selective uptake by vegetation | ⑨ | leaching of fertilisers, manure |
| ④ | oxidation/reduction | ⑩ | pollution |
| ⑤ | cation exchange | ⑪ | lake/sea biological processes |
| ⑥ | dissolution of minerals | | |

Figure 3.1. Schematic overview of the processes that affect the water quality in the hydrological cycle (source: Appelo & Postma 1996, p. 33).

capacity to sustain systems of aquatic biota in general (Meybeck 1998). This is so because we now realize that our well-being is inseparable from the well-being, say, of the aquatic biota. If frogs are dying, we would be next in the line!

### 3.2.1 *Areas of science and technology involved in water quality*

Chemical attributes of waters which have a bearing on water quality, are measured in several ways for different purposes:
- Analysis of trace compounds, and their speciation,
- Determination of rates, such as of denitrification or nutrient uptake,
- Variation in concentrations ($C_i$) in relation to discharges ($Q$), particularly during flood events,
- Temporal variation in concentrations, $C_i = f(t)$, or fluxes $F_i = f(t)$ at a given station, including abnormal variations due to extreme events, and

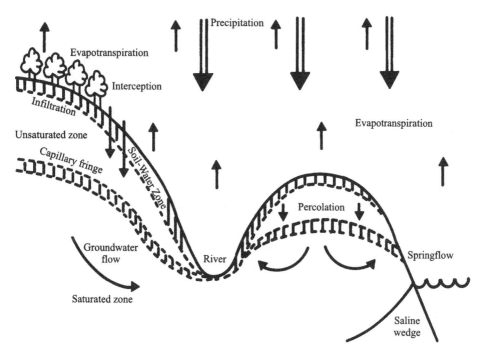

Figure 3.2. The 'Natural 'Water cycle (source: Petts & Eduljee 1994, p. 158).

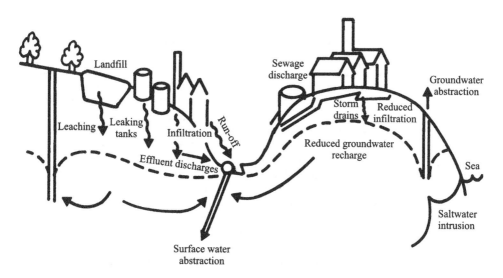

Figure 3.3. Influence of industrial and urban developments on the natural water cycle (source: Petts & Eduljee 1994, p. 158).

– Spatial variations, in terms of longitudinal variations, geological formations in the catchment, consideration of stream orders, and other forms of spatial hetereogeneities, etc. (Meybeck 1996). Table 3.1 (source: Meybeck 1996) summarizes the various areas of geosciences, biosciences and engineering which are involved in the measurement, interpretation and mitigation actions related to water quality.

The relationship between the rock types and the geochemistry of groundwater can be delineated using a Piper diagram. The diagram has three components: a trilinear plot of cations ($Ca^{2+}$, $Mg^{2+}$, $Na^+ + K^+$), a trilinear plot of anions ($HCO_3^-$, $SO_4^{2-}$, $Cl^-$).

### 3.2.2   *Evolution of water quality in the historical past*

It is possible to reconstruct the history of pollution of waters from the lake sediment cores. On the basis of such studies in Wales, UK, and in Latium, Italy, we now know that pollution of waters due to mines and cities has been occurring since Roman times. Lake sediment studies have been made use of to trace the deterioration of water quality during the nineteenth and twentieth centuries (vide review of Schmidtke 1988). The inorganic sedimentation rate in Lake Koltjarn, Sweden, has increased by more than five times during 1800-1920 due to rapid deforestation and consequent increase in the content of suspended solids in the tributary streams. Evidence from acidophilic diatoms in the lake sediments show that pH in Lake Gardsjon, Sweden, decreased drastically since 1960. Sharp increase in the contamination of sediments by

Table 3.1. Areas of science and technology involved in water quality.

| Subject area | Topic | Focus |
| --- | --- | --- |
| Geochemistry | Origins/processes | pH and redox influences; speciation of trace compounds; Dissolved vs particulates; ranges |
| Physical geography | Transport rates; chemical versus mechanical erosion | Major elements; annual fluxes; spatial distribution; transport regime |
| Hydrology | Hydrograph decomposition | Variations in chemical tracers |
| Ocean sciences | Inputs to oceans and seas | Nutrients; contaminants; organic matter; regional and global fluxes; trends |
| Biogeochemistry | Nutrient cycling; Biological interactions | C, N, P, Si uptake; organic matter decay; Fe, Mn, S cycles; organic tracers |
| Hydrobiology | Ambient quality; Extreme quality | Concentration ranges; Time/space structure |
| Ecotoxicology | Bioaccumulation; Biomagnification | Contaminants; speciation and uptake; Extreme concentrations |
| Environmental engineering | WQ standards; Pollutant inventory; Trends; Recovery rate | Time/space structure; Diffuse/point sources; Concentrations and fluxes; Longitudinal profiles |
| Hydraulic engineering | River bed erosion; Reservoir silting | TSS levels, fluxes, regimes |

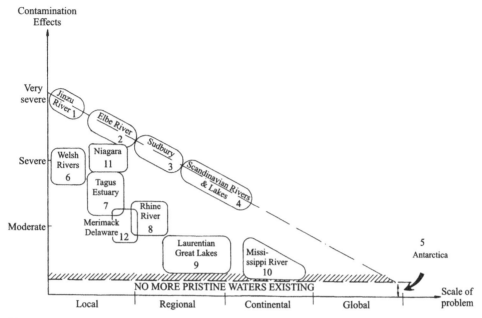

Figure 3.4. Evolution of water quality as evidenced from sediment records (source: Meybeck & Helmer 1989).

PCBs have been noted in the case of Lake Geneva, Switzerland. Since 1960, mercury levels in the southwestern Lake Ontario, Canada, near the mouth of Niagara River, went up by 100 times the natural background levels. There has been a drop in mercury levels in the 1970s, consequent upon the elimination of the mercury discharges into the Niagara River. Figure 3.4 (source: Meybeck & Helmer 1989) shows the evolution of water quality as evidenced from sediment records.

### 3.2.3 *Water quality descriptors*

The number of water quality descriptors have been rising exponentially. In 1890s, water quality was described in the form of a few simple parameters, namely, dissolved oxygen, pH, and faecal coliforms (incidentally, these are still the only parameters that are reported in most developing countries). With the development of industry, energy sector and the modern high-input agriculture, the factors that need to be monitored have risen exponentially (Fig. 3.5; source: Meybeck & Helmer 1989). Figure 3.6 (source: Meybeck & Helmer 1989) depicts the evolution in the number and variety of water quality descriptors. Thus, the number of water quality descriptors used by US EPA and the European Community has risen to 100, and more are being added. The analysis for such a large number of descriptors has been rendered possible because of the availability of modern automated analytical instruments (ICP – AES, ICP – MS, GC – MS, GC – ECD, gamma- ray spectrometer, etc.).

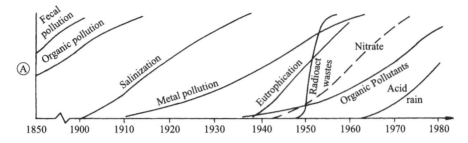

Figure 3.5. Diversification of the factors that need to be monitored, arising from the increasing complexity of pollutants (source: Meybeck & Helmer 1989).

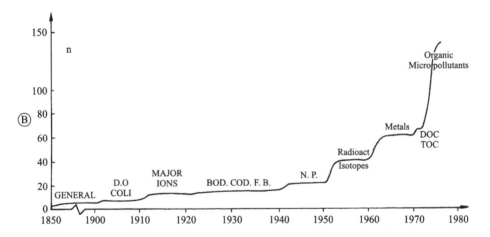

Figure 3.6. Evolution of the number and variety of water quality descriptors (source: Meybeck & Helmer 1989).

Currently, two levels of monitoring are employed. Most monitoring activities cover electrical conductivity, pH, temperature, total suspended solids (TSS), and major ions and nutrients (e.g. $PO_4^{3-}$, $NO_3^-$, $NH_4^+$), etc. Advanced monitoring covers organic and inorganic micropollutants. The costs of advanced monitoring per sample may range over two orders of magnitude relative to basic monitoring, and hence only industrialized countries can afford advanced monitoring. In the case of metals, particulate analyses have replaced the dissolved metal analyses. Since metals tend to be adsorbed on particulates than be present in solution, analysis of particulates gives more meaningful and reliable results.

### 3.2.4   *Chemical analysis for water quality*

Though a detailed description of the methods of sampling, chemical analysis, reporting formats, etc. of water in relation to water quality, is beyond the scope of the book, a summary is provided for the sake of completeness. The volume, 'Standard methods for the examination of water and wastewater' (1985) is a veritable bible in regard to water analysis. The hydrogeology and geochemistry of the waters have been dealt with in the classic works of Stumm & Morgan (1981), Hem (1985), Drever (1988), Domenico & Schwartz (1990).The common parameters used in water analysis are as follows (Table 3.2; source: Appelo & Postma 1996):

Water samples are usually stored in PVC bottles. Acidulation of the samples (by the addition of 0.7 ml of 65% $HNO_3$) is the time-tested method of conserving the water samples. It stops bacterial growth, inhibits oxidation and prevents adsorption or precipitation of cations. Filtration of the sample is important because experience has shown that heavy metals in water tend to be adsorbed on the suspended particles in water. Degassing of $CO_2$ is avoided by pressure filtration using an inert gas and employing a 0.1 μm filter. Field analyses involving electrode measurements should preferably be carried in a flow cell in order to prevent the entry of air.

The analytical data regarding various ions (obtained by the use of AAS, ICP – AES, ICP-MS, etc. in the laboratory or by portable kits with ion-specific electrodes in the field) are usually reported in mg $l^{-1}$. How the data is to be recalculated is illustrated with the help of an example (say, $SO_4^{2-}$):

The molecular weight of $SO_4$ is 96.06 (S: 32.06 + 4 O: 4 × 16 = 64)

mg $l^{-1}$ (say, 8.25) is converted to mmol $l^{-1}$ by dividing it by the molecular weight of $SO_4$ (96.06) = 8.25/96.06 = 0.0859 mmol $l^{-1}$

Avogadro number of molecules of $SO_4$ is obtained by multiplying mmol $l^{-1}$ with Avogadro number: $0.0859. 10^{-3} \times 6.10^{23} = 0.51. 10^{20} = 5.1. 10^{19}$.

Table 3.2. Common parameters used in water analyses.

| Parameter | Description |
|---|---|
| Hardness | Sum of $Ca^{2+}$, $Mg^{2+}$, and sometimes $Fe^{2+}$. Expressed in meq $l^{-1}$ or mg $CaCO_3$ $l^{-1}$, or in hardness degrees. 100 mg $CaCO_3$ $l^{-1}$ roughly equals 1mmol $Ca^{2+}$ $l^{-1}$ or 2 meq $Ca^{2+}$ $l^{-1}$. |
| Colour | Measured by a comparison with a solution of cobalt and platinum |
| Eh | Redox potential expressed in volts. Measured with platinum/reference electrode |
| pe (pE) | Analogue to pH. Redox potential expressed as – log ($e^-$), where $e^-$ is the activity of electrons. pE = Eh/0.059 at 25°C |
| Alkalinity (Alk) | Acid neutralizing capacity. Determined by titrating with acid down to pH of about 4.5. Equal to the total concentrations of carbonates and bicarbonates, expressed in mmol $l^{-1}$ |
| Acidity | Base neutralizing capacity. Determined by titrating upto a pH of about 8.3. Equal to $H_2CO_3$ concentration in most samples |
| COD | Chemical Oxygen Demand. Measured as chemical reduction of permanganate or dichromate solution, and expressed as oxygen equivalents |
| Electrical conductivity | μ S $cm^{-1}$ (= μ mho $cm^{-1}$). EC is roughly equivalent to 100 meq (anions or cations) per l. |
| pH | – log [$H^+$], log of $H^+$ activity |

A given ion may be positively or negatively charged. meq $l^{-1}$ is obtained by multiplying mg $l^{-1}$ by the charge of $SO_4^{2-}$ (i.e. –2): $0.0859 \times -2 = -0.1718$ meq $l^{-1}$.

The meq $l^{-1}$ of positively and negatively charged ions in water is calculated. As the solutions are electrically neutral, the sum of positive and negative charges in a solution should balance. This provides a simple method of assessing the reliability of a water analysis.

Stiff diagrams (Hem 1985) can provide a graphical overview of different water compositions. They are constructed by recalculating the water analyses into milliequivalents per liter, and then plotting anions and cations on three axes. The shape of the Stiff diagrams can be used as a finger-print to broadly identify the nature of the aquifer (rock) with which a given groundwater must have been in contact. Appelo & Postma (1996, p. 3-4) give an illustration of how the Stiff diagrams can be used to distinguish between groundwaters from (say) limestone, dolomite, serpentine rock, granite, etc. aquifers (Fig. 3.7; source: Appelo & Postma 1996, p. 4).

### 3.2.5   *Trends in water quality*

Meybeck & Helmer (1989) delineated four phases in water quality trends as a consequence of socioeconomic developments and water pollution control measures (Fig. 3.8).

*Phase I*: Pollution of water increases $(O - A)$ largely as a consequence of population growth. This is the kind of pattern that is currently in evidence in the developing countries, which are characterized by subsistence agriculture and lack of industrialization. Such a pattern existed in mid-nineteenth century in the countries which are presently highly industrialized.

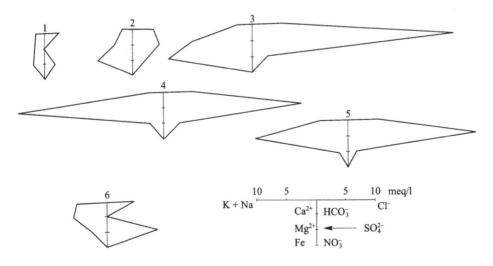

Figure 3.7. Stiff diagrams (source: Appelo & Postma 1996, p. 4).

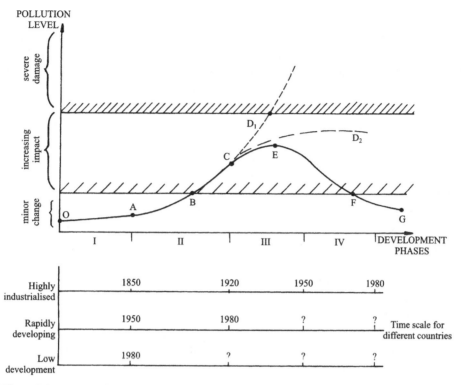

Figure 3.8. Water quality trends as a consequence of socio-economic developments and water pollution control measures (source: Meybeck & Helmer 1989).

*Phase II*: Pollution of water increases exponentially as a consequence of rapid increase in industrialization. The principal sources of pollution are: energy producers, industry, transportation, agriculture, and consumer services. Let us take energy production as a factor of environmental impact (the units of energy are MJ = $10^6$ joules; and GJ = $10^9$ joules). According to Goldemberg (1992), the per capita energy consumption of man has increased rapidly from 104 MJ d$^{-1}$ for the *advanced agricultural man* (corresponding to the present position in the developing countries), 308 MJ d$^{-1}$ for the *industrial man* who started using coal, to 1025 MJ d$^{-1}$ for the present-day *technological man* (corresponding to industrialized countries). In the beginning of the twentieth century, the pollution reached the threshold point (*B*) in the industrialized countries (e.g. North America, Western Europe, Japan, etc.). At this point, pollution level of water had a recognizable impact on the economy, as it impaired the use of water for some vital purposes (say, for drinking) and needed vast sums of money for treatment. The ensuing public outcry stirs the responsible water authorities and the governments to initiate pollution control measures (say, at the time point *C*). The period, *B − C*, therefore represents the time lag between the point when the threshold is reached (*B*) and the time when the control measures have been put in place (*C*). Figure 3.2 shows that control measures (corresponding to point *C*) got initiated in the industrialized countries in 1920s, and in the rapidly industrializing developing countries (e.g. Brazil, South Korea, Southeast Asian countries) in 1980s. Several of the Least Developed Countries (LDCs) (e.g. most African countries – Sub-Saharan Africa contains two-thirds of the low-income countries in the world) have yet to initiate effective control measurements, due to lack of infrastructure and trained personnel.

Phase III: Pollution can be contained by the implementation of effective pollution control measures. Three scenarios are possible:

- *C – D1*, where the pollution control measures have been ineffective, leading to further deterioration of water quality. The pollution reaches the second threshold point *D1*, with severe damage to the ecosystem and drastic impairment in water use – this happened in the Angonia province of Nigeria,
- *C – D2*, the pollution control had some beneficial effect, but was not adequate enough to curb the problem completely. As a consequence, pollution level still rises but at a lower rate, with the ever-present danger of serious damage to the ecosystem – Ganga water project in India is an example of this kind of scenario, and
- *C – E*, stabilization of pollution at the current levels.

Phase IV: Reversing the trend of pollution – bringing about steady reduction in pollution to a tolerable threshold level (*F*), with further improvement in quality (*G*). Though the original pristine stage (*O*) may never be reached, the trend *E – F – G* would be a reasonably satisfactory scenario. Many examples of phase IV can be found in the industrialized countries.

Rapidly industrializing countries are faced with the herculean task of curbing the exponential growth in pollution (*E – F*) within a short period of about 25 years, what the industrialized countries took 100 years to achieve (*E – F*).

The developing countries are faced with a cruel dilemma. Suppose a country uses modern fertilizers and pesticides to achieve high crop production. As should be expected, some of these substances end up in streams which are used for drinking water. The kind of analytical equipment and the skills needed to analyze the pesticide residues in water and food are beyond the reach of most of the developing countries. Under the circumstances, a sensible option open to them is to pool the analytical facilities on a regional basis (such as, SADC – Southern Africa Development Community), just as has been done for locust control.

## 3.3 WATER QUALITY IN RELATION TO WATER USE

Water quality for certain purposes (say, drinking) may be impaired by natural causes (e.g. highly fluorous spring and river waters in northern Tanzania) or anthropogenic causes (e.g. effluent discharges or acid rain).

Generally, water in the natural state is fit for most human uses, but this is not always so. The waters of the rivers in arid zones (e.g. Ethiopia) are unfit for drinking purposes, as high evaporation make them salty. The *As* and *F* contents of the hydrothermally influenced rivers (e.g. Firehole River in the Yellowstone National Park, Wyoming, USA) are three orders of magnitude higher than the most common natural concentrations. Some river waters are unsuitable for irrigation because of their high salt content, and high sodium to calcium ratio. High Dissolved Organic Carbon (DOC) values found in the plain rivers in the tropical regions render them unsuitable for use for some industrial purposes. Similarly, high level of hardness in carbonate-draining rivers, high content of silica in basalt-draining rivers, and high TSS in the arid zones with high slopes or in volcanic areas, impair the use of the river waters for particular purposes.

Impairment may be caused not only due to degradation in the chemical and physical attributes of water, but also due to aquatic organisms. In the recent decades, the zebra mussel (*Dreissena polymorpha*) which clogs water intakes, has spread all over Europe and North America. The thick and vast mats of water hyacinth (*Eichhornia crassipes*) in tropical waters is seriously impeding fluvial transport.

The various uses of water can be classified into three categories, depending upon the stringency of specifications for a particular use:
– Most demanding uses: A – Drinking water; B – Fisheries;
– Less demanding uses: C – Recreation, D – Industrial uses, E – Irrigation,
– Least demanding uses: F- Power and cooling; G – Transport.
Table 3.3 (source: Meybeck 1998) shows how the water quality contaminants constrain the use of water.

Table 3.3. Water quality constraints in water use.

| Pollutant | A | B | C | D | E | F | G |
|---|---|---|---|---|---|---|---|
| Pathogens | xx | 0 | xx | 0 to xx (1) | x | na | na |
| Suspended solids | xx | xx | xx | x | x | x (2) | xx (3) |
| Organic matter | xx | x | xx | x to xx (4) | + | x (5) | na |
| Nitrate | xx | x | na | 0 to xx (1) | + | na | na |
| Salinity (9) | xx | xx | na | x to xx(10) | xx | na | na |
| Inorganic toxics (11) | xx | xx | x | x | x | na | na |
| Organic toxics | xx | xx | x | x to xx (1) | x | na | na |
| Protons(acidification) | x | xx | x | x | na | x | na |
| Aquatic biota | x (5, 6) | x (7) | xx (12) | x to xx (4) | + | x (5, 13) | x (8) |

xx – Presence of the pollutant markedly impairs the use of water for the concerned purpose; can only be used for the purpose after major treatment; x – Minor impairment of use, needing minor treatment; 0 – No impairment of use is involved. na – Not applicable. + – Degradation of water quality in this way may be beneficial for this specific use.

1: Food industries; 2: Abrasion, 3: Sediment settling in channels, 4: Some industries (e.g. electronic industries) require low DOC, 5: phytoplankton may cause clogging of filters, 6: Bacterial activity may affect odor and taste, 7: Higher algal biomass is acceptable in fish ponds, 8: Development of water hyacinth (*Eichhornia crasspipes*), 9: Also includes B and F, 10: Ca, Fe, Mn in textile industries, 11: As, Cd, Cu, Hg, Pb, Se, Sb, Sn. Zn, etc. 12. Loss of transparency from algal development, 13. Development of zebra mussels in pipes (*Dreissena polymorpha*).

The information provided in Table 3.3 indicates the directions for proper use and reuse:
- The highest quality water (at the rate of, say, 2-4 l per capita per day) should be reserved for drinking purposes (as different from water for other domestic uses, such as bathing, washing clothes and utensils and sanitation, which can use slightly inferior water). In many Indian homes, the traditional practice has been to keep drinking water separately from water to be used for washing purposes. It is no doubt technologically feasible but prohibitively expensive to purify contaminated water (say, municipal waste water) to a level of purity acceptable for domestic purposes,
- Municipal waste water which contains (say) pathogens can be used in most industries, except food industries, and
- Sewage which contains organic matter and nitrates is not only acceptable but even be desirable for use in irrigation.

### 3.3.1 *Water quality criteria for various uses*

The water quality criteria for various uses of water have been spelt out under concerned sections: for irrigation, under Chapter 6.1, and for industrial, domestic, recreational and other uses, under Chapter 7.

### 3.3.2 *Water quality standards of potability*

Water quality criteria specify a concentration of constituents in water which, if not exceeded, are expected to result in an aquatic ecosystem suitable for higher use of water. Train (1979) gave the quality criteria for safe drinking water in terms of 'no effect' level. 'Safe' drinking water intake is taken as less than 1% of 'no effect' in terms of the most sensitive animal tested. It should be emphasized that the long term 'no effect' level (in terms of mg $kg^{-1}$ of body weight $d^{-1}$) of most of the pesticides are not precisely known.

In 1984, the World Health Organization (WHO) has prescribed norms for the quality of potable (i.e. drinking) water, and these have been updated periodically. For instance, on the basis of the studies on As toxicity, WHO has recently revised the permissible limit of As in drinking water from 50 µg $l^{-1}$ to 10 µg $l^{-1}$. Some countries (e.g. Nordic countries) have adopted more stringent criteria than those of WHO, whereas many developing countries have reconciled themselves to less stringent criteria due to operational difficulties (for instance, the permissible fluoride content in drinking water in Tanzania is 8 mg $l^{-1}$, as against the international norm of 2 mg $l^{-1}$).

The Federal Drinking Water standards, prescribed by the US Environmental Protection Agency in 1988 (Tables 3.4 and 3.5) has two categories:
1. Primary standards dealing with microbial, turbidity, inorganic and organic chemical contaminants which are expressed in terms of the Maximum Contaminant Level (MCL) which should not be exceeded, and

Table 3.4. Federal Primary Drinking Water Standards (US EPA).

| Contaminant | Source | Maximum Contaminant Level (mg $l^{-1}$) |
|---|---|---|
| Total coliform | Human and animal faecal matter | 1 per 100 ml |
| Turbidity | Erosion, runoff, discharges | 1-5 turbidity units |
| Arsenic | Geological, mining, smelting | 0.05 |
| Barium | | 1.0 |
| Cadmium | Geological, mining, smelting | 0.01 |
| Chromium | Plating, tanneries | 0.05 |
| Lead | Leaches from lead pipe, and lead-based solder pipe joints | 0.05 |
| Mercury | Paint, paper, vinyl chloride, fungicides, geological | 0.002 |
| Nitrate | Fertilizer, sewage, feedlots, geological | 10 |
| Selenium | Geological, mining | 0.01 |
| Silver | Geological, mining | 0.05 |
| Fluoride | Geological, additive to drinking water and toothpaste | 4 |
| Endrin (cancelled) | Insecticide used on cotton, small grains, orchards | 0.0002 |
| Lindane | Insecticide used on seed and soil treatment, foliage application, wood protection | 0.004 |
| Methoxychlor | Insecticide used on fruit trees, vegetables | 0.10 |
| 2,4 – D | Herbicide used in agriculture; forests,range pastures and aquatic environments | 0.1 |
| 2,4.5-TPsilvex (cancelled) | Herbicide | 0.01 |
| Toxaphene | Insecticide used on cotton, corn, grain | 0.005 |
| Benzene | Solvent used in chemical industries, pharmaceuticals, pesticides, paints and plastics | 0.005 |
| Carbon tetrachloride | Common in cleaning agents, industrial wastes for the manufacture of coolants | 0.005 |
| p-Dichlorobenzene | Insecticides, moth balls, air deodorizers | 0.075 |
| 1,2,- Dichloro ethane | Insecticides, gasoline | 0.005 |
| 1, 1-dichlomethylene | Plastics, dyes, perfumes, paints | 0.007 |
| 1,1,1 – Trichlomethane | Food wrappings, synthetic fibres | 0.02 |
| Trichlomethylene | Waste from dry cleaning materials, pesticides, Paints, varnishes | 0.005 |
| Vinyl chloride | Polyvinyl chloride (PVC) and waste from plastics and synthetic rubber | 0.002 |
| Total trihalomethanes (TTHM) | Forms when surface water containing organic material is treated with chlorine | 0.10 |
| Gross alpha particle activity | Radioactive wastes, uranium deposits | 15 pico- Curies $l^{-1}$ |
| Gross beta particle activity | Radioactive wastes, uranium deposits | 4 millirems $yr^{-1}$ |
| [226] Radium, & [228] Radium (total) | Radioactive waste, geological | 5 pico- Curies $l^{-1}$ |

Table 3.5. Federal Secondary Drinking Water Standards (US EPA).

| Contaminant | Level |
|---|---|
| pH | 6.5-8.5 |
| Chloride | 250 mg $l^{-1}$ |
| Copper | 1.0 mg $l^{-1}$ |
| Foaming agents | 0.5 mg $l^{-1}$ |
| Sulfate | 250 mg $l^{-1}$ |
| Total Dissolved Solids (TDS) | 500 mg $l^{-1}$ |
| Zinc | 5.0 mg $l^{-1}$ |
| Fluoride | 2.0 mg $l^{-1}$ |
| Color | 15 color units |
| Corrosovity | Non-corrosive |
| Iron | 0.3 mg $l^{-1}$ |
| Manganese | 0.05 mg $l^{-1}$ |
| Odour (Threshold Odour no.) | 3 |

2. Secondary standards dealing with parameters like pH, TDS, etc., whose recommended level is indicated. Health effects that could arise if the intake exceeds the maximum contaminant levels, are described in Chapter 10. It may be noted that pesticides such as endrin have been banned in the industrialized countries, though they continue to be used in the developing countries because they are cheap and potent.

Though the pH range for potable water could be 6.5 to 8.5, it should be remembered that acid pH tends to mobilize toxic elements, such as Pb, Cd, and Al, and is therefore undesirable. For instance, the median lead level (in $\mu g \ l^{-1}$) is 170 at pH of 6.8, 44 at pH of 6.9 to 7.5, 32 at pH of 7.6-8.3, and 21 at pH > 8.4. This indicates that even moderately acid drinking water can carry about four times more lead than water at the normal pH of 7.0, and that kind of acid pH is not uncommon in the areas affected by acid rain.

## 3.4   WATER QUALITY MONITORING SYSTEMS

The US Environmental Protection Agency, 401 M St., SW, Washington, DC 20460, USA, has brought out a computerized Environmental Monitoring Methods Index (EMMI) – 1994 (obtainable from National Technical Information Services, Tel: + 1-703-487 4650). It contains information on the recommended methods of analysis of various elements and compounds in different media, ways of reporting data, 'Permissible Exposure Limits' to toxic substances, etc.

### 3.4.1   *GEMS/WATER project*

The monitoring of water quality of rivers started around 1890 for some European rivers, such as the Thames and the Seine, which were getting highly polluted due to domestic sewage. In those days, only a few parameters, such as dissolved oxygen, pH and faecal coliforms, were measured. As the pollution from various sources became more and more, there has been an exponential rise in the number of water quality descriptors to more than 100 at present.

The GEMS/WATER Project is the most ambitious international effort to date to monitor the quality of surface waters (1983, 1988) through the voluntary efforts of the WHO member states. By 1983, GEMS/WATER network covered 255 river stations from 59 countries. The project suffers from the following inadequacies:

- While the bulk (about 90%) of the stations do report some principal water quality descriptors (e.g. faecal coliforms, Electrical conductivity, dissolved oxygen, pH, etc.), very few stations report highly toxic pollutants,
- The target of 24 measurements per year is rarely met. Basic descriptors are usually measured monthly. Most harmful inorganic and organic pollutants are measured only 4-6 times per year, and
- The present system does not take into account short episodes of heavy pollution or pollution loads under conditions of high-water discharges (as in the case of floods).

The percentage of river stations under the GEMS/WATER project which regularly report a given water quality descriptor, is given below (GEMS 1983):

*Basic descriptors:*
Faecal coli (74), TSS (63), Elec. conduc. (95), Diss. oxygen (90), pH (90)

*Major ions:*
$SiO_2$ (28), $Ca^{2+}$ (71), $Mg^{2+}$ (66), $Na^+$ (48), $K^+$ (33), $Cl^-$ (98), $SO_4^{2-}$ (59), $HCO_3^-$ (88)

*Nutrients and organic matter:*
$PO_4^{3-}$ (60), $P_{tot}$ (28), $NO_3^-$ (81), $NH_4^+$ (57), $N_k$ (50), TOC (21), BOD(73), COD (40)

*Pollutants and miscellaneous elements:*
F (40), B (15), CN (15), Phenol (21), Cu (36), Cr (23), Fe (18), Mn (37), Ni (27), Se (7), Zn (20)

*Highly toxic pollutants:*
As (26), Cd (30), Hg (33), Pb(36), DDT (20), Aldrin (22), Dieldrin (23), PCB (8)

Globally, the GEMS/WATER monitoring has a rating 4.1 out of the optimum of 10.0. The ratings vary from more than 6.0 for North America, 5.4 for Europe, 4.1 for Asia, 3.8 for Oceania, and 2.6 for Africa. The ratings are an indication of the status of the environmental monitoring infrastructure in different continents.

### 3.4.2  *Regional problems of water quality*

The size of a water body determines its effectiveness in diluting the pollution. Thus, on a local scale (10,000 km$^2$) pollution may reach very severe levels, as it happened in Japan in 1950s (Minamata disease due to mercury pollution and itai-itai disease due to cadmium pollution). The average contaminant levels decrease at regional scale (about 1 million km$^2$) and continental scale (about 50 million km$^2$). On a global scale, the contamination of the hydrosphere manifests itself in the form of higher contaminant levels of metals in the Antarctic surface ice, and presence of synthetic organics, such pesticide residues, in the biota (e.g. DDT has been reported in the eggs of the penguins).

Some issues of water degradation are global in scope. Increase in the salt concentrations of rivers and surface water bodies has been occurring due to anthropogenic activities, such as irrigation (e.g. the Colorado and Nile rivers), industrial activities (e.g. the Rhine), or mining (e.g. the Rhine, Weser, Vistuala). Deforestation increases the soil erosion and the sediment load (e.g. the Ganges carries a suspended load of $1670 \times 10^6$ t yr$^{-1}$, in its discharge of 971 km$^3$ yr$^{-1}$). All the sediment load does not reach the sea – good part of it settles down on the flood plains and behind the dams. Most of the major rivers in the world have been dammed and channelized, with the consequence that about $1.5 \times 10^{15}$ g yr$^{-1}$ of suspended material, or about 10% of the global TSS, is trapped behind the dams.

Contamination of water by synthetic organic pollutants has emerged as a serious problem, not only because such pollutants have become ubiquitous, but also because of the technical difficulties and expense in monitoring them. Such pollutants (such as, PCBs) have been reported in several south Asian rivers, Parana river in South America, the Seine and Loire rivers in France, etc. Some pesticides such as DDT are no longer used in the industrialized countries, and the concentration levels of DDT have declined sharply in those countries. But DDT and other pesticides continue to be used in the developing countries. Agarwal et al (1986) have reported the highest levels ever of DDT in the waters of the Jamuna river which flows through Delhi, India.

Some broad patterns of water contamination are discernible on a global scale. Many streams and rivers in South America, Africa and Asia (particularly, in the Indian sub-continent) show high levels of coliforms, BOD and nutrient levels, with consequent waterborne morbidity and mortality. In several countries, eutrophication of lakes and ponds has spread to slow moving rivers. For instance, hypereutrophic conditions have arisen in the Rhine and Loire rivers, with summer chlorophyll levels of 100 to 200 µg l$^{-1}$. Such a high level of biomass poses severe problems in water treatment, besides causing estuarine anoxia. Nitrate levels in some places in Western Europe and North America have already reached the critical levels of 10 mg N l$^{-1}$. Acid rain is known to cause the acidification of surface waters, as is well documented in Northeast America, Scandinavia and Central Europe. The process gets accentuated when such waters flow on acid crystalline rocks or non-carbonate sandstones.

## 3.5 QUALITY OF SURFACE WATERS

Surface water (in rivers, lakes and reservoirs) which is easily accessible for our economic needs and critically important for water ecosystems, constitute only 0.26% of the total amount of freshwater on earth. Thus only a small proportion of the freshwater is renewable (Shiklomanov 1998).

Ocean waters are subject to homogenization processes. Consequently, the chemistry of the ocean waters is more or less uniform globally. In contrast, the chemical composition of continental surface waters vary up to four orders of magnitude, depending upon the environmental conditions.

### 3.5.1 *River basins and runoff to the oceans*

The volume of discharge ($km^3 yr^{-1}$) of 37 major rivers in the world have been given by Shiklomanov (1998). Amazon alone produces about one-sixth of the annual global river runoff. Five largest rivers (the Amazon, Ganges with Brahmaputra, Congo, Yangtze and Orinoco) account for 27% of the world's water resources.

There are some drainless regions in the world where the rivers are not connected to the World Ocean. Such regions are called endorheic (e.g. North Africa, Middle East, Central Asia). Though the endorheic regions cover an area of about 30 M $km^2$ or about 20% of the land area, only 2.3% (about 1000 $km^3 yr^{-1}$) of the global runoff is formed in these regions, because they are occupied by deserts or semi-deserts.

In the regions of exorheic drainage (i.e. where the rivers are connected to the World Ocean), the water resources of a river basin are formed in mountainous areas with large precipitation. But in the process of a river moving towards its mouth, considerable amount of the river runoff is lost due to evaporation. This is particularly true of rivers in the tropical lands, such as the Ganges and Indus in Asia, and the Niger and Zambeze in Africa. The total amount of runoff in the exorheic regions of the world that does not reach the mouth because it has been lost due to evaporation, is estimated at 1100 $km^3 yr^{-1}$. Thus, the total water flow into the World Ocean is somewhat less than the total renewable water resources in the continental areas of the earth.

The Atlantic ocean receives about half of the total river water inflow into the World Ocean, because four of the six largest rivers in the world (namely, the Amazon, the Congo, the Orinoco and Parana) drain into it. Also, the river water flow into the world ocean is uneven – about 40% of the river runoff enters the ocean in the equatorial region (10°N to 10°S).

### 3.5.2 *Effect of environmental factors on surface water quality*

The material in this section is largely drawn from the extensive contributions of Meybeck and his associates.

The dissolved matter carried by the rivers is derived principally from three sources: inputs from the atmosphere, erosion of weathered surface rocks and soils, and degradation of terrestrial organic matter. When such dissolved matter passes through the soil or porous rocks en route to the river, it may get involved in a number of processes – it may get recycled through the terrestrial biota, or stored in soils, or exchanged between the dissolved and particulate matter. It may lose some volatile substances to the atmosphere, and it may take part in the production and degradation of aquatic plants. The complexity of the composition of the dissolved matter in the rivers is thus a consequence of the multiplicity of the sources from which it is derived and the numerous pathways through which it moves.

The concentrations of elements and compounds in the river waters are a function of physical factors (e.g. climate, relief), chemical factors (e.g. solubility of minerals in the context of the ambient Eh and pH) and biological factors (e.g. uptake of elements by vegetation, degradation by bacteria).

Meybeck & Helmer (1989) summarized the previous work on the most important factors controlling the chemistry of the river water, as follows:
– Occurrence of highly soluble (e.g. halite, gypsum) or easily weathered (e.g. calcite, dolomite, pyrite, olivine) minerals in the rocks or soils in the catchment area.
– Distance from the sea – the farther a given area is from the sea, the less will be the ocean aerosol input to land (in the form of $Na^+$, $Cl^-$, $SO_4^{2-}$, $Mg^{2+}$).
– Aridity of an area, which controls the precipitation/runoff ratio and has a bearing on the concentration of dissolved substances.
– Terrestrial primary productivity, and the release of nutrients (such as, C, N, Si, K).
– Ambient temperature, which controls the biological soil activity, and the rate of weathering linked to it.
– Geomorphic setting, arising from tectonism and uplift due to earth movements.
Meybeck updated the compilations of Livingstone (1963) and Alekin & Brazhnikova (1968) about the chemical composition of river discharges. Meybeck (1979) gave the following estimate of the natural average composition of surface waters (in mg $l^{-1}$) for exorheic drainage: $SiO_2$: 10.4; $Ca^{++}$: 13.4; $Mg^{++}$: 3.35; $Na^+$: 5.15; $K^+$: 1.3; $Cl^-$: 5.75; $SO_4^-$: 8.25; $HCO_3^-$: 52. The estimate is based on 60 big rivers which represent 63% of the total river discharge to the ocean. The chemistry of the surface waters for the rest 37% of the discharge has been estimated on the basis of the mean of the dissolved matter in the runoff, average watershed temperature, with relief correction. Lithological factors of the catchment area are not taken into account; they tend to get smoothened out when catchment areas of the order of $10^6$ $km^2$ are taken into account. It is found that other sources of dissolved material to the ocean, such as the glaciers, leaching of the fresh volcanic rocks and underground waters, are negligible.

Atmosphere contributes about 15% of the dissolved substances in the surface waters. Humid mountainous regions which are exposed to intense chemical denudation (30 mm/1000 years) account for 45% of the dissolved inputs to the ocean. About 74% of the dissolved silica in the surface waters comes from the tropical regions (as is well known, the solubility of silica in water increases with temperature).

The distribution of the major rock types on the surface of the earth is as follows: volcanic rocks: 7.9%, plutonic rocks: 11%, metamorphic rocks: 15%, carbonate rocks: 15.9%, shales: 33%, sandstones: 15.8%, evaporites: 1.25% (Meybeck 1987). Thus, not only are the sedimentary rocks present in greater abundance on earth, but the relative rates of dissolution of the sedimentary rocks are far higher than those of crystalline rocks: evaporites – about 80%, carbonate rocks – about 20%, crystalline rocks – about 1.2%, shales and sandstones – about 1%. Consequently 89% of the chemical erosion products are derived from sedimentary rocks, of which two-thirds are derived from carbonate rocks.

About 57% of $HCO_3^-$ in surface waters is derived from atmospheric $CO_2$. Rest is derived from the present-day chemical erosion of the continents (9 mm/1000 yrs).As a consequence of the operation of the above processes, 95% of the surface waters are of the calcium carbonate type. Two major varieties could be recognized: sedimentary regions are characterized by magnesium sulfate waters (43% of waters), and the crystalline regions by sodium chloride waters (34%).

Meybeck & Helmer (1989) summarized the data about the distribution of dissolved elements in waters (Table 3.6) and the enormous geographic variability in regard to the dissolved major elements in pristine (i.e. not affected by anthropogenic pollution due to habitations, roads, farming, mining, etc. and located away from sources of atmospheric pollution) waters (Table 3.7).

Three categories are considered:

1. Pristine streams draining the most common rock types, such as, granite, gneiss, volcanic rocks, sandstones, shales and carbonate rocks,
2. Pristine streams draining rock types (such as, coal shale or salt rock) or in geological settings, which are rare (e.g. equatorial forest waters of the Amazon, or the juvenile waters of the Yellowstone National Park in USA),
3. Pristine streams influenced by intense evapotranspiration (e.g. as in Ethiopia).

On the basis of the statistical analysis of the content of the dissolved matter in streams (1-100 km$^2$) and rivers (100,000 km$^2$), Meybeck & Helmer (1989) derived two global averages: Discharge Weighted Natural Concentrations (DWNC) and Most Common Natural Concentrations (MCNC). As is to be expected, 97% of all the water discharged to the oceans is of the $Ca^{2+} - HCO_3^-$ type.

### 3.5.3 *Geographic variability of dissolved matter in river waters*

Natural concentrations in small watersheds (1-10 km$^2$) have a much larger range (2-3 orders of magnitude) than in the case of large rivers (100,000 km$^2$) (1-2 orders of magnitude). In both the cases, some parameters such as, $H^+$, $Na^+$, $Cl^-$, $SO_4^{2-}$ and Total Suspended Solids (TSS), are most variable.

The great range of variation in the chemistry of stream waters can be illustrated on the basis of the sulfate content and pH in some unpolluted rivers and streams

Table 3.6. Distribution of dissolved elements (source: Meybeck & Helmer 1989).

| | Unit | Streams * (1-100 km$^2$) | Rivers * (1-100,000 km$^2$) | DWNC | MCNC |
|---|---|---|---|---|---|
| SiO$_2$ | µmol l$^{-1}$ | 10 – 830 | 40 – 330 | 170 | 180 |
| Ca$^{2+}$ | µ eq l$^{-1}$ | 3 – 10,500 | 100 – 2500 | 670 | 400 |
| Mg$^{2+}$ | µ eq l$^{-1}$ | 4 – 6600 | 70 – 1000 | 275 | 200 |
| Na$^+$ | µ eq l$^{-1}$ | 2.6 – 15,000 | 55 – 1100 | 225 | 160 |
| K$^+$ | µ eq l$^{-1}$ | 3 – 160 | 13 – 100 | 33 | 27 |
| C$^-$ | µ eq l$^{-1}$ | 2.5 – 15,000 | 17 – 700 | 162 | 110 |
| SO$_4^{2-}$ | µ eq l$^{-1}$ | 2.9 – 15,000 | 45 – 1200 | 172 | 100 |
| HCO$_3^-$ | µ eq l$^{-1}$ | 0 – 5750 | 165 – 2800 | 850 | 500 |
| TZ$^+$ | µ eq l$^{-1}$ | 45 – 20,000 | 340 – 4000 | 1200 | 800 |
| pH | | 4.7 – 8.5 | | | |
| TSS | mg l$^{-1}$ | 3 – 15,000 | 10 – 1700 | 450 | 150 |
| DOC | mg l$^{-1}$ | 0.5 – 40 | 2.5 – 8.5 | 5.75 | 4.2 |
| POC | mg l$^{-1}$ | 0.5 – 75 | | 4.80 | 3.0 |
| POC$^0$ | | 0.5 – 20 | | 1.0 | 2.0 |
| TOC | mg l$^{-1}$ | 1.5 – 25 | | 10.5 | |
| N-NH$^{4+}$ | mg l$^{-1}$ | | 0.005 – 0.04 | | 0.015 |
| N- NO$_3^-$ | mg l$^{-1}$ | | 0.05 – 0.2 | | 0.010 |
| N$_{org}$ | mg l$^{-1}$ | | 0.05 – 1.0 | | 0.26 |
| P- PO$_3^-$ | mg l$^{-1}$ | | 0.002 – 0.025 | | 0.010 |

Minimum and maximum values, corresponding to 2% and 98% of the distribution. In the case of nutrient elements, the corresponding figures are 10% and 90%. DWNC – Global Discharge Weighted Natural Concentrations; MCNC – Most Common Natural Concentrations TSS – Total Suspended Solids; DOC – Dissolved Organic Carbon; POC – Particulate Organic Carbon; POC$^0$ Percentage of organic carbon in TSS.

(Fig. 3.9; source: Meybeck & Helmer 1989; A. Lower Amazon tributaries; B. French streams, C. Mackenzie River of Canada). The sulfate levels in the Canadian and French streams have a range of about three orders of magnitude, and the high values are caused by the weathering of evaporite deposits. Vegetation may be responsible for the extremely low levels of sulfate in the Amazon waters, and their log-normal distribution indicates a point source. The H$^+$ distribution also shows a great variability, of four orders of magnitude. The lowest pH values (3.7) are to be found in the black, organic acid-rich waters of the Amazon, whereas waters draining the carbonate rocks are characterized by the highest pH. The nutrient load of rivers is schematically shown in Figure 3.10 (source: LOICZ Newsletter).

The great global variability in the dissolved matter content of rivers precludes the possibility of any river being designated as a typical/reference river, to serve as a baseline for purposes of determining the extent of pollution. For the same reason, averages for continents are also meaningless. According to Meybeck & Helmer (1989), it would be more meaningful to set up baselines on the basis of regional reference values, taking into account the lithology for major ions, climate, drainage

Table 3.7. Geographic variability of dissolved elements in pristine waters (source: Meybeck & Helmer 1989).

| | Elec. conduc. | pH | TZ+ | SiO₂ | Ca²⁺ | Mg²⁺ | Na⁺ | K⁺ | Cl⁻ | SO₄²⁻ | HCO₃⁻ |
|---|---|---|---|---|---|---|---|---|---|---|---|
| 1. Granite | 35 | 6.6 | 166 | 150 | 39 | 31 | 88 | 8 | 0 | 31 | 128 |
| 2. Gneiss | 35 | 6.6 | 207 | 130 | 60 | 57 | 80 | 10 | 0 | 56 | 136 |
| 3. Volcanic rocks | 50 | 7.2 | 435 | 200 | 154 | 161 | 105 | 14 | 0 | 10 | 425 |
| 4. Sandstone | 60 | 6.8 | 223 | 150 | 88 | 63 | 51 | 21 | 0 | 95 | 125 |
| 5. Shale | – | – | 770 | 150 | 404 | 240 | 105 | 20 | 20 | 143 | 580 |
| 6. Carbonate rock | 400 | 7.9 | 3247 | 100 | 2560 | 640 | 34 | 13 | 0 | 85 | 3195 |
| 7. Amazonian clearwaters | 5.7 | 5.1 | 111 | 31 | Tr | 10.7 | 51 | 35.5 | 19.7 | N.D. | N.D. |
| 8. Amazonian black waters | 29.1 | 3.7 | 212 | 9.8 | Tr | 1.6 | N.D. | N.D. | 33.8 | N.D. | N.D. |
| 9. Batholith | | | 2700 | 332 | 150 | 2490 | 52 | 5 | 93 | 472 | 2020 |
| 10. Coal shale | | | 40,700 | 116 | 4350 | 10,000 | 26,100 | 260 | 420 | 29,000 | 10,700 |
| 11. Salt rock | | | 312,000 | 20 | 30,350 | 5640 | 276,000 | 197 | 266,000 | 27,700 | 3000 |
| 12. Hydrothermal | | 8.0 | 4130 | 1660 | 245 | 40 | 3830 | 184 | 1653 | 290 | 2230 |
| 13. Evapotranspiration rivers | 2230 | 9.2 | 21,800 | 1310 | 110 | 120 | 20,900 | 640 | 5500 | 1350 | 16,000 |

Units: Electrical conductivity in μS cm⁻¹; TZ⁺ is the sum of ions and cations in μeq l⁻¹; Silica in μmole l⁻¹. 1-6: Pristine streams draining the most common rock types; 7-12: Pristine streams draining rare rock types or in rare geological setting; 13. Rivers influenced by evapotranspiration in Ethiopia.

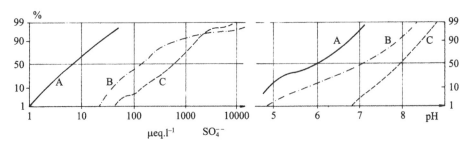

Figure 3.9. Sulfate content and pH of some streams (source: Meybeck & Helmer 1989).

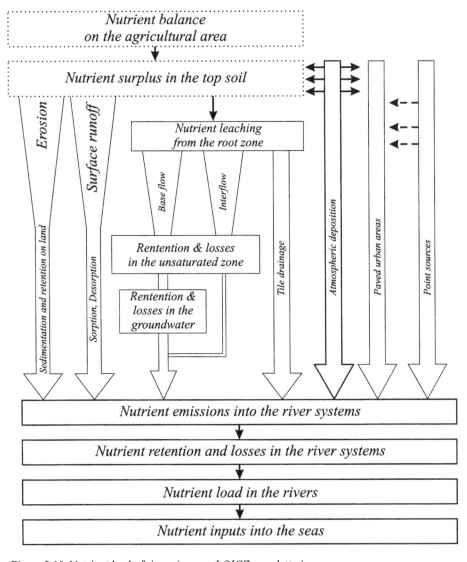

Figure 3.10. Nutrient load of rivers (source: LOICZ newsletter).

pattern, vegetation, etc. DWNC (Discharge Weighted Natural Content) averages can be used to determine the order of abundance of various components from major elements to trace elements. MCNC (Most Common Natural Content) averages are more useful for ecological studies. Both DWNC and MCNC estimates are based on pristine or least contaminated rivers. They differ from the averages under the GEMS – Water program, which cover 255 rivers, but which include the rivers of Western Europe and North America, almost all of which are undoubtedly contaminated.

### 3.5.4 *Trace element composition*

Trace element composition of the pristine river waters is poorly known, and there are several reasons for this unsatisfactory situation – samples may easily be contaminated, the actual level of concentration of a given element may be below the detection limit of the instrument for that element, it is extremely expensive to determine the inorganic and organic micro-pollutants, and thus far, few pristine rivers have been studied for trace elements, etc.

Globally, the trace element composition of the river waters is still poorly known, though new data have been reported for Amazon, Zaire, Yang Tse, Huang He, Ganges, etc.

The percentages of the natural dissolved component in the total elemental river transport are as follows:

> 99-90%: Cl, Br;
> 90-50%: S, Na, Sr, C, Ca, Li
> 50-10%: Sb, Mg, N, B, Mo, As, F, Ba, K
> 10-5%: Cu, P
> 5-1%: Ni, Si, Rb, U, Co, Cd, Mn, Th, V, Cs
> 1-0.5%: Ga, Pb, Lu
> 0.5-0.1%: Ti, Gd, La, Ho, Yb, Tb, Er, Sm, Cr, Fe, Eu, Ce, Zn, Al
> 0.1-0.05%: Sc, Hg

The above data indicates that most metals (e.g. Pb, Cd, Hg, Zn, Cr, etc.) do not go into solution. They tend to be adsorbed on the suspended particles. It therefore follows that trace elements should be determined in the particulate phase rather than from the dissolved matter.

## 3.6  QUALITY OF GROUNDWATER

Generally, the quantity of renewable water resources is estimated from the total river runoff. The runoff which enters the groundwater feeds the rivers. But some groundwater runoff goes directly to the sea and may get evaporated. Such groundwater constitutes a renewable water resource.

In Africa, where the arid and semi-arid regions occupy half of the continent, groundwater resources are of great importance. According to the estimates of FAO made in 1995, the total volume of renewable water resources, unrelated to river runoff, is 188 km$^3$ yr$^{-1}$ for the continent as a whole or about 5% of the total runoff volume (Shiklomanov 1998).

It is important to bear in mind that the quantity of fresh groundwater (about 30% of all freshwater) is about 100 times more than the total amount of fresh surface water (about 0.3% of all freshwater).

The geochemistry of groundwater is useful not only to evaluate the water quality but also to obtain information (such as, residence times, flowpaths and aquifer characteristics) about the environments through which the water has passed, and the extent of exposure to anthropogenic pollution.

Table 3.8 gives a comparative study of the chemical composition of rainwater, groundwater and surface water.

A comparison of the relative concentrations of the various ions in the rainwater, groundwater and surface water, leads to the following conclusions:

- With the exception of NH$_4^+$ which is derived from the atmosphere, all other ions are enriched in the groundwater and surface waters due to contact with rocks and soils. The enrichment is maximum in the case of Ca$^{2+}$,
- Surface water is more enriched than groundwater in the case of K$^+$ and Mg$^{2+}$, and
- Groundwater is more enriched than surface water in the case of Na$^+$, Ca$^{2+}$, Cl$^-$ and SO$_4^{2-}$.

Chloride is a conservative parameter, and hence the chloride concentration ratio between the groundwater and rainwater could be made use of to estimate the an-

Table 3.8. Comparison of the chemical composition of waters.

| Parameter | Rainwater (mmol l$^{-1}$) RW | Groundwater (mmol l$^{-1}$) GW | Surface waters (mmol l$^{-1}$) # SW | GW/RW | SW/RW |
|---|---|---|---|---|---|
| Na$^+$ | 0.070 | 0.30 | 0.224 | 4.29 | 3.20 |
| K$^+$ | 0.004 | 0.025 | 0.033 | 6.25 | 8.25 |
| Mg$^{2+}$ | 0.009 | 0.089 | 0.138 | 9.89 | 15.33 |
| Ca$^{2+}$ | 0.015 | 0.49 | 0.334 | 32.66 | 22.27 |
| NH$_4^+$ | 0.12 | 0.005 | | 0.04 | |
| Cl$^-$ | 0.078 | 0.30 | 0.162 | 3.85 | 2.08 |
| HCO$_3^-$ | 0 | 1.04 | 0.852 | | |
| SO$_4^{2-}$ | 0.074 | 0.15 | 0.086 | 2.02 | 1.16 |
| NO$_3^-$ | 0.064 | 0.063 | | | |

# mg l$^{-1}$ data of Meybeck (1979) has been converted to mmol l$^{-1}$ by dividing by the weight of the element or molecule. Thus, 8.25 mg l$^{-1}$ of SO$_4^{2-}$ is equivalent to 8.25/96.06 = 0.086 mmol l$^{-1}$.
RW: Average of rainwater at Deelen airport, 1978-1980 (Appelo & Postma 1996, p. 2). GW: Average composition of groundwater in Veluwe area (Netherlands), from depths upto 192 m below surface (Appelo & Postma 1996, p. 2).

nual recharge of groundwater. If the ratio is 3.85 (as in the above table) and if the precipitation is 1000 mm, the annual recharge of groundwater in that area is $1000/3.85 = 260$ mm.

### 3.6.1 *Groundwater chemistry in unsaturated zones*

The chemistry of groundwater in a given setting is a function of chemistry of rainwater, dry deposition (i.e. atmospheric fallout during dry periods) and seasonal variation in infiltration, etc. These effects are best seen in the unsaturated zone, particularly where it does not contain reactive minerals like carbonates.

Figure 3.11 (source: Hansen & Postma 1995) is a depiction of the chemistry of groundwater in the unsaturated zone of a sandy, carbonate-free sediment under a coniferous forest in Denmark. The concentrations are plotted cumulatively. The dissolved ion content is dominated by $Na^+$ and $Cl^-$ which are of marine origin. The profile (about 5 m deep) corresponds to one year, as the average annual water transport rate through the unsaturated zone is about 5 m yr$^{-1}$. The net infiltration occurs only during May to October. The salt content in the groundwater is attributable to two sources: washing down into the unsaturated zone of salts that accumulated in the soil in the previous period, and large amounts of sea salt deposited during the westerly storms. The maxima of $Na^+$ and $Cl^-$ observed at the depth of 3 m, may be a cumulative effect of both the processes.

As explained earlier (Section 1.5.5), forests play a beneficial role in feeding groundwater and the streams. There is a sharp difference in the evapotranspiration between grassland and forested areas. Forested areas are characterized by high interception (evaporation from wetted surface), which may account for about 40% of

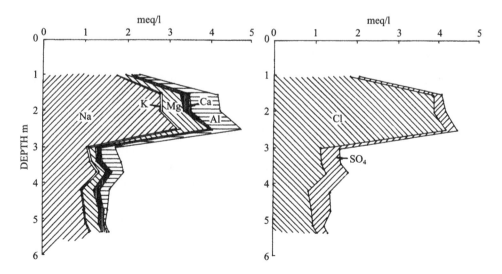

Figure 3.11. Groundwater chemistry in the unsaturated zone of a sandy aquifer under a coniferous forest in Denmark (source: Hansen & Postma 1995).

the annual precipitation. Thus, the total concentration of ions derived from precipitation in the groundwater downstream of forests is much higher than below grassland.

### 3.6.2   *Groundwater chemistry in carbonate aquifers*

*Kinetics of carbonate reactions*
Karst phenomena, such as solution flutes, sinkholes (dolines) and caves, are manifestations of the high susceptibility of carbonate rocks to rapid dissolution.

The rate of dissolution of calcite is a function of pH and partial pressure of $CO_2$. Plummer et al. (1978) delineated three regions in the dissolution kinetics of calcite.

First region, below pH of 3.5, where the proton attack is the dominating process, with the rate being controlled by the transport of $H^+$ to the surface of calcite:

$$CaCO_3 + H^+ \rightarrow Ca^{2+} + HCO_3^- \qquad (3.1)$$

Second region, pH 3.5-5.5, where the reaction becomes increasingly dependent upon $P_{CO_2}$. Here the rate is controlled both by transport and surface reaction.

$$CaCO_3 + H_2CO_3^* \rightarrow Ca^{2+} + 2\ HCO_3^- \qquad (3.2)$$

Third region, pH 5.5 to ca. 7.0, when there is a sharp drop in the dissolution rate.
The hydrolysis of calcite is governed by the following reaction:

$$CaCO_3 + H_2O \rightarrow Ca^{2+} + HCO_3^- + OH^- \qquad (3.3)$$

The backward precipitation of calcite takes place as per the following reaction:

$$Ca^{2+} + HCO_3^- \rightarrow CaCO_3 + H^+ \qquad (3.4)$$

### 3.6.3   *Groundwater chemistry of Bermuda aquifer*

When aragonite-rich coastal marine sediments undergo upliftment (of the order of hundreds of meters) either due to sealevel oscillations or due to tectonic movements, they get elevated above the sealevel and enter the freshwater zone. Such carbonate aquifers are common in the tropics, notably in the Caribbean and the Pacific. The Bermuda aquifer is one of the best studied aquifers of this type (Plummer et al. 1976, as summarized by Appelo & Postma 1996, p. 135-136). The sediments are oldest in the central part of Bermuda, and become younger towards the edges.

When the aragonite-rich carbonates enter the freshwater environment, aragonite will recrystallize into a low Mg-calcite by the following equation:

$$CaCO_{3\ aragonite} \rightarrow Ca^{2+} + CO_3^{2-} \rightarrow CaCO_{3\ calcite}$$
$$\text{dissolution} \qquad\qquad \text{precipitation} \qquad (3.5)$$

Recent sediments contain about 35% aragonite, 50% high Mg-calcite, and 14% of low Mg-calcite. With time, there is steady shift to low Mg-calcite. Thus one million year old limestones consist almost wholly of low Mg-calcite (Fig. 3.12; source: Plummer et al. 1976). This massive recrystallization has a profound influence on the groundwater geochemistry, leading to groundwater supersaturated with low Mg-calcite.

Figure 3.13 (source: Plummer et al. 1976) gives a description of the groundwater geochemistry of the Bermuda aquifer (*SI* is saturation index; it is given by log IAP/K, where IAP is the Ion Activity Product, and K is solubility product).

A and B) Groundwater in the central part of Bermuda is subsaturated for both calcite and aragonite. The presence of marshes in the central part of the island produces $CO_2$ – rich waters, which bring about extensive dissolution of limestone in the central part of the aquifer.

C) The highest values of $P\,CO_2$ are to be found in the central part of the island, which is characterized by the subsaturation for calcite and aragonite.

D) Aragonite accepts a small amount of Sr into its lattice, but not calcite. So when aragonite recrystallizes to calcite, $Sr^{2+}$ remains in the groundwater.

E and F) The part of $Ca^{2+}$ and $Mg^{2+}$ which is derived from limestone dissolution can be obtained by deducting the seawater contribution (on the basis of cation/Cl$^-$ ratio) from the total $Ca^{2+}$ and $Mg^{2+}$. The contribution from precipitation

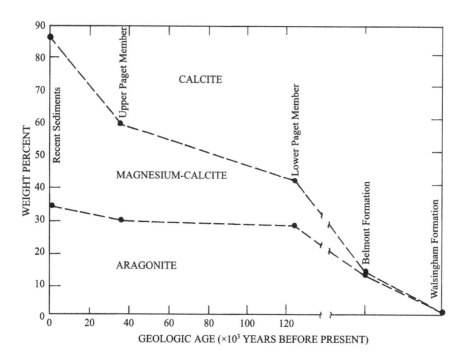

Figure 3.12. Change in the mineral composition of Bermuda Limestone as a function of age (source: Plummer et al. 1976).

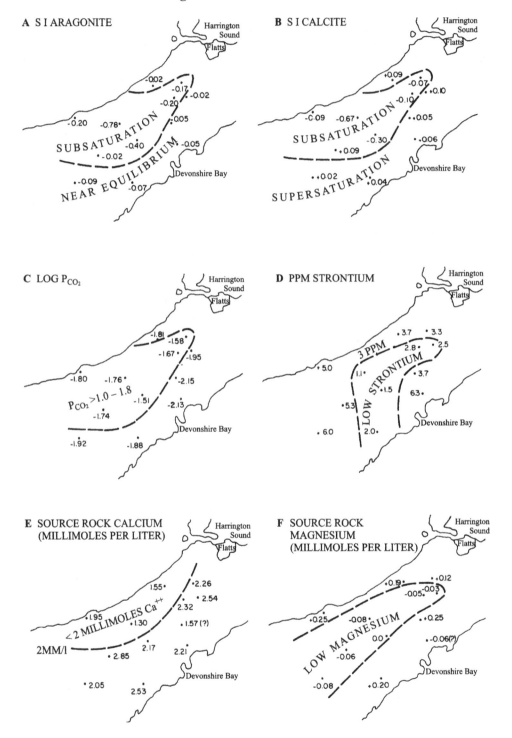

Figure 3.13. Groundwater geochemistry of Bermuda aquifer (source: Plummer et al. 1976).

is negligible. It has been found that much of $Ca^{2+}$ is derived from the dissolution of the rock, but very little $Mg^{2+}$. This may be because the process of dissolution of high Mg-calcite is much slower than the residence time of groundwater in the Bermuda aquifer (which is about 100 years).

### 3.6.4 *Quality of spring water*

Springs constitute the link between the groundwater and surface water. Most often they are the starting points for streams.

Appelo & Postma (1996, p. 116-119) brought out the relationship between the water quality of springs and the lithologies (micaschists, limestones, dolomites, gypsiferous layers, and serpentine rock) in Ahrntal Alps in northern Italy. It has been found that only some *reactive* minerals, such as gypsum, sulfides, calcite and dolomite, talc, serpentine and chrysotile, actively contribute to water composition. The low and uniform levels of $Na^+$ and $K^+$ in the waters indicate that the contribution from feldspar and mica are minimal.

Figure 3.14 (source: Appelo & Postma 1996, p. 118) is a plot of the difference $(Mg - Ca_c)$ versus electrical conductivity ($Ca_c$ corresponds to calcium concentration corrected for dissolution of gypsum). It shows how the rock types and mineral reactions influence the chemical characteristics of the spring water.

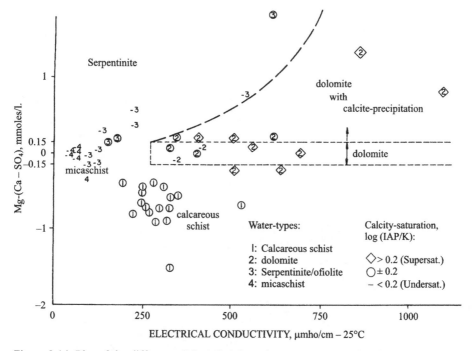

Figure 3.14. Plot of the difference $(Mg - Ca_c)$ in spring waters versus electrical conductivity (Appelo & Postma 1976).

The following four types of water have been delineated:

1. $CaCO_3$ water: This is derived from calcareous schists. The low EC is a consequence of the low $SO_4^{2-}$ content. It is only slightly supersaturated with respect to calcite.

2. Dolomite water: Its (Mg – $Ca_c$) numbers are around 0. The high EC is a reflection of high $SO_4^{2-}$ content, presumably derived from gypsum. Part of the $Ca^{2+}$ released from gypsum precipitates as thick travertine deposits of calcite at the spring hollow. With the precipitation of calcite, there is an increase in (Mg – $Ca_c$). The water is strongly supersaturated with respect to calcite.

3. Serpentine water: Relative to dolomite water, the serpentine water has lower EC, and a similar and occasionally higher (Mg – $Ca_c$). This water is always subsaturated with respect to calcite.

4. Micaschist water: This water has low EC of less than 100 $\mu$S cm$^{-1}$. It is strongly subsaturated with respect to calcite and dolomite.

The detailed chemical composition of the spring waters does not fully match the hard rock geology, however. This may be a consequence of the following causes:

– The surface deposits (moraines, slumps, slope deposits), rather than the underlying hard rocks, may be controlling the water chemistry, or

– the spring originates from a fault or fissure zone which transects the lithological units (Appelo & Postma 1996, p. 118-119).

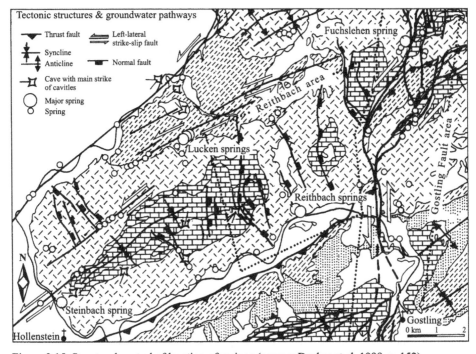

Figure 3.15. Structural control of location of springs (source: Decker et al. 1998, p. 152).

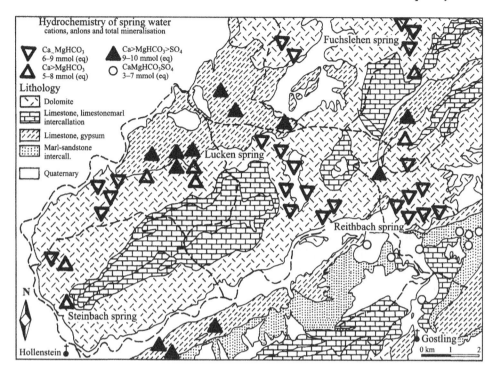

Figure 3.16. Relationship between lithology and hydrogeochemistry of spring water (source: Decker et al. 1998, p. 151).

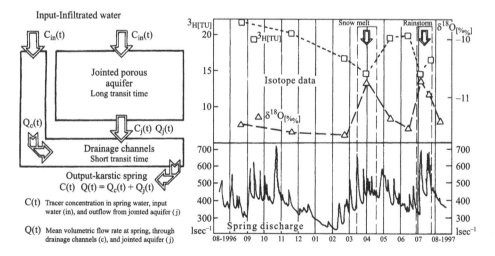

Figure 3.17. Two kinds of water movement in the Reithbach spring, as indicated by isotopic studies (source: Decker et al. 1998, p. 155).

That the locations of springs in the Reithbach area of the Alps are strongly subject to structural control is clearly evident from Figure 3.15 (source: Decker et al. 1998). Major springs (e.g. Lucken springs) are associated with strike-slip faults.

Figure 3.16 depicts the relationship between lithology and the hydrochemistry of the spring water (in terms of cations, anions and total mineralisation):

1. $Ca \sim Mg\ HCO_3$; 6-9 mmol (eq) spring waters are associated with dolomites.
2. $Ca > Mg\ HCO_3$; 5-8 mmol (eq) spring waters are associated with dolomites.
3. $Ca > Mg\ HCO_3 > SO_4$; 9-10 mmol (eq) spring waters are associated with limestone, gypsum.
4. $Ca\ Mg\ HCO_3\ SO_4$; 3-7 mmol (eq) spring waters are associated with Quaternary deposits.

A study of spring discharge ($1\ s^{-1}$) versus isotopic studies (using $^3H$ and $\delta^{18}O$) of the Reithbach tubular spring indicate two kinds of water movement: a slow movement through the jointed aquifer (residence time of 6-10 yrs), and fast movement through open fissures and drainage channels (residence time of a few days to few weeks) (Fig. 3.17: Source: Decker et al. 1998, p. 155).

## REFERENCES

*Suggested reading

Agarwal, H.C. et al. 1986. DDT residues in the River Jamuna in Delhi, India. *Water, Air, Soil Pollut.* 28: 89-104.

Alekin, O.A. & L.V. Braznikova 1968. Contribution à la connaissance de l' écoulement des substances dissoutes de la surface terrestre du globe. *Gidrokhimiceskie materialy* (in Russian) 32: 12-24.

*Appelo, C.A.J. & D. Postma 1996. *Geochemistry, Groundwater and Pollution*. Rotterdam: A.A. Balkema.

*Decker, K. et al. 1998. Karst springs, groundwater and surface runoff in the calcareous Alps: assessing quality and reliance of long-term supply. IAHS Publ. no. 248: 149-156.

Domenico, p. A. & F.W. Schwartz 1990. *Physical and chemical hydrogeology*. New York: Wiley.

Drever, J.I. 1988. *The geochemistry of natural waters*. 2nd. ed. New York: Prentice-Hall.

GEMS 1983. *Data Evaluation Report. GEMS/WATER programme*. WHO, Geneva. Doc EPF/83.55.

GEMS 1988. *The Quality of the Environment: A Health-based Global Assessment*. WHO, Geneva. WHO /PEP/88.15.

Goldemberg, J. 1992. Energy, Technology and Development. *Ambio*, 21(1): 14-17.

Hansen, B.K. & D. Postma 1995. Acidification, buffering and salt effects in the unsaturated zone of a sandy aquifer, Klosterhede, Denmark. *Water Resour. Res.* 31: 2795-2809.

*Hem, J.D. 1985. Study and interpretation of the chemical characteristics of natural water. 3rd. ed. US Geol. Surv. Water Supply Paper 2254.

Livingstone, D.A. 1963. Chemical composition of rivers and lakes. Data of Geochemistry. Chapter G, US Geol. Surv. Prof. Paper, 440-G, p. G1-G64.

Meybeck, M. 1979. Concentrations des eaux fluviales en éléments majeurs et apports en solution aux océans. *Rev. Geogr. Phys. Geol. Dyn.* 21: 215-246.

Meybeck, M. 1987. Global chemical weathering of surficial rocks estimated from river dissolved loads. *Amer. J. Sci.* 287: 401-428.

Meybeck, M. 1996. River water quality, global ranges, time and space variabilities. *Verh. Int. Verein. Limnol.*, 26, p. 81-96.

\* Meybeck, M. 1998. Surface water quality: Global assessment and perspectives. In H. Zebidi (ed.), *Water: A looming crisis?* Paris: Unesco, p. 173-185.

\* Meybeck, M. & R. Helmer 1989. The quality of rivers: from pristine stage to global pollution. *Global. Planet. Change*, 1: 283-309.

Meybeck, M. & R. Helmer 1992. An introduction to water quality. In D. Chapman (ed.), *Assessment of the Quality of the Aquatic Environment through Water, Biota and Sediment*. London: Chapman & Hall, p. 1-17.

Petts, P. & G. Eduljee 1996. *Environmental Impact Assessment for waste treatment and disposal facilities*. Chichester: John Wiley.

Plummer, L.N. et al. 1976. Hydrogeochemistry of Bermuda: a case history of groundwater diagenesis of biocalcarenites. *Geol. Soc. Amer. Bull.*, 87, p. 1301-1316.

Plummer, L.N., T.M.L. Wigley & D.L. Parkhurst 1978. The kinetics of calcite dissolution in $CO_2$ water systems at 5° to 60° C, and 0.0 to 1.0 atm $CO_2$. *Am. J.Sci.*, 278: 179-216.

Schmidtke, N.W. (ed.) 1988. *Contamination in large lakes*. I-IV. Boca Raton: Lewis.

*Standard methods for the examination of water and wastewater*. 1985. Joint Publication of APHA, AWWA, WPCF, 16 th. ed., Washington, D.C. Am. Publ. Health Assn.

Stumm, W. & J.J. Morgan 1981. *Aquatic Chemistry*. 2nd. ed. New York: Wiley.

Shiklomanov, I.A. 1998. *World Water Resources – A new appraisal and assessment for the 21st. century*. Paris: UNESCO.

Train, R.E. 1979. *Quality criteria for water*. London: Castlehouse Publications.

CHAPTER 4

# Augmentation and conservation of water resources

## 4.1 RAINWATER HARVESTING

Any method that saves water is a conservation method. Since ancient times, farmers and herders living in areas of limited rainfall, practiced a variety of water conservation measures. This allowed population densities of 10-500 persons $km^{-2}$ in areas with annual precipitation of 100-1500 mm. Prinz & Wolfer (1998) described the traditional techniques of water conservation under three heads:
- Making better use of rainfall – by minimizing runoff losses (in combination with increased infiltration), collection and storage of rainfall, minimizing evapotranspirational losses, and improving rainwater usage by plants,
- Collection of fog drip and dew, and
- Making use of groundwater without water lifting, through quanats, artesian wells and horizontal wells. These water conservation measures are low-cost, labor-intensive operations and are hence particularly suitable for the Developing countries. It should, however, be borne in mind that rainwater can be harvested only if there is rain – no rain, no rainwater harvesting!

### 4.1.1 *Rainwater harvesting from rooftops*

The following account is largely drawn from UNEP (1983) and Pacey & Cullis (1986).

In many regions of the world, heavy rainfall is confined to a three to four months in a year. During the rest of the time, it is mostly dry, with meager or no rainfall. The traditional practice in such regions has been to harvest and store rainwater and storm runoff during the rainy season for use during the dry season. Intercepting the rain before it reaches the ground has the advantage that water can be collected with minimum of contamination, and is therefore suitable for domestic use.

From time immemorial, many cultures practiced the collection, storage, treatment, distribution and use of rainwater and other forms of precipitation. Sophisticated examples of rainwater harvesting has been found in the ruins of the Palace of Knossos (1700 BC), the centre of Minoan Crete. In the Alhambra Palace in Granada, Spain, the harvested rainwater was used to operate the fountains and to keep the palace cool. In Mexico, two types of rainwater harvesting was practiced – slope wash was trapped through contour terraces (locally known as *trincheras*, or *metlepantli*) which were constructed by placing long rows of stones along contours of slopes. Apart from trapping the runoff, such an arrangement provides for the retention of the soil, and facilitates erosion control. Such a system was used widely in Rio Gavilan (Chihuahua), Tehuacan Valley (Publa) and Nochixtlan valley (Oaxaca) in Mexico.

Rainwater harvesting can be done on any kind of roof. Roofs of corrugated galvanized iron sheets are common in most Developing countries – they are cheap and durable and have low maintenance costs. They tend to rust in the coastal regions, however. The poorer people live in mudhuts with thatched roofs. In the case of thatched roofs, the rainwater would tend get colored, and therefore unattractive. Also thatched roofs are less durable than corrugated iron sheets or tiles. One way to get over the problem is to cover the thatched roof by PVC sheet during the rainy season of about four months. Alternatively, bituminous paper or sisal or coir-cement corrugated sheets could be used. Metal or bamboo guttering may be used. The first flush of rainwater is not collected, as it may be carrying dust, worms or unwanted debris. The harvested rainwater is made to pass through a baffle or filter to remove any impurities. The harvested water may be stored in a variety of containers, depending upon what a family can afford – 'ghala' tank, brick structure, or ferrocement tank, aluminum tank, etc. A ghala tank is a large granary basket. It is made by molding cement inside and outside of a bamboo framework. If the cement wall thickness is three cm, the ghala tank could have a capacity of 2500 l. The usual practice is to keep a tap to draw water about 15 cms above the bottom of the container to avoid mud that may accumulate at the bottom (Fig. 4.1; source: UNEP 1983).

A family can be assured of clean water for domestic use at the rate of about 60 l d$^{-1}$, assuming a rooftop with an area of 50 m$^2$, rainfall of 1000 mm, and collection efficiency of 50%. Bermuda is an excellent example of efficient application of rainwater harvesting technology. Bermuda depends for its supplies of water (80 l d$^{-1}$ per capita) almost exclusively on harvesting of precipitation (1430 mm). The rainwater is collected and stored in cisterns beneath the houses. Necessity, tradition, financial incentives and administrative procedures have been effectively integrated to bring about this highly desirable situation. The Portuguese colonizers promoted rainwater harvesting in some towns in Mozambique (e.g. Pemba) by incorporating the facility in the building design. Rainwater harvesting is practiced in Israel and in the Rajasthan province in India.

There are some simple methods of conserving water in homes. Bathrooms can be so designed that the wastewater from bathing and washing passes through a filter to a cistern from where it can be used to flush the lavatories by gravity flow. Indian-style, pour-flush lavatories with stainless-steel commodes require just 2.5 l of water per flush, as against 15 l required for normal WCs. In California, USA, some municipalities promote conservation of water by providing water-efficient

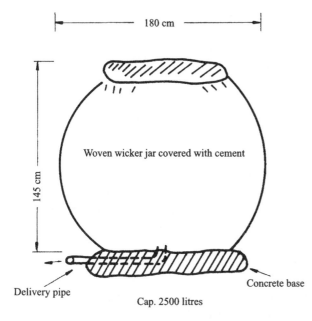

Figure 4.1. 'Ghala' tank (source: UNEP 1983).

WCs free to the families. Besides, wastewater from bathing and washing is often used to water the gardens which tend to have a preponderance of low water need plants (such as, bougainvillea, cactus, roses, etc.).

Farming communities in arid lands (e.g. South Africa) routinely save wash water and give it to animals, such as pigs.

Micro-enterprises are ideally suited to promote rainwater harvesting by individual families in the Developing countries.

## 4.2  HARVESTING OF SURFACE RUNOFF

As in the case of rainwater harvesting, collection, storage and distribution of storm runoff, have been practiced by ancient cultures from time immemorial. Between 305 and 311 AD, Emperor Galerius of Rome built a structure to control the outflow of Lake Balton in the Roman province of Pannonia (the present day Hungary). In South India, thousands of irrigation tanks were built by the ancient kingdoms, several of which continue to be in use to this day (currently, minor irrigation tanks in India have a storage capacity of about 14.3 M ha m).

In Negev desert, rainwater from rooftops, courtyards and streets is collected into settling tanks. Prior to storage, water collected from ground catchments can be cleaned by the use of silt traps and sand filters. Clear water is then channeled into underground cisterns in the house. The capacity of the cistern, which may be in the form of well-sealed, brick and mortar construction, may be 5-10 m$^3$. The water requirements of a family are usually computed at the rate of 1.5 m$^3$ per person, and one m$^3$ per camel. Above-ground storage tanks with a capacity of 1000 l could be built with galvanized iron sheets, but when full, such a tank may get deformed, unless reinforced by steel or wooden framework.

Diverting the flood for warping disperses the runoff and moderates the flood peak. Thus warping reduces the flood hazard, besides converting a barren gully or flood plain into a fertile cropland. A good example of this application is Mawan Creek in Jingbian county in China, with a catchment area of about 100 km$^2$. The usual practice is to construct a dam at the mouth of the gully to divert the muddy water to the flats on either bank of the stream. In some localities in India, flood water is diverted for leaching saline soil.

The usual practice is to construct a dam at the mouth of the gully to divert the muddy water to the flats on either bank of the stream. Alternately, intercepting ditches and canals are constructed along contour lines to intercept the overland flow on hill slopes, and the water thus intercepted is used for warping river flats and terraced fields (Fig. 4.2; source: UNEP 1983)

Harvested rainwater is protected from contamination by a simple device, namely, a baffle tank. The runoff passes through a baffle and sieve, before going over a stand-pipe. Sediment settles down in the bottom of the box, which needs to be periodically cleaned (Fig. 4.3; source: UNEP 1983).

In Mali and Burkina Faso in West Africa, a 'Zay' system of water conservation is followed. The Zay system consists of small pits, 5-15 cm in depth and 10-30 cm in diameter. Soil mixed with manure and grasses is put into the Zay to increase the water-holding capacity of the pits. Plants or bushes are grown in the pits after rains. Zays are often combined with bunds to conserve soil moisture and improve soil fertility. In Tunisia, a micro-catchment system (called '*Meskat*') provides fruit tree plantations with about 2000 m$^3$ of extra water during the rainy season. Studies made in Kayes province, Mali (with annual precipitation of 550 mm) show that the construction of earth dikes around a field, could increase the yield of sorghum from 0.8 t ha$^{-1}$ to 1.8 t ha$^{-1}$, even in dry years (Klemm 1990, quoted by Prinz & Wolfer 1998).

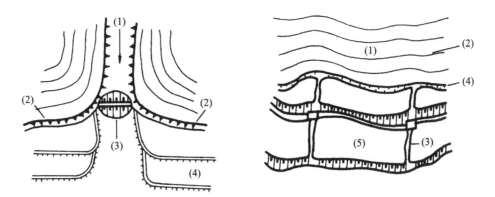

Figure 4.2. Construction of ditches and canals along contour lines for intercepting surface runoff (source: UNEP 1983).

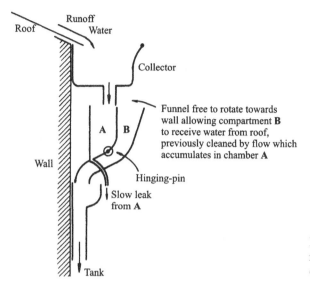

Figure 4.3a. Design of collector for rainwater harvesting from roof (source: UNEP 1983).

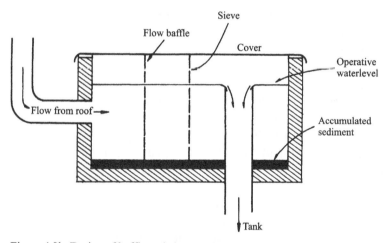

Figure 4.3b. Design of baffle tank (source: UNEP 1983).

Many areas in the world, including arid areas, are often covered with low clouds. In some countries (e.g. Mexico, Chile, Colombia, Sudan, Yemen, Oman), fog drip is collected in coastal and high mountain areas. In Crimea, there exists a system of earthenware pipes up to 10 m high, which might have served as condensation promoters.

The addition of organic matter, such as compost, manure, peat, paper, crushed coal, etc. leads to higher moisture retention and improved soil fertility. Some synthetic compounds, such as starch copolymers or granular polymers (such as Polyacrylamid or PAM) have been successfully applied to achieve the same purpose. The choice of the compound would depend upon the kind of root system (fine roots/deep roots) that the crop plant concerned has.

### 4.2.1  *Ways and means of increasing the runoff from an area*

All precipitation does not lead to runoff – in sandy areas or after a long spell of dry weather, rainwater may simply percolate down or evaporate. The UNEP document on 'Rain and Stormwater Harvesting' (1983, p. 43) suggests the following procedures to increase the runoff from an area:

1. The sloping surfaces may be cleared of vegetation and loose material. After doing this, the surface may be contoured and compacted. This will lead to greater efficiency in the runoff collection. In Australia and USA, the usual practice is to develop a V-shaped contour producing a 'roaded' catchment.
2. Vegetation management: Developing runoff enhancing vegetation in the areas immediately adjacent to the stream channel. Shrub cover may be cleared, but in a manner that will not cause erosion.
3. Mechanical treatment: Smoothing and compacting of surface. Where the run-on area is minimum, and contains a vertical mulched slot, there will be deep penetration of water, and evaporation losses may be reduced by about 50%.
4. Chemical application to reduce soil permeability: Colloidal dispersion and/or hydrophobic treatment with appropriate chemicals can be made use of to reduce soil permeability and enhance the runoff. Silicones are easy to apply and are relatively inexpensive. Silicone treatment would, however, be ineffective if the soil contains swelling clays like bentonite. When sodium methyl silanclate is applied, it penetrates into the soil to form an inert, hydrophobic resin, which is not biodegradable. A runoff efficiency of 94% have been reported for soils treated in this way, as against 35% runoff in the case of untreated, smooth beds. Repellancy drops off to 40% after a few years, but can be improved to 85% after re-treatment. Significant increases in runoff have been achieved by the compaction of low-clay, sodium-treated soils. Additionally, salt is a herbicide.
5. Surface binding treatment: Surface can be treated with impermeable suspensions in order to increase the runoff. For instance, asphalt emulsions can lead to virtually 100% runoff. Paraffin can be used in the form of granules or flakes which are then allowed to melt and spread. Other possibilities are: black PVC, polyvinyl fluoride, and butyl film.
6. Rigid surface coverings: Corrugated galvanized iron sheets can be used to collect the runoff, as is the case in Gibraltar.
7. Flexible surface coverings: Butyl rubber is commonly used for the purpose because of its low cost. In Hawaii and the Pacific Islands, butyl rubber sheeting is used to collect runoff from large catchments of 1-7 ha, on slopes of 40%.

### 4.2.2  *Storage and utilization of rainwater from ground catchments*

Rainwater storage ponds are best located where there is:
- A natural, saucer-shaped depression, or
- A shallow ravine with a small flowing stream which can be impounded, and

– The ground has high water retention capacity. In Australia, self-contained butyl bag installations with a capacity of 228,000 l are commercially available for providing water to the livestock.

Consider, for instance, the drinking water requirements of a sheep farm. At the rate of about 10 l d$^{-1}$ per animal, a farm with 500 sheep would need to have a pond which can provide at least 10$^6$ l per annum. Having spent considerable sums of money to harvest the rainwater, it is necessary to ensure that stored rainwater is not lost through seepage or evaporation.

If the soil porosity is high, the bottom and the sides of tank have to be sealed with a` suitable sealant such as an embedded membrane of plastic sheeting or bentonite. Powdered sodium bentonite is spread over the dry surface at a minimum rate of about 5 kg m$^{-2}$. Soils with high sand content need to be compacted, before bentonite is applied at the rate of 19.6 kg m$^{-2}$. In Hawaii, the usual practice of sealing is to use top and bottom layers of butyl laminated with 0.8 mm nylon.

The various techniques of reducing evaporation by covering the surface of rainwater storage basins, are summarized below (Table 4.1) (source: UNEP 1983):

A low-cost, ecologically-sound option would be to line the pond with clay (to reduce seepage), and cover the pond with bamboo or wooden poles and train viny vegetables with large leaves to cover the pond. The vines would not only reduce evaporation but would also provide vegetables.

It is well-known that water table slopes from a higher terrain in the direction of the plains. Groundwater flows horizontally along the slope. In Iran, gently sloping tunnels (called '*qanats*' or '*quanats*') were constructed thousands of years ago to bring groundwater from mountainous areas to arid plains. Qanats may have a slope of 1-2%, lengths up to 30 km, and could yield water in the range of 5-60 l s$^{-1}$ (in some cases, up to 270 l s$^{-1}$) (Fig. 4.4).Vertical dug wells were constructed to take the water to the surface. The qanat system was in vogue not only in Iran

Table 4.1. Techniques of reducing evaporation.

| Technique | Reduction in evaporation (%) |
|---|---|
| 1. *Changing of water color* | |
| Dye in the water | 6-9 |
| Shallow colored pans | 35-50 |
| 2. *Using wind barriers* | |
| Baffles | 11 |
| 3. *Shading the water surface* | |
| Plastic mesh | 44 |
| Blue, polylaminated plastic sheeting | 90 |
| 4. *Floating reflective covers* | |
| White butyl sheets | 77 |
| Polystyrene rafts | 95 |
| Continuous wax | 87 |
| Foamed butyl rubber | 90 |

but also in Afghanistan, India and China (Megasthanes who was travelling in India in 300 BC recorded the use of qanats for water supply in Baluchistan which is now a province in Pakistan).

n northern Yemen, there exists a system dating back to 1000 BC, whereby flood waters are diverted to irrigate 20,000 ha of land. The agricultural products from this system must have supported a population of about 300,000 people.

Sometimes, groundwater may exist under artesian conditions, and may be able to flow out under its own pressure (i.e. without pumping) (Fig. 4.5: source: Prinz & Wolfer 1998). There may be situations where the groundwater is trapped by a vertical impervious geological barrier. Such groundwater can be tapped by boring a horizontal hole, and inserting a steel pipe casing. Horizontal wells are inexpensive to construct, and no pumping is needed. There is no possibility of contamination by rubbish or animals (Fig. 4.6; source: Prinz & Wolfer 1998).

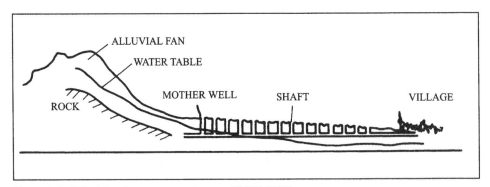

Figure 4.4a. Principle of qanat system (source: UNEP 1983).

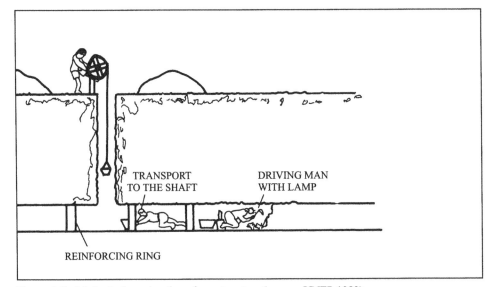

Figure 4.4b. Method of construction of qanat system (source: UNEP 1983).

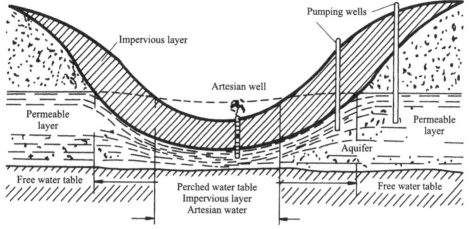

Figure 4.5. Principle of the artesian well (source: Prinz & Wolfer 1998).

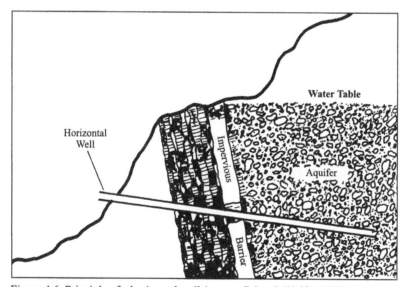

Figure 4.6. Principle of a horizontal well (source: Prinz & Wolfer 1998).

There are 500,000 tanks in peninsular India for providing drinking water and irrigation. In the province of Karnataka alone, tanks have a storage capacity of 200,000 m$^3$. But soil erosion and consequent silting reduces the storage capacity of the reservoirs at the rate of 0.5% per year. It is therefore critically important to minimize silting of reservoirs through practices such as the afforestation of the catchment, banning of cultivation of the foreshore lands, construction of soil traps in the catchment, etc.

Suspended sediment constitutes the chief visible impurity in the case water impounded by earth dams. Where water is especially precious, as in Eyre Peninsula in Australia, livestock is not allowed direct access to reservoirs, but is allowed to drink only from the troughs. Figure 4.7 shows the diagram of the earthen hillside dam, in which a simple float-controlled outflow system taps water just below the surface but away from the muddy sediments.

Pot filters are used in homes to remove turbidity. In most cultures, indigenous plant products or natural products are used to remove turbidity from drinking water.

## 4.3  GROUNDWATER RECHARGE

In the recent years, there has been much controversy about the construction of large dams (e.g. Sardar Sarovar dam over the Narmada River in India and the Three Gorges project over Yangtze river in China). Apart from the huge expenses involved, large populations may have to be moved and resettled away from the areas that are going to be flooded, silting reduces the life-span of the reservoirs, and buildings, cultural monuments, soils, forests, mineral deposits, etc. may be irretrievably lost, etc. But a case can always be made that the benefits (such as, flood control, power generation, water for irrigation, industries, domestic supply, and navigation) far outweigh the disadvantages. While it is true that some advantages of the high dams can still be achieved through a number of small dams, but large dams may still necessary to secure some benefits (such as, power generation and navigation).

Along with surface reservoirs, underground reservoirs created by artificial recharging of groundwater are becoming increasingly common (Asano 1985). Presently, the total volume of the world's surface reservoirs is about 6000 km$^3$, and the total surface area is up to 500,000 km$^2$ (Shiklomanov 1998, p. 21). Even 20 yrs ago,

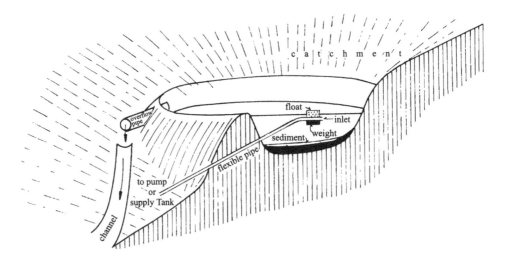

Figure 4.7. Float-controlled outflow system (source: UNEP 1983).

L'vovich (1979, p. 343) had the vision to propose the artificial conversion of surface runoff to underground runoff of the order of 3000 to 5000 $km^3 yr^{-1}$ – in other words, the capacity of the underground reservoirs would roughly have to be of the same order as surface reservoirs.

There are several advantages in storing water in underground reservoirs:
– Water in an underground reservoir is better protected against pollution than water in the surface reservoirs,
– Underground reservoirs do not interfere with the land use, as the surface area above the underground reservoir can still be used for economic purposes, such as agriculture,
– Water stored in the underground reservoirs is not lost due to evaporation (which may correspond to a layer of a few metres for surface water bodies in tropical countries),
– Though the recovery rate of the water stored in an underground reservoir is less than unity because of effluent seepage of groundwater, that is beneficial as it will feed the rivers and makes for an increase in stable runoff.

Groundwater recharging allows the withdrawal of more water from the aquifers than is possible under the normal circumstances. Several cities in the world presently get their supplies of water from such underground reservoirs. Since ancient times, the Turkmens living in the Kara Kum desert drained the surface runoff from the clayey areas of the desert into wells. Such water formed underground lenses of fresh water which were protected from evaporation. Men and animals (sheep) made use of the water thus stored.

Geopurification or Soil-Aquifer Treatment (SAT) of wastewater for being used for non-potable purposes is a kind of groundwater recharge with infiltration basin (Bouwer 1994). This subject has been dealt with in detail under Section 5.5.1.

In some townships (e.g. Long Island, New York), excessive withdrawal of fresh water from the coastal aquifers has resulted in the salinisation of fresh water due to the incursion of sea water. This process has been reversed by the recharge of storm water runoff into the aquifers. Though the runoff water is not exactly pure, it served the useful purpose of displacing the saline water, and mitigating the salinization of coastal aquifers.

When water is charged into highly permeable soils and rocks, it percolates deep down, and forms underground reservoirs. This method is simple, but it may need the flooding of large areas, and the quantities of storable water may not be much. The usual practice is the artificial injection of water into the aquifers by means of drilled wells. Normally we draw water *from* a well – in the case of artificial recharge, we pump the water *into* the well. An issue to be borne in mind in the case of artificial recharge is the possibility that it may trigger low-magnitude earthquakes. In 1962 when toxic fluid wastes were pumped under pressure into a borehole in the Denver Basin in USA, low-magnitude earthquakes got triggered. It has been found that nearly critical stress already existed in the rock mass, and the additional water pressure triggered the earthquakes.

A sensible solution is to go in for a combination of surface reservoirs and artificial injection, particularly where the recharging has to be done with flood water. The spring flood flow in the temperate regions usually lasts for one month, and rarely as long as three months. The monsoon flood flow in the tropical lands usually lasts 3 to 4 months. Thus the underground storage of the flood water needs to be accomplished in a few months' time. By having a combination of above-ground and below-ground reservoirs, groundwater recharge could be done throughout the year.

An excessive use of runoff water will reduce the groundwater runoff and consequently the stable part of the runoff. Groundwater moves slowly, and therefore has a high regulating capacity. Thus, a good practice would be to charge the groundwater during wet seasons and years, to be used when needed.

## 4.4   TAPPING OF GROUNDWATER INFLOW INTO THE SEA

Shiklomanov (1998) estimates that the direct groundwater runoff to the ocean is about 2100 km$^3$ yr$^{-1}$, amounting to about 5% of the total runoff of the Earth's rivers (42,700 km$^3$ yr$^{-1}$). Two aspects about the groundwater inflow into the sea need special mention: it is quite substantial, and very little of it is being presently tapped.

### 4.4.1   *Methodology of estimating the groundwater inflow into the sea*

Chapelle in his evocative book *The Hidden Sea* (1997) drew attention to the significance of observations made by Mark Twain *(Life on the Mississippi)* about the change in the level of Mississippi River from droughts to floods:

The rise is tolerably uniform down to Natchez – about 50 ft. But at Bayou La Fourche, the river rises to only 24 ft.; at New Orleans, only 15, and just above the mouth only two and half.

Chapelle (1997) argued that since the river channel becomes narrower towards the mouth, the behavior to be expected is exactly the opposite of the actual observation. He says that the conclusion is inescapable that Mississippi must be discharging large quantities of water into the sea through subterranean channels.

Moore (1999) coined the term, *subterranean estuary*, to describe a coastal aquifer where groundwater derived from land drainage, discharges into the sea.

Moore and his colleagues developed the strategy to use tracers (principally radium quartet) in the coastal ocean to assess fluid inputs from subterranean estuaries (Rama & Moore 1996; Moore 1996, 1999). The procedure involves the following steps:
– Identify tracers derived from fluid advection that are not recycled in the coastal ocean,
– Map the distribution of the tracer in the coastal ocean and evaluate other sources,

– Determine the exchange rate with the open ocean,
– calculate the tracer flux to the ocean, and hence the tracer flux from fluid advection,
– Measure the average tracer concentration in subsurface fluids to calculate fluid flux, and
– Measure the other components of interest in the fluids to calculate fluxes of reactive species to the coastal ocean due to submarine fluid input '(Moore 1999, p. 115).

The radium quartet comprises of four isotopes of radium derived from thorium parents, and having widely varying half-lives: $^{224}$Ra (t$_½$: 3.6 days), $^{223}$Ra (t$_½$: 11.4 days), $^{228}$Ra (t$_½$: 5.7 yrs), $^{226}$Ra (t$_½$: 1600 yrs) (Moore 1998). The use of the radium quartet to estimate the coastal mixing rates and the groundwater discharge to the oceans is based on two considerations:

– There is a profound difference in the geochemical behavior of thorium parent and radium daughter. Thorium is a hydrolyzate element, and tends to be tightly bound to particles, whereas radium gets mobilized in the marine environment – in other words, the daughters and parents follow different geochemical pathways, and
– Though the four radium isotopes have a coherent geochemical behavior, they have different rates of generation from the parent – the short-lived isotopes are generated on a time scale of a few days or few tens of days, whereas the long-lived Ra isotopes can be considered essentially stable in the context of period of observation. Thus, by the measurement of Th and Ra isotopes in sediments and waters, it is possible to quantify the chemical exchange in the coastal ocean.

### 4.4.2  *Case studies*

Moore (1998) gave a case study of the application of this methodology in the case of Winyah Bay Transect in the coastal waters of South Carolina, USA. The different radium isotopes were measured in the waters offshore upto a distance of 100 km. Plots were made of activity (dpm/100 l) of the individual radium isotopes vs. the distance from the coast (km). The horizontal eddy diffusion coefficient ($K_h$) was computed from $^{224}$Ra activity plot. The $^{226}$Ra plot showed that the concentrations decrease in a roughly linear fashion (from 17 to 9 dpm/100 l), 15 to 50 km offshore. The offshore gradient has demonstrably been caused by dilution but not due to chemical or biological removal. On the basis of the comparison of the $^{226}$Ra fluxes offshore and in the submarine groundwater, it was surmised that the submarine groundwater inflow to the sea is of the order of 40% of the surface water flow. It does not, however, follow that all the groundwater flux is fresh water, as the groundwaters responsible for $^{226}$Ra flux have variable salinity (Moore 1998).

The strategy used in this study can be extended to make an estimate the fluxes of nutrients, trace metals and organic compounds. There is little doubt that

Table 4.2. $^{226}$Ra fluxes in different locations (source: Moore 1999).

| Region | Setting | Size (km$^2$) | $^{226}$Ra flux (dpm d$^{-1}$) | $^{226}$Ra flux (dpm m$^{-2}$ d$^{-1}$) |
|---|---|---|---|---|
| North Inlet, SC, USA | Salt marsh | 32 | $3 \times 10^9$ | 100 |
| FSUML, Gulf of Mexico, USA | Near shore | 620 | $8 \times 10^{10}$ | 130 |
| South Atlantic Bight | Inner shelf | 6 400 | $2 \times 10^{11}$ | 30 |
| South Atlantic Bight | Section of outer shelf | 2 600 | $3 \times 10^{11}$ | 100 |
| Bay of Bengal | Inner shelf, low river discharge | 18,000 | $1 \times 10^{13}$ | 500 |
| Amazon River | Total shelf annual average | 130,000 | $3 \times 10^{12}$ | 20 |

groundwater discharge to the oceans takes place all over the world. Using Ba and other tracers, Tsunogai et al. (1996) identified fluid seepage in Sagami Bay, Japan, at a depth of 900-1100 m. According to them, such discharges of fluids into the ocean play an important role in controlling ocean chemistry.

A comparison of $^{226}$Ra fluxes in different case studies are given in Table 4.2.

When the total tracer fluxes (dpm d$^{-1}$) are normalized to the area (dpm m$^{-2}$ d$^{-1}$), most of the $^{226}$Ra fluxes are found to be of the order of 100 dpm m$^{-2}$ d$^{-1}$.

The exception is the Bay of Bengal, where the normalized flux is about five times greater. This calls for an explanation. The annual flux of the sediment from the Ganges and the Brahmaputra rivers into the Bay of Bengal is among the highest in the world (the Ganges carries a suspended load of $1760 \times 10^6$ t yr$^{-1}$, in its discharge of 971 km$^3$ yr$^{-1}$). The desorption of $^{226}$Ra and Ba from the sediment provides point sources of these elements into the sea. As is well known, the highest discharges of water and sediment in these rivers occur during the monsoon season (June-September). Moore (1997) made a significant observation that during March 1991 when the discharge of the Ganges and Brahmaputra rivers was at their lowest, the fluxes of $^{226}$Ra and Ba in the northern Bay of Bengal were still comparable to the river-derived fluxes at the time of the peak discharge during the monsoon. Such an observation can only be explained by invoking a large non-riverine source of Ra and Ba. Moore therefore concluded that submarine fluids must be responsible for discharging these fluxes of $^{226}$Ra and Ba.

At the other end of the scale, the Amazon shelf contributes only 20 dpm m$^{-2}$ d$^{-1}$ though the water and sediment discharges are enormous.

## 4.5  AMELIORATION OF MINE WATER

All aspects of the mitigation of the environmental impact from mining waste, have been comprehensively dealt with in a state-of-the-art report by MiMi (1998), a Swedish organization devoted to the study. The following account draws heavily from this report.

### 4.5.1 *What is Acid Mine Drainage?*

Acid Mine Drainage (AMD) gets generated due to the oxidative dissolution of the iron-containing sulfide minerals, such as pyrite. Both purely chemical reactions as well as microbially catalyzed reactions are involved. AMD may arise from the mining of coal, lignite, metallic sulfides, uranium, etc. Under oxidizing conditions, and in the presence of catalytic bacteria, such as *Thiobacillus ferrooxidans*, sulfides are oxidized into sulfuric acid, as per the following equation:

$$4\,FeS_2 + 15\,O_2 + 2\,H_2O = 2\,Fe_2\,(SO_4)_3 + 2\,H_2SO_4$$

Pyrite + Oxygen + Water = Iron sulfate + Sulfuric acid

(4.1)

Surface runoff and groundwater seepages with waste piles tend to be highly acidic, and corrosive, and contain high concentrations of iron, aluminum, manganese, copper, lead, nickel and zinc, etc. in solution. The discharge of such waters into streams destroys the aquatic life, and the stream water is rendered non-potable.

An understanding of the physical, chemical and biological processes that lead to the production of AMD is necessary for the following purposes:
– To minimize the production of AMD,
– To dispose of AMD from the operating mines or for the decommissioning of waste piles, as required by law, and
– To ameliorate AMD to allow it to be used for beneficial purposes, such as irrigation, industrial and domestic purposes.

Irrespective of whatever technology is used to mitigate the problem of acid mine drainage, it is necessary to study the focuses of oxidation and flow-pattern of waters in the mine, identification of sources of acid mine water, and how the mine water spreads. Whole-rock analyses and leaching tests can be used to predict the nature and extent of AMD that could develop in a given mine or from a waste pile. Rock samples are leached with water, and the leachate is analyzed for parameters which indicate the pathways of weathering, namely, pH, specific conductance and sulfate. The mineralogical and chemical composition of the rock, the pyrite content and the presence or absence of calcareous material are the determining factors (Table 4.3)

Table 4.3. Guidelines for the choice of tests for acid mine drainage.

| Pyrite content | Leachate characteristic | Method recommended |
|---|---|---|
| < 1% (C.M. absent) | Slightly acidic, low SC | WR or LT |
| 1-1.5% (C.M. absent) | Acidic | WR or LT |
| > 1.5% (C.M absent) | Acidic | WR or LT |
| < 1% (C.M. present) | Alkaline, low SC | WR or LT |
| 1-1.5% (C.M. present) | Alkaline, high SC | LT |
| > 1.5% (C.M. present) | Slightly acidic | LT |

C.M. = Calcareous Material; SC = Specific conductance, WR = Whole rock; LT = Leaching tests

Several countries have prescribed the allowable concentrations in mine efflu-ents: pH: 7, SS (Suspended Solids): 30 mg $l^{-1}$; BOD (Biochemical Oxygen De-mand): 30 mg $l^{-1}$; Pb: 0.2 mg $l^{-1}$; Fe: 0.1 mg $l^{-1}$; Cu: 0.1 mg $l^{-1}$. In other words, the mining companies are expected to treat the mine water in such a manner that the discharge stays within the prescribed limits.

In many countries, a company responsible for noncomplying discharges is is-sued a Notice of Violation by the Environmental Agency of the government con-cerned. If the company does not take prompt remedial action, it is penalized.

The US Bureau of Mines has developed a simple, low-cost, portable and highly efficient system to neutralize the acid mine drainage on site. The only drawback of the system is that it requires at least 130 kPa water pressure, and may not be able to remove manganese if the iron content is low. Apatite can be used to ame-liorate AMD. Apatite is soluble only in acid conditions. So it will act only when the AMD becomes sufficiently acid. The phosphate ion can sequester and precipi-tate $Fe^{3+}$, $Al^{3+}$, $Mn^{2+}$, etc.

### 4.5.2    Mitigation of Acid Mine Drainage (AMD)

Since the supply of both oxygen and water is necessary for the generation of AMD, an obvious way to prevent the formation of AMD is to block the entry of oxygen and water to the mine or waste pile. This is easier said than done, for the simple reason that Fe (III) that may be present in the partly oxidized waste, could serve as an oxidant and still generate AMD. Also, if the pore water in the mine waste is acidic, the mobility of heavy metals gets strongly increased due to their higher solubility and lower tendency for sorption. Thus, if the waste dump contains buffering substances such as calcite, or if lime is added, the develop-ment of acid drainage, and the release of heavy metals could be substantially mitigated.

The methodology and techniques for the prevention and control of AMD are summarized as follows (source: MiMi 1998):

1. Restricting the transport of oxygen and/or water into the waste dump: Dry cov-ering of the waste is an effective way to accomplish this.
   - Oxygen diffusion barriers to limit the diffusion of oxygen into the waste. Experience in Sweden shows that a single layer cover of thickness of 1.0 m results in the reduction of pyrite weathering rate and metal release, in the re-gion of 80-90%.
   - Oxygen consuming barriers: to consume oxygen before it could penetrate the waste. A cover of 2 m of organic waste or lime stabilized sewage sludge can be used.
   - Low permeability barriers: to act as a barrier against the infiltration of the precipitation. These can be constructed of fine-grained soils, clays, geotex-tile or bentonite liners or geomembranes.

– Reaction inhibiting barriers: to create a chemical environment favorable for limiting the reaction rates and metal release.

Flooding of the waste to limit the transport of oxygen into the waste has been tried in some cases in Sweden.

2. Changing the chemical properties of the waste (such as, separation of pyrite or addition of a buffering substance, such as lime) or physical properties of the waste (such as, compaction to reduce porosity and permeability). This is expensive.

3. Treatment of the leachate with the objective of reducing the metal concentrations in the water that is discharged from the waste pile.

### 4.5.3 *Use of mine water*

There is severe scarcity of drinking water in the coalfield areas of eastern India. On one hand, the water-table has gone down to 200-250 m due to mining activities, thus making the tapping of groundwater prohibitively expensive. On the other hand, there is abundance of mine water which, however, is not potable because of its high acidity, and the high content of metals, such as iron. The Central Mining Research Institute, (CMRI), Dhanbad, Bihar, 826 001, India, has developed a treatment process which is claimed to render the mine water potable (item 6.2.13, CSIR Rural Technologies, New Delhi, India 1995). Filtration is done adjacent to the settling pond. Two filter beds are used to work alternatively at the time of changing the bed. A slow or rapid filtration may be employed depending upon the situation. A disinfectant is incorporated in the treatment process to destroy the pathogens.

The presence of high iron content in groundwater is objectionable because of discoloration, turbidity, bad taste and tendency to form deposits in the distribution mains. The National Environmental Engineering Research Institute, Nagpur 440 020, India, developed a simple plant to remove iron from groundwater by precipitating the iron impurity as a ferric sludge (item 6.2.4, CSIR Rural Technologies 1995). The plant is to be attached to a hand pump. It has a capacity of 2500 l $d^{-1}$ (10-hr operation) and costs about USD 500. The plant has three chambers. 'The water from the hand pump is sprayed over an oxidation chamber. The aerated water flows over baffle plates to a flocculation chamber and then to sedimentation chamber. The water then passes through plate settlers and to the filter from where the filtered water is drawn through a tap after chlorination'. The ferric sludge need to be scoured out twice a month.

### 4.5.4 *Passive treatment of acid rock drainage*

Gusek (1995) gave a lucid review of the techno-economic aspects of passive treatment of acid rock drainage.

The conventional method of amelioration of acid rock drainage is the liming of the runoff. Liming neutralizes the water and chemically precipitates the metals.

However, liming is expensive, leaves behind large quantities of sludge, and has to be continued long after the mine ceased operating. For this reason, much R&D effort has been concentrated in developing low-cost, low-maintenance, passive treatments of AMD. These involve the utilization of vegetation and sediment microbial communities found in wetlands to reduce the acidity and precipitate the metals. The techno-economic viability of the passive treatment is now well established. For instance, the Tennessee Valley Authority (TVA)'s Fabius Mine in Alabama, USA, replaced an earlier lime-treatment plant by a large, passive treatment system. The latter treats 126 l s$^{-1}$ (about 2000 gpm) of coal mine drainage. It has been operating for several years and discharging compliant effluent.

Interestingly, wetlands established for water quality improvement have been found to provide habitat for abundant development of herptofaunal wildlife (Lacki et al. 1992).

It has been known that wetlands are capable of improving the water quality by reducing the contaminants through the precipitation of metal hydroxides, sulfides and carbonates and pH adjustments. Whether these reactions would occur under oxidizing (aerobic) conditions or reducing (anaerobic) conditions would depend on the Eh of the environment, and the chemistries of soil and water. Where natural wetlands are not available, wetlands are constructed. The latter are engineered so as to optimize the biogeochemical processes that take place in the natural wetlands. Figure 4.8 (source: Kolbash & Romanovski 1989) shows the design of a constructed wetland. The wetland plants that are most commonly used are *Typha*, *Schoenoplectus*, *Phragmites* or *Cyperus*.

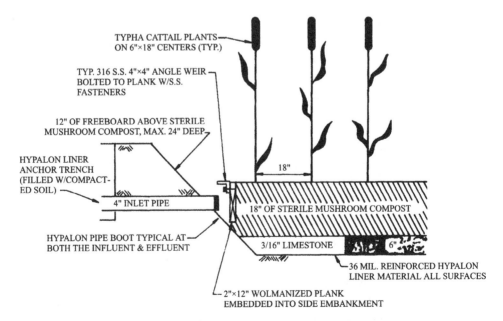

Figure 4.8. Design of a constructed wetland (source: Kolbash & Romanovski 1989).

The important physical, chemical and biological mechanisms that operate in the passive wetland treatment are as follows:
- Hydroxide precipitation catalyzed by bacteria in the aerobic zones,
- Sulfide and carbonate precipitation catalyzed by bacteria in anaerobic zones,
- Filtering of suspended material,
- Metal uptake into live roots and leaves,
- Ammonia-generated neutralization and precipitation, and
- Adsorption and exchange with plant, soil and other biological material.

The predominant mechanisms by which microorganisms remove soluble metals from solution are as follows:
- Volatilization – whereby microorganisms methylate metals,
- Extracellular precipitation – whereby metals are immobilized by the metabolic products produced by microorganisms. Sulfate-reducing bacteria reduce $H_2SO_4$ to $H_2S$, which would readily react with soluble metals to form insoluble metal sulfide minerals,
- Extracellular complexing and subsequent application – whereby chelating agents (known as siderophores) synthesized by microorganisms have a high binding efficiency for some metals, resulting in the generation of metal-binding polymers,
- Binding to bacterial, fungal and algal cell walls, and
- Intra-cellular accumulation (Brierley et al. 1989). Studies made by White & Gadd (1996) showed that the most efficient nutrient regime for bioremediation using sulfate-reducing bacteria required both ethanol as a carbon source and cornsteep as a complex nitrogen source.

Brierly (1990) gave a detailed review of the techniques of bioremediation of metal-contaminated surface and groundwaters. Advances in biotechnology have made it possible to make use of nonliving microorganisms immobilized in polymer matrices to remove low concentrations (<1 to about 20 mg $l^{-1}$) of heavy metal cations in the presence of high concentrations of alkaline earth metals ($Ca^{2+}$ and $Mg^{2+}$) and organic contaminants. The removal process is so effective that the effluent more than satisfies the requirements of US National Drinking Water Standards.

Davison (1993) describes a proprietary Lambda Bio-Carb Process which is an *in situ* bioremediation system utilizing site-indigenous, mixatrophic cultures hybridized for maximum effectiveness. Lambda has catalogued about 6000 microorganisms suitable for the purpose. The system is utilizable in conjunction with wetlands, and is capable of self-adjustment in response to influent changes. It has been successfully used to treat sites contaminated by heavy metals, hydrocarbons, organics, agricultural wastes and other hazardous compounds.

The economics of the passive treatment can be illustrated with a case (Eger & Lapakko 1989). Drainage from the Dunka mine in the mineralized Duluth complex in northern Minnesota has increased the concentration of metals (Ni, Cu, Co and Zn) in the creeks in the proximity, up to 400 times. This was naturally unacceptable to Minnesota Pollution Control Agency, and the company concerned had to give an undertaking to achieve the water quality goals. A feasibility study was made of the options for treating $6 \times 10^8$ l yr$^{-1}$ of mine water:

- A full-scale treatment plant (lime precipitation with reverse osmosis): capital cost: USD 8.5 million, and annual operating cost: USD 1.2 million.
- Passive treatment (combining infiltration reduction, alkaline treatment and wetland treatment): capital cost: USD 4 million; annual operating cost: USD 40,000.

The passive treatment of AMD did not work well in some parts of Canada, presumably because of low temperature (ca. 4°C).

## 4.6 DESALINIZATION OF SALT WATER

Desalinization of seawater in coastal areas or brackish water in inland areas is technically feasible, but expensive (about USD 1 per 1000 m$^3$ for every 10 mg l$^{-1}$ salt removed by reverse osmosis, or about USD 2000-3000 per 1000 m$^3$ for seawater) (Bouwer 1994). Desalinization of seawater is practiced largely in energy-rich countries of the Middle East. According to Al-A'Ame & Nakhle (1995), the cost of desalinization of seawater ranges from USD 1500 to 4500 per 1000 m$^3$, with the cost in the case of Jubail in Saudi Arabia being about USD 2700 per 1000 m$^3$.

In the case of inland areas, one has to reckon with the problem of disposal of the reject water (brine). The latter may have to be flash evaporated with the resulting salts being disposed of in landfills or other suitable repositories. In some situations, it may be an attractive proposition to use the brines to make commercial table-salt or the production of chemicals, such as sodium, potassium, chlorine, etc. compounds.

Reverse osmosis (RO) is the preferred process for desalinization of brackish water. The process is based on the principle that when water is forced through a semi-permeable membrane under pressure, the dissolved salts are held back. The Central Salt and Marine Chemicals Research Institute (CSMCRI), Bhavnagar 364 002, India, developed the RO process to provide drinking water to rural communities which have only brackish water (item 6.2.12, CSIR Rural Technologies, New Delhi 1995). In this process, brackish feed water is first passed through a pressure sand filter to remove most of the suspended particles. The pH is adjusted, and the feed water is then passed under a pressure of 30-35 kg cm$^{-2}$ through a spiral wound cellulose acetate membrane in the RO unit. The membrane rejects 85-90% of dissolved salts.

Clean water is produced at the rate of 300-400 l m$^{-2}$ of membrane surface. CSMCRI designed stationary plants with a capacity of 10-15 m$^3$ d$^{-1}$ and a mobile plant with a capacity of 15 m$^3$ d$^{-1}$ to cater to the needs of remote villages.

The choice of membranes depends upon the water chemistry, such as the TDS and the ratio of monovalent to multivalent cations. With improvements in membrane technology, the RO process is steadily becoming more affordable.

# REFERENCES

Al-A'Ame, M.S. & G.F. Nakhle 1995. Wastewater reuse in Jubail, Saudi Arabia. *Water Res.* 29(6): 1579-1584.

Asano, T. (ed.) 1985. *Artificial Recharge of Groundwater*. Boston: Butterworth Publishers.

Bouwer, H. 1994. Role of geopurification in future water management. In *Soil and Water Science: Key to Our Understanding Our Global Environment*. Soil. Soc. Amer. Sp. Publ., 41: 73-81.

Brierley, C.L. 1990. Bioremediation of metal-contaminated surface and groundwaters. *Geomicrobiology J.* 8: 202-233.

Brierley, C.L., J.A. Brierley & M.S. Davidson 1989. Applied microbial processes for metals recovery and removal from wastewater. In Beveridge, T.J. & R.J. Doyle (eds), *Metal Ions and Bacteria*.

Chapelle, F.H. 1997. *The Hidden Sea*. Tucson, Arizona: Geoscience Press.

Davison, J. 1993. Successful acid mine drainage and heavy metal site bioremediation. In Moshiri, G.A. (ed.), *Constructed Wetlands for Water Quality Improvement*, Chap. 16, Boca Raton: CRC Press.

Eger, P. & K.Lapakko 1989. Use of wetlands to remove nickel and copper from mine drainage. In Hammer, O.A. (ed.), *Constructed Wetlands for wastewater treatment – municipal, industrial and agricultural*, Chap. 42 e, Lewis Publishers.

Gusek, J.A. 1995. Passive-treatment of acid rock drainage: what is the potential bottom line? *Mining Engg.* 97: 250-253.

Kolbash, R.L. & T.L. Romanovski 1989. Windsor Coal Company Wetland: An overview. In Hammer, O.A. (ed.), *Constructed Wetlands for wastewater treatment – municipal, industrial and agricultural*, Chap. 42 f, Lewis Publishers.

Lacki, M.J., J.W. Hummer & H.J. Webster 1992. Mine-drainage treatment as habitat for herptofaunal wildlife. *Environ. Management* 16(4): 513-420.

L'vovich, M.I. 1979. *World Water Resources and their Future*. Washington, DC: Amer. Geophys. Uni.

MiMi.1998. *Prevention and control of pollution from mining waste products*. State-of-the art-report.

Moore, W.S. 1996. Large groundwater inputs to coastal waters revealed by $^{226}$Ra enrichments. *Nature* 380: 612-614.

Moore, W.S. 1997. The effects of groundwater input at the mouth of the Ganges – Brahmaputra Rivers on barium and radium fluxes to the Bay of Bengal. *Earth Planet. Sci. Lett.* 150: 141-150.

Moore, W.S. 1998. Application of $^{226}$Ra, $^{228}$Ra, $^{223}$Ra, and $^{224}$Ra in coastal waters to assessing coastal mixing rates and groundwater discharge to the oceans. *Proc. Ind. Acad. Sci. (Earth Planet Sci.)* 107(4): 343-349.

Moore, W.S. 1999. The Subterranean estuary: A reaction zone of groundwater and seawater. *Marine Chem.* 65: 111-125.

Pacey, A. & A. Cullis 1986. Rainwater Harvesting: The Collection of Rainfall and Runoff in Rural Areas. London: I.T. Publications.

Prinz, D. & S. Wolfer 1998. Opportunities to ease water scarcity (Water conservation techniques and approaches: Added values and limits). In H. Zebidi (ed.), *Water: A looming Crisis?* IHP-V 18: 521-530. Paris: UNESCO.

Rama & W.S. Moore 1996. Using the radium quartet to estimate water exchange and groundwater input in salt marshes. *Geochim. Cosmochim. Acta* 60: 4645-4652.

Shiklomanov, I.A. 1998. World Water Resources: A new appraisal and assessment for the 21st century. Paris: UNESCO.

Tsunogai, U. et al. 1996. Fresh water seepage and pore water recycling on the seafloor: Sagami Trough subduction zone, Japan. *Earth Planet. Sci. Lett.* 138: 157-168.

UNEP 1983. Rain and Stormwater Harvesting in Rural Areas. Dublin: Tycooly.

White, C. & G.M. Gadd 1996. A comparison of carbon/energy and complex nitrogen sources for bacterial sulphate-reduction: potential applications to bioprecipitation of toxic metals as sulphides. *J. Industr. Microbiology* 17: 116-123.

CHAPTER 5

# Wastewater reuse systems

## 5.1 WHY REUSE WASTEWATER?

In Europe in medieval times, waste water was simply thrown out of the window. As a courtesy to the passersby, a housewife would shout 'gardez l'eau' (watch out for the water!). This was anglicized to *gardyloo* (presently, the British ship which dumps wastes in the ocean is named gardyloo) (Bouwer 1993). In the city of York in England, there is a street called Shambles, with sides sloping towards the centre. It was originally a street of the butchers and one could easily see even now how the liquid wastes from the butchers shops must have flown along the centre of the street.

When it became evident that the contamination of drinking water by sewage was responsible for the outbreaks of diseases such as dysentry, cholera, and typhoid, municipalities organized elaborate sewage treatments for wastewater before letting it into a river or lake or sea. As the removal of nitrogen, phosphorus, heavy metals and toxic organic compounds from the sewage became more and more expensive, reuse of wastewater, for non-potable uses, particularly in water-short areas, became a highly sensible proposition.

Wastewater means discarded water. Wastewater is just a euphemism for sewage. Wastewater is generally of two kinds:

1. Sanitary or foul sewage: composed of sanitary or domestic wastewater, liquid material collected from residences, buildings and institutions, industrial or trade wastes arising from manufacturing, municipal wastes, industrial effluents from dairies, bakeries, breweries, etc.
2. Storm water: Runoff of rainwater.

After a dry spell, the runoff from urban areas may contain organic carbon, pathogens, suspended solids, hydrocarbons, lead washed from the highways, acid rainfall, etc. Hence it is necessary to treat such water.

There are few places on earth where potable water is so plentiful and so freely available that there is no necessity to reuse the waste water. There is little doubt that in future the reuse of wastewater will become routine all over the world, for the following reasons:

It has been estimated that in 1995 the volume of waste water was 326 km$^3$ yr$^{-1}$ in Europe, 431 km$^3$ yr$^{-1}$ in North America, 590 km$^3$ yr$^{-1}$ in Asia, and 55 km$^3$ yr$^{-1}$ in

Africa (Shiklomanov 1998, p. 37). Comprehensive treatment of wastewater before discharging into a water body is expensive, and is practiced only in Industrialized countries, and even there, the coverage is not 100%. Most countries tend to discharge the wastewater either untreated or partly treated, into the hydrological system. This has disastrous consequences. *Every cubic metre of waste water discharged into water bodies or water courses contaminates and degrades 8 to 10 cubic metres of good water.* Hence, prime waters get polluted and become non-potable. Thus most parts of the world are already facing the problem of degradation of quality of water because of the inherently wrong practice.

It is said that the Chinese term for Crisis consists of two ideograms representing Danger and Opportunity. The waste water problem has indeed a win-win solution. Ecological treatment of waste water has a number of merits: besides removing the pollutants effectively, it is low-cost, energy saving, resource recovery, easy operation and maintenance. The reuse of wastewater not only avoids contamination of the hydrological system but the treated water can be put to beneficial uses such as irrigation, industry, recreation, etc.

Where municipal waste water containing human faeces is discharged untreated into the sea, the seawater gets contaminated with pathogens, as indicated by the presence of thousands of E.Coli/100 ml. The pathogens get into the fish and other kinds of seafood, and when people consume such seafood, cholera and other gastro-intestinal epidemics develop, as has happened in Peru some years ago. In 1997-98 when the El Niño – induced flooding along the East Coast of Africa washed the human and animal wastes into the sea, the cholera epidemic resulting from the consumption of pathogen-contaminated seafood killed an estimated 5000 people in the coastal belts of Somalia, Kenya, Tanzania and Mozambique.

Thus, neither Industrialized countries nor Developing countries can afford not to reuse waste water. The precise method of treatment of wastewater and its reuse to be adopted would depend upon the techno-socio-economic situation obtaining in each case.

## 5.2 CONVENTIONAL WASTEWATER TREATMENT

The various aspects that need to be taken into account in designing systems of wastewater reuse are summarized in Table 5.1 (source: Wilson 1985, p. 1659-1662).

The treatment processes needed to meet the health criteria (prescribed by WHO 1973) for wastewater reuse are summarized in Table 5.2.

Waste water is generally subjected to a two-phase treatment: Mechanical or primary treatment and Biological or chemical treatment. Tertiary treatment is dealt with later under Advanced waste water treatment.

Pre-treatment: Sewage may sometimes be highly foul smelling because of the presence of hydrogen sulfide, mercaptans and amines. Chlorine/hydrogen per-

Table 5.1. Aspects to be considered in waste water reuse (source: Wilson 1985, p. 1659-1662).

|  | Biological contaminants | Inorganic chemicals and physical parameters | Organic chemicals |
|---|---|---|---|
| Character | Bacteria, viruses and parasites (including cysts and eggs) | Heavy metals; nutrients (N,P), TDS, specific ions such as Na, hardness, color, odor, taste, temperature, turbidity and SS, pH, radioactivity | More than 100 compounds have been identified |
| Sources | Human and animal wastes are the major source | Heavy metals arise from industry, hospitals, or corrosion of water pipes; Sewage is generally nutrient rich ($\times$ 100 ppm TDS); Turbid, relatively warm; radioactivity is usually low | Natural substances in surface water; industrial (e.g. petrochemical) discharges; chlorination creates compounds, such as chloroform |
| Levels in raw waste water | Millions of Bacteria, thousands of viruses, abundant parasites | Domestic sewage normally contains < 0.5 mg l$^{-1}$ of heavy metals; contains more (in mg L$^{-1}$) P (10), N (40), SS (200), TDS (300) than original raw water | Generally at ppb levels; could be much higher where industrial discharges are important |
| Desired levels in drinking water | Less than one coliform/100 ml; virus standard of < 1 plaque-forming unit/ l of water | US EPA standards exist for drinking water in regard to toxic heavy metals, nitrate – N, TDS, SS, radioactivity, etc. (Tables 3.4 and 3.5). | Some organics are hazardous even at very low concentrations. Standard for Trihalomethanes is 100 ppb. |
| Feasibility of treatment | Complete disinfection requires larger chorine doses and/or use of ozonation, lime coagulation, etc. High turbidity impedes virus removal | Lime coagulation (softening) removes most metals, besides reducing hardness, TDS, etc. Some metals (As, Se, Mo) difficult to remove. N & P removal needs advanced treatment; Reverse osmosis can be used for demineralization | Activated carbon effectively removes large amounts of organics.Ozonation may improve performance. Further Treatment can reduce total organic C levels to 1-2 mg l$^{-1}$ |
| Effects of Recharge | Movement through soil largely purifies biological contaminants | Heavy metal content is reduced, but TDS and Na may increase. Good P removal, but minimal N removal | May reduce organic content. |

oxide/chlorine dioxide/ferric salts may be used to deodorize the sewage. Solids more than one or two cms across are removed by screening. Quartz grains larger than 0.2 mm diameter are got rid off through settling tanks, called grit chambers.

Primary (Mechanical) treatment: Primary treatment consists of settling. Sewage is kept in settling tanks 'with a detention time at design flow of about 2 hours and calculated settling velocity of 0.4 mm s$^{-1}$' (Wilson, quoted in Dean & Ebba Lund 1981, p. 150). In the course of primary treatment, about half of BOD (Biological Oxygen Demand) and SS (Suspended Solids) are removed, and get concentrated

Table 5.2. Treatment processes needed to meet the health criteria for waste water use (source: WHO 1973).

| | Ir. 1 | Ir. 2 | Ir. 3 | R1 | R2 | Ind. | M.1 | M.2 |
|---|---|---|---|---|---|---|---|---|
| Health criteria | A+F | B+F/D+F | D+F | B | D+G | C/D | C | E |
| Primary treatment | 000 | 000 | 000 | 000 | 000 | 000 | 000 | 000 |
| Secondary treatment | | 000 | 000 | 000 | 000 | 000 | 000 | 000 |
| Sand filtration or equivalent polishing methods | | 0 | 0 | | 000 | 0 | 000 | 00 |
| Nitrification | | | | | | | 0 | 000 |
| Denitrification | | | | | | | 0 | 00 |
| Chemical clarification | | | | | | | 0 | 00 |
| Carbon adsorption | | | | | | | | 00 |
| Ion exchange or other means of removing ions | | | | | | | 0 | 00 |
| Disinfection | | 0 | 000 | 0 | 000 | 0 | 000 | 000* |

Ir. – Irrigation; Ir 1: Crops not for direct human consumption, Ir 2: Crops eaten cooked; fish culture, Ir.3: Crops eaten raw, R – Recreation; R 1: No contact, R 2: Contact, Ind.: Industrial use, M – Municipal use; M 1: Non – potable, M 2 – Potable.

Health criteria: A – Freedom from gross solids; significant removal of parasite eggs, B – As A, plus significant removal of bacteria, C – As A, plus more effective removal of bacteria, plus removal of viruses, D – Not more than 100 coliform organisms per 100 ml in 80% of the samples, E – No faecal coliform organisms in 100 ml, plus no virus particles in 1000 ml, plus no toxic effects on man, and other drinking water criteria, F – No chemicals that lead to undesirable residues in crops or fish, G – No chemicals that lead to irritation of mucous membranes or skin. In order to meet the given health criteria, processes marked 000 are Essential. In addition, one or more processes marked 00 will also be essential., and further processes marked 0 may some times be essential. *Free chlorine after one hour.

in a sludge that contains upto 5% solids. In some cases, lime or ferric salts are added to raise the pH to about 11 in order to precipitate calcium carbonate or calcium phosphate. Oil, grease or low density goods like rubber float to the surface, and are removed by skimming.

Secondary (Biological) treatment: The purpose of the biological treatment is to remove organic matter making use of the bacteria, protozoa and fungi present in the sewage. Primary effluent is made to pass through trickling or percolating filters, which are similar to slow sand filters. Most of BOD is removed by biological action. The residual solids settle out in the clarifier and are removed as stable sludge. Another variety is called Suspended Growth reactors. In these reactors, wild microorganisms form flocs which will aggregate further and settle, leaving a clear supernatant which is low in BOD and SS. The settled sludge is called activated sludge, because it is not a chemical precipitate, but owes its activity to biological action. In lagoons and oxidation ditches, organic matter is destroyed by bacterial oxidation. Stirring of lagoons can make the aeration process more efficient. Lagoons require lot of land, but the operating expenses are very low.

The final step in the biological treatment is clarification to remove bacterial floc from the effluent.

Sludge collected from primary and secondary clarifiers will still be mostly water. It may be concentrated by filtering or centrifuging to raise the solid content to 20-25%. If sludge is kept without exposure to oxygen, anaerobic bacteria will produce methane. Anaerobic digesters can convert half of the organic matter in the sludge to methane, which can be collected and used as fuel. Switzenbaum (1991) gave a good summary of the anaerobic treatment technology for municipal and industrial wastewaters. Digested sludge makes good fertilizer.

### 5.2.1 *Advanced wastewater treatment*

An Advanced Waste Treatment (AWT) Program was established by an Act of US Congress in 1961 with the objective of finding better ways of purifying waste water.

AWT processes are designed to treat waste water, but they could be used to treat any polluted river water, with minor modifications. The following summary is largely drawn from Wilson (1985).

### *Chemical precipitation*

*Lime treatment*: When sewage is flocculated with lime, orthophosphates as well as carbonates get precipitated. Tertiary lime clarification can keep the phosphate below one mg $l^{-1}$, and reduce the suspended solids below 20 mg $l^{-1}$, and even lower if filtration is added.

*Alum or iron flocculation*: Alum flocculation is capable of reducing phosphates to one mg $l^{-1}$. Ferric salts, $FeCl_3$ or $Fe(SO_4)_3$, have been successfully used to remove phosphates from the secondary effluent, and to reduce turbidity.

*Activated carbon*: The excellent adsorptive capabilities of activated carbon arise from its high porosity and aromatic nature. Columns of granular activated carbon (GAC) are used to remove organic matter in the secondary effluent to below one mg $l^{-1}$. Activated carbon cannot, however, adsorb alcohol, ethanol, sugars, amino acids, fruit acids, and several carbohydrates. It so happens that most of the oxygenated compounds that are not adsorbed on carbon, are biodegradable. The dense flora of bacteria associated with carbon columns act as a biological reactor and consume the BOD (Biological Oxygen Demand), so long as oxygen or its equivalent, such as nitrogen, is available. But this is not an unmixed blessing. The bacterial action produces slimes which clog the pore spaces in GAC, and reduce its efficiency. Also, when the BOD exceeds DO (Dissolved Oxygen), the bacteria would produce anaerobic conditions, leading to the production of hydrogen sulfide and foul smells.

Porous ion exchange resins can be used as substitutes for activated carbon. Barring some specialized applications, synthetic resins are not used to the same extent as activated carbon for two reasons: synthetic resins are expensive, and they cannot adsorb the wide range of organic compounds that activated carbon can.

*Salt removal*: Removal of salts, such as sodium chloride, is difficult and expensive. Removal of sodium chloride by distillation is technically feasible, but very expensive because of its high energy needs (it requires about 100 kg of coal, equivalent to 2.89 GJ, to evaporate one m$^3$ of water). Reverse osmosis (RO) is being increasingly used to effect the removal of salt. Operating pressures of the order of 50 atmospheres are used. The rate of flow through the membrane is slow (10 gal/day/sq.ft. = 3.6 l d$^{-1}$ m$^{-2}$), and hence large areas of membrane are needed. The membrane may be fouled by the deposition of calcium carbonate and sulfate scales, and accumulation of organic matter. Elaborate cleaning programs have been developed to deal with the fouling of the membrane. One way to avoid the fouling problems is to pre-treat the water before subjecting it to reverse osmosis. It is not necessary to treat all the flow by R.O. If for instance, the waste water contains 800 mg l$^{-1}$ of dissolved salts, a blend of equal parts of R.O. processed water (i.e. without salt) and treated feed stream would contain about 400 mg l$^{-1}$ of dissolved salts, which is perfectly acceptable.

*Treatment systems*: There is no ready-made treatment system which is applicable to any kind of waste water. Invariably, a treatment has to be custom-deigned depending upon the kind of waste water that is available for renovation, government regulations, affordability, social acceptability, etc. Biodegradable substances are best removed by activated sludge. Some pathogens in waste water can have serious health consequences – so much so that several steps may have to be taken to reduce the concentration of bacteria by factors of two to one million. Primary treatment reduces the pathogenic bacteria by about 50%. Secondary treatment with activated sludge reduces them to 10%. Coagulation with lime at high pH, or filtration with alum or iron will bring down the bacterial concentrations to 0.1-1%. Further disinfection with chlorine or ozone will destroy the bacteria completely.

In the customary treatment cycle, mechanical treatment and primary sedimentation precedes other processes. But in some cases, lime addition in the course of primary treatment proved most effective. This is followed by the secondary or biological treatment involving activated sludge. Chemical treatment to remove nutrients, such as phosphates, can be combined with biological treatment. Since activated carbon cannot remove low molecular weight, chlorinated organic compounds, a sensible practice would be to reduce the precursors of these compounds before final chlorination. When ammonia is removed through aeration, volatile chlorinated hydrocarbons are also largely eliminated.

It is a good practice to store the product water for a period of several weeks in a reservoir or underground, before distribution.

Pumpel Thomas (Abstract, ICOBTE V, Vienna, July 1999) designed a biofilm reactor for the removal of trace elements from wastewater (rinsing water, effluent from conventional precipitation, groundwater, drainage or leachate water). It is based on the bioabsorption of metals through the process of inoculation with efficient bacteria, supply of missing nutrients and adjustment of microenvironments. The following bacteria are made use of: AS 302 *Pseudomonas mendocina*, BP 7/26 *Arthrobacter* sp., CH 34 *Alcaligenes eutrophus*, K 1/8 A *Pseudomonas fluorescens*, MB 127 *Methylobacillus* sp. The feed rate ($m^3 h^{-1}$) used depends upon the kinds of water: 2.5 for wastewater from plating industries, five for mine drainage water (with As, U, etc.). At a feed rate of 30 $m^3 h^{-1}$, the following removal efficiencies have been achieved:

| Element | Concentration (mg $l^{-1}$) | Removal efficiency (%) |
|---|---|---|
| Co | 1.8 | 80 – 90 |
| Ni | 0.8 | 80 – 100 |
| Zn | 0.4 | 100 |
| Cu | 0.2 | 100 |

The sludge produced (Co – 15%; Ni – 10%; Fe – 6%; Cu – 2%; Zn – 1%) is economically valuable, as it is feasible to recycle it pyrometallurgically.

Thus the new technique not only removes organic and inorganic pollutants, but also allows metal recycling. The most attractive feature of the new technology is its low-cost, as would be evident from the following cost estimates (Eu/$m^3$): Biologically active sand filter: 0.15; Ion exchange: 0.44; Reverse Osmosis: 0.48.

The new technology holds great promise.

### 5.2.2 *Treatment of waste water through marshing*

This issue has been covered in paper 3.6 in Laconte & Haimes (1982).

Where the availability of land is not a constraint, waste water can be treated by letting it into a marsh (either natural or man-made). This process is called marshing or psuedo-marshing. Marshing treatment brings about the following benefits:

1. The large surface area of the marsh in contact with air would ensure good aeration of the waste water. Aeration also helps the movement of semi-aquatic plants, and in clear weather, to the oxygenation of submerged vegetation,
2. Waste water would get thoroughly decanted, when its speed of movement decreases on entering the marsh, and water undergoes filtration by the submerged parts of the vegetation,
3. When the running waste water gets into contact with sediment, the latter may return to suspension, and get digested by higher plants,
4. Exposure to the sun brings about bacteriological purification,

5. A bacterial bed of high capacity (a reed-bed may have 35-250 shoots per $m^2$) may come into existence because of the large contact surface between the effluent and the biological covering fixed on submerged living or dead.

Marshing has several advantages:
- Minimum installation costs,
- Very low working costs, and easy maintenance,
- Good efficiency,
- Resistance to accidental toxic discharges, and
- Tertiary treatment is ensured during the hot, and critical periods. The only catch in the method is that it needs a large area.

## 5.3  BIO-POND TREATMENT OF WASTE WATER

Wang (1991) gave a detailed analysis of the Chinese experience in the biological methods of treatment of waste water. In 1988, China had 86 conventional sewage treatment works with a total capacity of three M $m^3$ $yr^{-1}$, accounting for about 11% of the municipal sewage flow in the whole country. As the country did not have the requisite financial and technical resources to build conventional sewage treatments for the rest 89% of the municipal sewage flow, they went in a big way for eco-pond technologies.

### 5.3.1   *Land treatment*

The acute water shortage in northern China necessitated the reuse of domestic and industrial waste water for irrigation and other purposes. Sewage irrigation increased from 42,000 ha in 1963, to 1.4 million ha in 1983. For instance, Tianjin city has 153,000 ha of sewage irrigation farmlands. Land treatment of wastewaters has been able to achieve high pollutant removal efficiencies as follows: SS, BOD, COD, total phosphorus and bacteria: 95%; trace metals, phenols and cyanide: 90%, COD, total nitrogen and potassium: 80%. The effluent from the irrigated fields is used to charge the depleted aquifers. On an average, one $m^3$ of sewage applied on a farm increased the grain yield by 0.5 kg.

There have been failures also. In the case of some irrigation projects (less than 10% of the total area), the toxic and harmful substances in the industrial wastewater have resulted in decreasing yields and pollution of crops, soil mantle and groundwater. It is therefore crucial that toxic substances be removed from wastewater before it is used in irrigation.

### 5.3.2   *Eco-ponds*

The ecological and biochemical processes that go on in the eco-ponds under aerobic, facultative and anaerobic conditions, are shown in Figure 5.1 (source: Wang

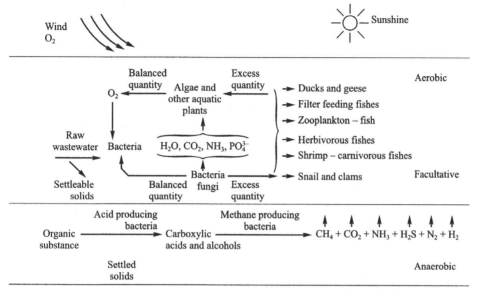

Figure 5.1. Ecological and biochemical processes in eco-ponds (source: Wang 1991).

1991). This diagram helps us to understand what kind of ecological conditions are necessary for what kind of ecological production. For instance, aerobic conditions have to be maintained in an eco-pond if the objective of developing a eco-pond is to grow aquatic plants for human and animal food. On the other hand, if the objective is the production of methane to be collected and used as fuel, anaerobic conditions are necessary.

The design of the flowsheet for treatment of waste water would take into consideration different local situations:

Where sewage is produced all the year round, whereas irrigational needs for reusing the waste water are seasonal, the following flowsheet is made use of:

Raw waste water → Preliminary treatment → Facultative ponds → Storage lagoons → Farmland irrigation

In areas where adequate land is not available and where the weather permits irrigation during most of the year, storage lagoons could be avoided and the ponds may directly supply water for irrigation. Then the following flowsheet can be adopted:

Raw waste water → Preliminary treatment → Facultative ponds → Year-round irrigation

The eco-pond systems are designed in the following ways, depending upon the climatic conditions and the composition of the sewage water (Wang 1991):

*In arid or semi-arid, cooler areas:*

Domestic sewage → Preliminary treatment → Facultative ponds → Storage ponds or lagoons (fish farming/duck or geese raising in warm season) → Irrigation of farm land → Effluent recharging ground water

High strength organic wastewaters → Anaerobic ponds → Facultative ponds → Storage lagoons (fish farming/duck or geese raising in warm season) → Effluent recharging ground water

*In humid, warmer areas:*

Organic wastewaters → Preliminary treatment → (Anaerobic ponds) → Facultative ponds (with growth of floating plants) → Fish farming in paddy fields → Lotus or reed ponds → Effluent to receiving waters

High strength organic wastewaters, such as those from sugar mills, wineries, food processing industries, are sent to anaerobic digesters after pre-treatment. Methane formed in the digesters is removed to be used as fuel in the houses of farmers (China produces 720 M m$^3$ of methane gas annually, and some 15 million farmers use methane gas as domestic fuel). The supernatant from the anaerobic digesters is then sent to hydrophyte-growing facultative ponds or duck/geese ponds or lotus/reed ponds. The sewage sludge may be applied as fertilizer on farmland, grassland or forest, or it can be used to grow mushrooms or earthworms (fish food).

### 5.3.3 Fish farming

Using the pretreated wastewater for fish farming has become lucrative in all parts of China. The high rate of fish production (4000 to 16,000 kg ha$^{-1}$) with very low cultivation cost, makes fish farming more attractive financially than raising rice or wheat or vegetables. The case history of fish farming ponds in Changsha city, Huanan province of China, illustrates the benefits of ecological wastewater treatment: About 25,000 m$^3$ d$^{-1}$ of municipal sewage and 50,000 m$^3$ d$^{-1}$ of organic waste waters from industries and animal farms, are sent to the fish farms. The total area of sewage fish ponds in the city is 1430 ha. Under conditions of sewage hydraulic loading of 400-500 m$^3$ ha$^{-1}$ d$^{-1}$, BOD$_5$ loadings of 20-30 kg ha$^{-1}$ d$^{-1}$, and pond water temperature of 15-25°C, the ponds achieved the following removal efficiencies: SS 74-83%, BOD$_5$: 75-91%, Total nitrogen: 70%. The fish production ranged from 4500-6000 kg ha$^{-1}$ yr$^{-1}$.

In macrohydrophyte ponds, water hyacinth (*Eichhornia crassipes*), water peanuts (*Alternathera phioloxeroides*) and water lotus (*pistia stratiotes*) grow fast for the following reasons:
– Their utilization of solar energy (3-5%) is much higher than crop plants (0.5-1.0%),

– The hydrophytes make use of nitrogen and phosphorus nutrients in the waste-water, and
– The bacterial number and biomass in a water hyacinth pond is 10-20 times greater than in an ordinary pond of equal volume. The hydrophytes are used as food by fish, poultry, pigs and other animals.

Relative to water hyacinth (*Eichhornia crassipes*) ponds, water peanut (*Alternathera philoxeroides*) ponds are more efficient in removing $BOD_5$, COD, TN and $NH_3$-N and less efficient in removing TP and $PO_4^{3-}$. The high pollutant ability of water peanut pond is attributed to the bacteria, *Bacillus*, *Psecomonas*, and *Alcaligenes*. These adhere to the roots of the plants and bring about the degradation of material containing C, N and P into $CO_2$, $NH_4$-N, and $PO_4$-P which are then taken up by the roots of the plants. Thus, N and P in waste water are effectively removed, while benefiting the plant (Hang Xu et al. 1991).

Shi & Wang (1991) studied the purifying efficiency and the mechanism of the aquatic plants in the bio-ponds. They found that the increasing peroxidase activity in the plant body promoted the phytophysiological metabolic activity resulting in the acceleration of the pollutant removal rate.

*'Rice – fish' symbiotic system*:

The practice of raising paddy and fish together has a long tradition in China dating back to Han Dynasty (206 BC to 220 AD). The rice plant benefits the fish by the intake of substances (e.g. excrement) that are harmful to the fish. As the fish (grass carp) swim through the water in search of food, this activity promotes aeration by breaking up the algae film and stirring up the upper soil layer. The grass carp excrete 72% of its diet – so much so, 6800 grass carp can produce 1650 kg of manure per ha, which is more than enough to fertilize the paddy field. When young, the grass carp eat the abundant zooplankton that is present in the sewage sludge or supernatant from the digesters. As they mature, they eat weeds and the lower leaves of the rice plants. Also, the grass carp can kill more than 90% of the mosquito hatchings in the paddy field.

The practice of raising rice and fish together is very much in vogue in Eastern India and Bangladesh.

The 'Rice – duckweed – fish' system is a modification of the 'rice-fish' system. Duckweed is an excellent food for the fish. Under this system, fish and duckweed are grown in the ditches, and rice is grown on the ridges. In this system, not only grass carp but also *Nile Tilapia*, common carp and silver carp can also be grown.

### 5.3.4  Comprehensive recycling of wastewater

Figure 5.2 gives the schematic chart of the comprehensive recycling of the waste water (Source: Wang 1991, p. 18).

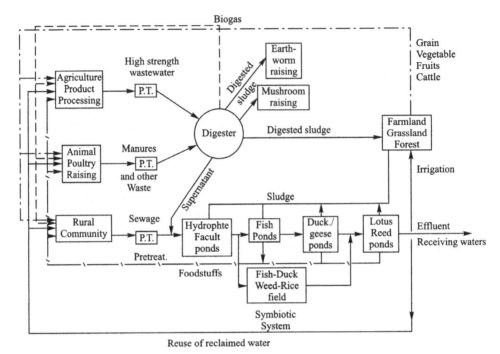

Figure 5.2. Comprehensive recycling of wastewater (source: Wang 1991).

Unlike the conventional waste water treatment plants which are energy-intensive (for pumping, stirring, filtration, reverse osmosis, etc.), eco-ponds are powered by solar energy which is free. Eco-ponds serve the following useful purposes:
– They are able to remove SS, BOD, COD, nutrients such as N and P, refractory organic compounds, bacteria, viruses and mineral salts,
– They use the food webs involving producers (such as algae and aquatic plants), and decomposers (such as, bacteria and fungi) to provide food to the consumers (such as, fish, clams, duck and geese), biogas fuel and organic fertilizer, and
– The effluent from the eco-ponds is sufficiently pure for it to charge the groundwater, or for non-potable uses (such as, irrigation, industrial, recreation, etc. uses).

## 5.4  TYPES OF WASTEWATER REUSE

The various issues involved in wastewater reuse have been discussed in a number of studies (National Academy of Sciences 1972, Ayers 1975, AWWA 1985, Kramer 1985, WHO 1989).

Figure 5.3 (source: *Future of Water Reuse* – Proc. of the Water Reuse Symposium, San Diego, CA., USA, Aug. 84, v.3, p. 1738) shows the various types of

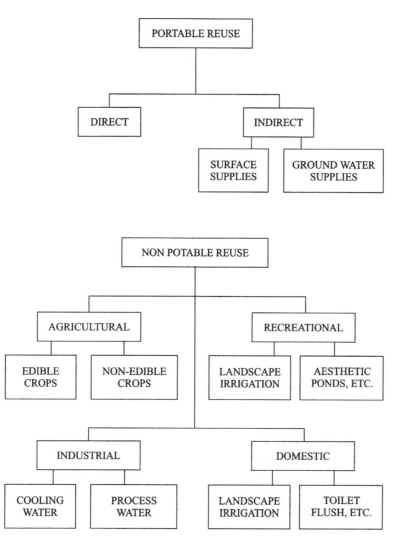

Figure 5.3. Types of wastewater reuse (source: AWWA 1985, p. 1738).

water reuse. The following case histories have been drawn from Dean and Lund (1981, p. 237, 241, 243, 245).

*Potable reuse:*

*A.1: Direct reuse*: Highly treated, reclaimed waste water is piped directly into the drinking water system,

   *Case history*: Windhoek, Namibia, South West Africa, lies in an extremely dry area, with the nearest river 720 km. away. Waste water (4500 m$^3$ d$^{-1}$) is treated in the following sequence: primary, secondary, maturation ponds, high pH lime,

$NH_3$ stripping, $CO_2$, clarification, $CO_2$, sand filtration, break-point chlorination, activated carbon, final chlorination. The microbiological quality of reused water is as good as fresh water supplies in the neighboring South Africa.

*A.2: Indirect reuse*: Highly treated waste water is discharged to the environment for dilution, natural purification and subsequent withdrawal for water supply (e.g. discharge to river upstream of water plant intake, or replenishment of groundwater supplies either through infiltration from the ground surface or by direct injection into the aquifers).

*Case history*: Thames River basin, London, England. About 1.2 million $m^3$ $d^{-1}$ of sewage is treated as follows: Primary, trickling filters or activated sludge, nitrification and denitrification when required. Water quality meets the public health requirements including nitrates below 10 mg $l^{-1}$ of N.

*Non-potable reuse:*

*B.1:* Agricultural irrigation, including irrigation for edible and non-edible crops, pasture irrigation, and livestock watering.

*Case history*: Idelovitch & Michail (1985) gave the case history of the groundwater recharge for wastewater reuse in the Dan Region project. The wastewater from the cities of Tel Aviv, Jaffa and the neighboring municipalities (involving about 400,000 people) is treated and recharged to groundwater. After undergoing the Soil – Aquifer Treatment (SAT), the reclaimed water is mainly used for non-potable uses (mainly unrestricted irrigation of agricultural crops). The project involves the intentional recharge of municipal effluent to a potable groundwater aquifer. The bulk of the recharged effluent is extracted by recovery wells, and is used for purposes compatible with the quality of water. Systematic monitoring of the effluent is crucial for the success of this kind of project. The recharge facilities involve four basins and occupy an area of about 30 ha. The volume of effluent recharged in the four basins during 1977-1981 was about 60 M $m^3$. The flow diagram of the Dan region sewage reclamation project is given in Figure 5.4 (source: Idelovitch & Michail 1985).

That the Soil – Aquifer treatment of the recharge effluent can be highly effective in removing the pollutants should be evident from the data given in Table 5.3 (source: Idelovitch & Michail 1985). The chemistry of the water in the observation wells shows that when the recharged water passes through the Soil – Aquifer system, the concentrations of Cd, Cr, Se, Cu, Mn, K and phenols are reduced virtually to the same levels as in the native groundwater (Table 5.4; source: Idelovitch & Michail 1985, p. 505).

In 1970, a cholera epidemic broke out in Jerusalem when greedy farmers illegally used raw sewage for watering lettuce supplied to the city.

*B.2: Industrial uses*: Power plant cooling, processing plants and construction.

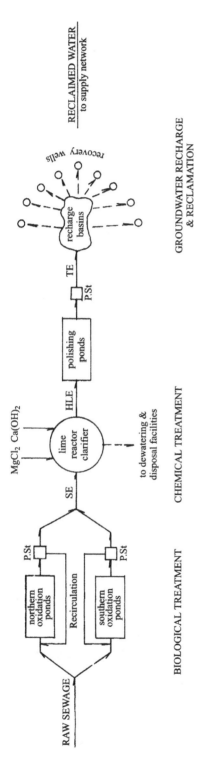

Figure 5.4. Flow diagram of the Dan region sewage reclamation project, Israel (Idelovitch & Michail 1985).

Table 5.3. Extent of purification of sewage effluent by soil-aquifer treatment (source: Idelovitch & Michail 1985).

| Parameter | Unit | Recharge effluent | Reclaimed (well) water |
|---|---|---|---|
| COD | mg l$^{-1}$ | 60-90 | 5-15 |
| KMNO$_4$ consumption as O$_2$ | mg l$^{-1}$ | 10-15 | 1-2 |
| UV$_{254}$ absorbance | m$^{-1}$ | 20-25 | 2-6 |
| Detergents | Mg l$^{-1}$ | 0.5-1.5 | 0.1-0.3 |
| Kjeldahl N | mg l$^{-1}$ | 7-14 | 0.2-1.5 |
| Ammonia as N | mg l$^{-1}$ | 1-8 | < 0.02 |
| Nitrate as N | mg l$^{-1}$ | < 0.01-0.5 | 2-10 |
| Total N | mg l$^{-1}$ | 7-15 | 2-12 |
| Phosphorus | mg l$^{-1}$ | 1-3 | 0.02-0.04 |
| Boron | mg l$^{-1}$ | 0.3-0.5 | 0.1-0.3 |
| Sodium | mg l$^{-1}$ | 120-160 | 100-150 |
| SAR | – | 4-7 | 3-4 |

Table 5.4. Extent of removal of toxic components of the recharged water during passage through the soil – aquifer system (source: Idelovitch & Michail 1985).

| Parameter | Units | Recharge water | Natural groundwater | Water in the observation wells |
|---|---|---|---|---|
| Cadmium | µg l$^{-1}$ | < 4 | < 2 | < 2.3 |
| Chromium | µg l$^{-1}$ | 15 | 7 | < 4.5 |
| Selenium | µg l$^{-1}$ | 9 | 1 | < 1.5 |
| Copper | µg l$^{-1}$ | 18 | 7 | < 6.4 |
| Manganese | µg l$^{-1}$ | 27 | 12 | 19 |
| Potassium | µg l$^{-1}$ | 25 | 0.6 | 3.2 |
| Phenol | µg l$^{-1}$ | 4 | < 1 | < 1.5 |

*Case history*: Contra Costa County, California, USA. Waste water (114,000 m$^3$ d$^{-1}$) is treated as follows: lime in primary, activated sludge with nitrification, rapid sand filtration, ion exchange, softening, final chlorination, before being used for industrial cooling.

A 3810 MW nuclear power plant in Phoenix, Arizona, USA, is cooled with sewage water at a design flow rate of 3.2 m$^3$ s$^{-1}$ (about 276,000 m$^3$ d$^{-1}$). The wastewater is first subjected to primary and secondary treatments in the sewage plant. It is then transported to the nuclear power plant. On site, the secondary effluent is further treated with lime precipitation, trickling filters and sand filters, to minimize the formation of scales in the pipe system. The cooling water is recycled about 15 times. By this time, the salt concentrations would have reached a level of about 15,000 mg l$^{-1}$ (salinity level roughly half of seawater). The brines are then discharged into a 196 ha lake, from which salts are periodically removed and disposed in a landfill (Bouwer 1994).

*B.3: Landscape irrigation and recreation*: including irrigation of turf and ornamental plants in golf courses, parks; playgrounds, landscaping, use of reclaimed water to fill artificial lakes for recreational and aesthetic purposes.

*Case history*: Santee, California, USA. Waste water (5000 m³ d⁻¹) is treated as follows: biological treatment, slow sand filtration through natural sand beds. The treated water was used to fill a series of recreational lakes. To prevent heavy growths of algae, phosphates were removed from the biologically treated sewage.

*B.4: Industrial process water:* water use in manufacturing.

*B.5: Municipal non-potable use:* use of water for toilet flushing, fire fighting, and air-conditioning.

*B-6: Environmental applications:* wildlife refuges, or in-stream benefits.


5.5  USE OF WASTEWATER IN IRRIGATION

The standards of water quality for the use of wastewater in irrigation have been given by Matters (1981). Kramer (1985, p. 1669) listed the allowable limits for wastewater for specific uses (Table 5.5). He gave a summary of the regulations regarding the treatment requirements required for the use of wastewater in the state of Arizona, USA.

The land application of waste water has to take into consideration the removal of pathogenic microorganisms. The bacterial survival depends upon moisture, temperature, organic matter and the presence of antagonistic soil microflora. The removal of the viruses by adsorption depends upon the salt concentration, pH, soil composition, organic matter and the electronegativity of the virus.

Waste water is applied to land through (a) spray irrigation, (b) overland flow, and (c) infiltration percolation (Jones 1982)

*Spray irrigation*: Waste water is sprayed on to land at some pre-arranged rates (cms/week). The material then infiltrates and percolates within the boundaries of the disposal site. In USA, spraying is used on flat lands, or on slopes upto 20%, or in wooded sites (as in California).

*Overland runoff*: In this arrangement, waste water moves along a flow path as a sheet flow down the slope. Slopes are usually less than 10%. The waste water infiltrates the top two feet of the soil. Vegetation usually picks up P and N from waste water. The water flows through the furrows, and plants are grown on the ridges between the furrows. The application rate may be 20-30 times of spray irrigation.

*Rapid infiltration*: The principal purpose of this kind of application is groundwater recharge. Normally, the liquid applied in this method is the disinfected effluent which has undergone secondary treatment.

As fresh water becomes scarcer, the need and justification for the reuse of wastewater is getting strengthened. The place of wastewater treatment and reuse in meeting the urban demands for water, could be appreciated from the following scheme:

The sources of fresh water are:
- stream flow,
- groundwater,
- agricultural tradeoff, and
- interbasin transfer.

These waters are subjected to raw water treatment before being supplied to the urban areas. Wastewater gets produced in the cities when the fresh water is used for domestic, municipal and industrial purposes.

The wastewater has to be treated in the following ways before being reused:

Primary,
$\downarrow$

Secondary  $\rightarrow$ wastes from the secondary treatment end up as urban effluent,
$\downarrow$

Tertiary   $\rightarrow$ wastes from the tertiary treatment end up as urban effluent.

In most cases, water would be ready for reuse after the tertiary treatment. In a few cases, desalting may have to be undertaken after the tertiary treatment before the water is reused.

Particulars regarding the use of wastewater are shown in Table 5.5 (p. 177), and Fig. 5.5 (Jones, 1982). Climate is a critically important factor in deciding about the method of application.

In Phoenix, Arizona, USA, large quantities (225 million gallons per year = 910 $\times$ $10^3$ m$^3$ yr$^{-1}$) of wastewater is applied.

### 5.5.1  *Geopurification*

Geopurification or Soil – Aquifer – Treatment (SAT) is a low-technology treatment system, whereby the sewage effluent is subjected to groundwater recharge with infiltration basins. After SAT, the water will have very low suspended solids content, essentially zero BOD, much reduced concentrations of N, P, organic compounds and heavy metals, and almost zero levels of pathogens.

Four types of Soil-Aquifer Treatments are possible:
1. Natural drainage of renovated water into stream, lake or low-lying area,
2. Collection of renovated water by subsurface drain,
3. Infiltration areas in two parallel rows and lines midway between, and
4. Infiltration areas in centre surrounded by circle of wells (Fig. 5.6; source: Bouwer 1994).

Table 5.5. Allowable limits for wastewater for specific uses (source: Kramer 1985, p. 1669).

| Description | 1 | 2 | 3 | 4 | 5 | 6 | 7 | 8 | 9 | 10 |
|---|---|---|---|---|---|---|---|---|---|---|
| pH | 4.5-9 | 4.5-9 | 4.5-9 | 4.5-9 | 4.5-9 | 4.5-9 | 4.5-9 | 4.5-9 | 6.5-9 | 6.5-9 |
| CFU/100 ml | | | | | | | | | | |
| Geometric mean * | 1000 | 1000 | 1000 | 1000 | 1000 | 200 | 25 | 2.2 | 1000 | 200 |
| Single sample not to exceed | 4000 | 4000 | 4000 | 4000 | 2500 | 1000 | 75 | 25 | 4000 | 800 |
| Turbidity (NTU) | – | – | – | – | – | – | 5 | 1 | 5 | 1 |
| Enteric virus (PFU) | – | – | – | – | – | – | 125/401 | 1/401 | 125/401 | 1/401 |
| Entamoeba histolytica | – | – | – | – | – | – | – | N.D. | – | N.D. |
| Giardia lamblia | – | – | – | – | – | – | – | N.D. | – | N.D. |
| Ascaris lumbricoides | – | – | – | – | – | – | N.D. | N.D. | N.D. | N.D. |
| Common large tapeworm | – | – | N.D. | N.D. | – | – | – | – | – | – |
| Trace substances | R 9-21-209 | R 9-21-209 | R 9-21-209 | R 9-21-209 | R 9-21-209 | R 9-21-209 | R 9-21-209 | R 9-21-209 | R 9-21-209 | R 9-21-209 |
| Organic chemicals | – | – | – | – | – | – | – | – | – | – |
| Radiochemicals | – | – | – | – | – | – | – | – | – | – |
| Treatment likely to achieve standards | A | A | A | A | A | A | B | C | A | A |

CFU – Colony Forming Units; NTU – Nephelometric Turbidity Units; PFU – Plaque Forming Units; N.D. – None Detectable (i.e. should not be present); A – Secondary disinfection; B – Secondary disinfection, infiltration; C – Soil aquifer treatment of secondary disinfection, coupled with chemical coagulation. * Geometric mean of minimum of five samples. 1 – Orchards; 2 – Fiber, seed and forage; 3 – Pastures; 4 – Livestock watering; 5 – Processed foods; 6 – Landscape area, restricted access; 7- Landscape area – open access; 8 – Food to be consumed raw; 9 – Partial body; 10 – Whole body.

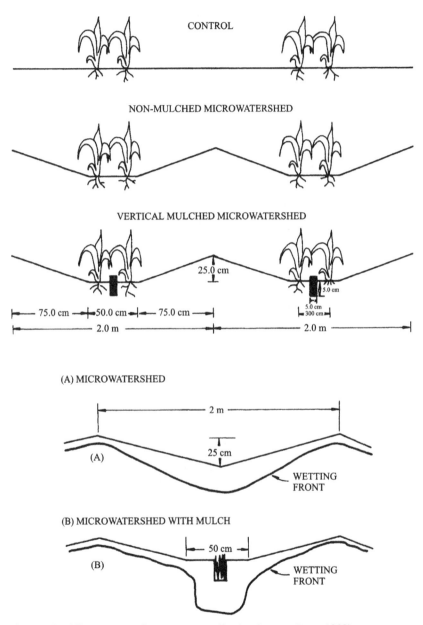

Figure 5.5. Different types of wastewater application (source: Jones 1982).

Table 5.6 (source: Bouwer 1994) shows the improvement in quality when mildly chlorinated secondary effluent (activated sludge) is let into infiltration basins, and the treated water then picked up through the recovery wells. Such water can be used for unrestricted irrigation.

Studies in the Dan region sewage reclamation project of Israel show that the long-term (20 yr) use of SAT reduces the trace element sorption capacity of the

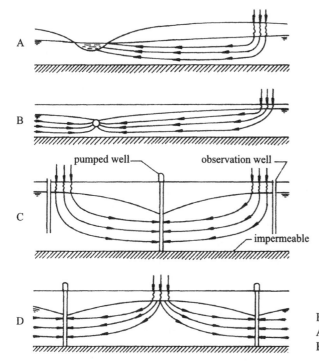

Figure 5.6. Four types of Soil – Aquifer Treatment (SAT) (source: Bouwer 1994).

Table 5.6. Extent of purification of secondary effluent effected by Soil – Aquifer Treatment (SAT) in Phoenix, Arizona (Source: Bouwer 1994).

| | Secondary effluent (mg l$^{-1}$) | Recovery well samples (mg l$^{-1}$) |
|---|---|---|
| Total Dissolved Solids (TDS) | 750 | 790 |
| Suspended Solids (SS) | 11 | 1 |
| Ammonium $NO_3 – N$ | 16 | 0.1 |
| Nitrate $NO_3 – N$ | 0.5 | 5.3 |
| Organic $NO_3 – N$ | 1.5 | 0.1 |
| Phosphate P | 5.5 | 0.4 |
| F | 1.2 | 0.7 |
| Br | 0.6 | 0.6 |
| Biochemical $O_2$ demand | 12 | 0 |
| Total organic C | 12 | 1.9 |
| Z | 0.036 | |
| Co | 0.008 | |
| Cd | 0.0001 | |
| Pb | 0.002 | |
| Fecal coliforms/100 ml | 3500 | 0.3 |
| Viruses, PFU*/100 l | 2118 | 0 |

* Plaque Forming Units.

soils (Roehl & Banin, Abstract, ICOBTE V, Vienna, July 1999). The soils undergo leaching of carbonates and manganese oxides and the accumulation of organic matter. The recharged soils have been found to have drastically reduced sorption capacity for (say) copper. This has been attributed to the acidolytic dissolution of the carbonates and manganese oxides, both processes being driven by the decomposition of the additional organic matter arising from recharge.

### 5.5.2 *Regulations regarding wastewater reuse*

Regulations regarding wastewater reuse may either specify water quality requirements or prescribe the treatments required for specific reuse. The second approach is generally preferred, as it is easy to enforce.

Secondary treatment required for (vide Table 5.5):
– Irrigation for fibrous or forage crops not intended for human consumption,
– Irrigation of orchard crops by methods which do not result in direct application of water on fruit or foliage,
– Watering of farm animals other than producing dairy animals.

Secondary treatment and disinfection required for (vide Table 5.5):
– Irrigation for any food crops where the product is subjected to physical or chemical processing sufficient to destroy pathogenic organisms,
– Irrigation of orchard crops by methods which involve direct application of water to fruit or foliage,
– Irrigation of golf courses, cemeteries, etc.,
– Watering of producing dairy animals,
– Water stored in impoundments intended for partial body contact recreation.

Tertiary treatment and disinfection required for (vide Table 5.5):
– Waters stored in impoundments intended for full body recreation,
– Irrigation of school grounds, playgrounds, lawns, etc. where children congregate,
– Irrigation of food crops which are consumed in their raw or natural state.

Several countries have prescribed regulations about the use of waste water in irrigation (Sepp 1976). Waste water is sought to be used in a manner which while providing irrigation to plants, will not have adverse health consequences to the consumers of plant-derived foods. Two cases are cited:

*Israel*: Settled sewage (primary effluent) may be used on industrial and fodder crops, pastures and hay, vegetables eaten in a cooked state, fruit trees, ornamental plants or seed plants. Irrigation should be stopped one month before the harvest in the case of apples, pears and plums, broccoli and cauliflower when furrow irrigated, and tomatoes used for canning if furrow irrigated.

*California* (USA): It is forbidden to use raw sewage to irrigate crops. Settled but undisinfected sewage may be used only on industrial, grain and fodder crops, and on vegetables grown for seed purposes. Such sewage may not be used on water growing vegetables, berries, low-growing fruits and vineyards and orchards during fruit growth. Completely treated, well oxidized, and reliably disinfected effluent conforming to the bacterial requirements of US Public Health Service (PHS) for drinking water standards, is allowed to be used on all crops.

## 5.6  ECONOMICS OF WASTEWATER REUSE

Technology is available for the primary and secondary and tertiary (coagulation, sand filtration and chlorination) treatment of wastewater to bring down the faecal coliform concentrations to essentially zero, but that is expensive (about USD 200-500/1000 $m^3$; Richard et al. 1992). Desalination of water is equally expensive – about one USD per 1000 $m^3$ for every 10 mg $l^{-1}$ salt removed with reverse osmosis, or about 2000-3000 per 1000 $m^3$ for seawater (Bouwer 1994).

Al-A'Ame & Nakhle (1995) gave a case study of the relative economics of wastewater reuse versus desalination. In Jubail, Saudi Arabia, there is a municipal wastewater plant of 19 MGD (million gallons per day) (73,625 $m^3$ $d^{-1}$) capacity, and an industrial wastewater plant of 11 MGD capacity (42,625 $m^3$ $d^{-1}$).

The wastewaters are subjected to biological treatment followed by pressure filtration. The effluents have the following characteristics:

|            | TDS | TSS | BOD | SAR  |
|------------|-----|-----|-----|------|
| Municipal  | 936 | 4.4 | 2.4 | 7.4  |
| Industrial | 762 | 2.1 | 2.4 | 10.5 |

Treatment of these wastewaters costs less (USD 2000 per 1000 $m^3$) than desalination (about USD 2670 per 1000 $m^3$ in Jubail). The treated effluent is largely used for landscape irrigation. Thus, wastewater reuse in Jubail is not only cost-effective but is environmentally beneficial.

Desalination is a viable option only for the energy-rich countries of the Middle East.

Developing countries have neither the capital nor human resources needed to build, maintain and operate plants for either Advanced Wastewater Treatment or desalination. For those countries, the WHO (1989) guidelines could be used, namely, maximum faecal coliform concentrations of 1000 per 100 ml and upto one helminthic egg per liter. Low technology treatments like lagooning and bioponds with detention times of more than a month are effective in the removal of most of the pathogens in the wastewater, after which the effluent can be recharged to groundwater.

Primary and secondary treatments typically cost USD 100/1000 $m^3$. Costs of post-treatment technologies (carbon filtration, reverse osmosis on half the flow and disinfection) for a 0.4 million $m^3/d$ plant, work out to about USD 230/1000 $m^3$. The costs would increase to USD 300/1000 $m^3$ if the carbon absorption were deleted, and the entire flow went through reverse osmosis followed by disinfection. As membrane technologies are improving and becoming cheaper, the latter may turn out to be the most cost-effective option in future. The total cost of converting raw sewage into potable water is about USD 400/1000 $m^3$ (Bouwer 1993).

Soil – Aquifer Treatment (SAT) by itself is inexpensive, since only pumping costs from recovery wells are involved – about USD 5/1000 $m^3$ if the groundwater is shallow, and USD 50/1000 $m^3$ 1000 if it is deep (50 m) (Bouwer 1993). The preferred option is a sequence which avoids the most expensive component, namely, the tertiary treatment component (AWT), and substitutes it by SAT.

## 5.7　HEALTH HAZARDS IN WASTEWATER REUSE

It is known that waste waters from urban areas contain a variety of faecally-excreted human pathogens including helminths, protozoans, bacteria and viruses. Waste water may contain high concentration of pathogens ($10^7$-$10^4$ organisms per gm of faeces) with long persistence times (weeks to months) (Fig. 5.7; source:

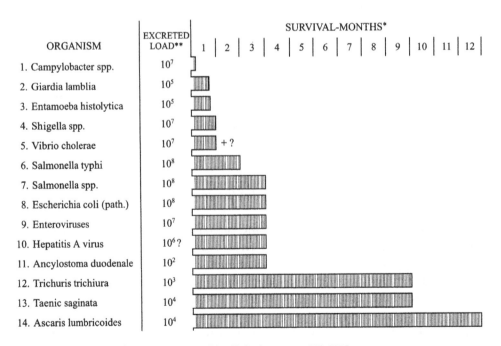

\* Estimated average life of infective stage at 20°–30°C
\*\* Typical avg. number of organism/gm feces

Figure 5.7. Abundances and persistence of pathogens in human faeces (source: Gunnerson et al. 1985).

Gunnerson et al. 1985). Thus, when waste water is used for irrigation, the concentrations of the pathogens in the irrigated soil and irrigated crops is sufficiently high and their persistence sufficiently long as to create a health hazard to the exposed population groups (WHO 1989).

Gunnerson et al. (1985) gave the following summary of health hazards arising from the use of raw or only partially treated waste water in the Developing countries:

- To the general public consuming salad or vegetable crops irrigated with raw waste water: ascariasis, trichuriasis, cholera, and possibly tapeworm (from eating meat of the cattle grazing on waste water irrigated pasture) infections, leading to cholera, typhoid fever, giardiasis, etc.
- To workers engaged in waste water irrigation: ancyclostomiasis (hookworm), ascariasis, possibly cholera, and to a much lesser extent, infection caused by some other bacteria or viruses.
- To the general public living in the proximity of waste water irrigation projects, particularly sprinkler irrigation projects using raw or poorly treated waste water: Minor transmission of diseases, particularly to children, caused by enteric viruses; also possibly limited transmission of shigellosis and other bacterial diseases due to contact.

Gunnerson et al. (1985) ranked the pathogenic agents (connected with waste water irrigation) in the following order:

- High risk (high excess incidence of infection): Helminths (*Ancylostoma, Ascaris, Trichuria* and *Taenia*),
- Medium risk (medium excess incidence of infection): Enteric bacteria (cholera, typhoid, *shigella*, etc.),
- Low risk (low incidence of excess infection): Enteric viruses.

## 5.8 USE OF SEWAGE SLUDGE AS FERTILIZER

Sewage sludge is a byproduct of wastewater treatment. Agricultural utilization of the sewage sludge not only constitutes a sustainable outlet for the disposal of the sludge, but it also enhances crop productivity as the sludge contains valuable micro-nutrients, trace elements and organic matter. Sometimes the sludge may contain toxic elements and pathogens which may enter the food chain and cause diseases in the humans and animals.

Sewage sludge is used in agriculture as a fertilizer and soil conditioner. It has been observed that sludge application reduces the surface runoff, and gives some protection against soil erosion.

The composition of sewage sludge varies greatly, depending on the source of sewage (domestic, industrial, etc.). The mean concentration level and range (percent content of dry matter) of the fertilizer content are as follows: N – 2.2 (1.5-3.5), P – 1.7 (1.0-2.5), K – 0.15 (0.05-0.3). The undesirable consequences of sludge application arise from the heavy metal content of the sludge, which may be taken up by the plants.

Table 5.7 compares the metal concentrations in dry sewage sludge and soil (in mg kg$^{-1}$) (source: O'Neill 1985, p. 207).

The metals present in sewage sludge may be divided into three categories on the basis of their availability to plants:
- Unavailable forms, such as insoluble compounds (oxides or sulfides);
- Potentially available forms, such as insoluble complexes, metals linked to ligands, or forms attached to clays and organic matter; and
- Mobile and available forms, such as hydrated ions or soluble complexes. A soil which has high pH (i.e. alkaline) and high cation exchange capacity, will be able to immobilize the metals added to it via the sewage sludge. A soil which is presently alkaline may not always remain that way. Thus, if at a later stage the pH drops (i.e. becomes acidic), the metals may get released.

Elements such as iron, zinc and copper are essential elements needed by man. The same elements may become toxic at high concentrations, however. Cadmium is a highly toxic element. Because of its geochemical affinity with zinc, it enters the plants along with zinc.

In a study made in Norway, sludge application at rates of 60-120 t (dry matter) per ha increased the Ni and Zn contents of plants, but had no significant effect on the concentration levels of Cd and Pb. Excessive application of sludge could increase the content of heavy metals in the crops to levels which render them unfit for human consumption.

Liming (1.5-6 t of CaCO$_3$ ha$^{-1}$) reduces the heavy metal uptake by plants *in the short run*. The heavy metals will continue to persist in the top soil, however. It has been reported that sludge-amended soils almost invariably contain higher levels of Cd than garden soils. Potential increase in the cadmium content of plants due to the application of sewage sludge constitutes the most important health hazard, which can be mitigated by increasing the available content of zinc in the soil. Sludge should not be applied to soils which are used for growing vegetables.

### 5.8.1 *Case study of sewage sludge in Cairo, Egypt*

As the Greater Cairo Wastewater project becomes fully operational, substantial quantities of sewage sludge are being produced. The sludge was treated in different ways to understand the chemical changes arising therefrom. Experiments were conducted to determine to what extent the application of treated sludges affect the heavy metal concentrations in plants.

Table 5.8 (source: Abou Seeda, Abstract, ICOBTE V, Vienna, July 1999) gives the chemical characteristics of the sludge after being treated in different ways.

The sludges thus treated are used to fertilize the spinach plants. The effect of application of variously treated sludges on the heavy metal concentrations (in mg kg$^{-1}$ of dry matter) in the tissues of the growing plants. was then studied. The

Table 5.7. Comparison of metal concentrations in dry sewage sludge and soil (in mg kg$^{-1}$); source: O'Neill 1985, p. 207).

| Metal | Mean conc. in sewage sludge | Mean conc. in soil | Conc. proportion, sludge/soil |
|---|---|---|---|
| Hg | 2 | 0.25 | 8 |
| Cd | 20 | 0.4 | 50 |
| Cu | 250 | 50 | 5 |
| Cr | 500 | 50 | 10 |
| Pb | 700 | 25 | 28 |
| Zn | 3000 | 100 | 30 |
| Fe | 16,000 | 30,000 | 0.5 |

Table 5.8. Chemical characterization of the sludge treatments (source: M. Abou Seeda 1999).

| Chemical properties | Air dried sludge | Aerobic digested sludge | Anaerobic digested sludge | Ash sludge |
|---|---|---|---|---|
| pH (1:5) | 6.91 | 7.30 | 6.35 | 9.42 |
| EC (1:5) m S$^{-1}$ | 3.55 | 5.10 | 9.98 | 0.79 |
| Total O.M. (%) | 51.20 | 43.12 | 47.15 | – |
| Soluble O.M. (%) | 2.24 | 3.59 | 7.89 | – |
| HA (% of total) | 0.07 | 0.14 | 0.70 | – |
| FA (% of total) | 3.42 | 2.45 | 5.34 | – |
| Total N (%) | 2.73 | 1.93 | 2.32 | – |
| Total P (%) | 1.31 | 1.29 | 1.69 | 2.89 |
| Total K (%) | 1.12 | 1.18 | 1.52 | 2.93 |
| Soluble – N (mg kg$^{-1}$) | 731 | 1179 | 6760 | – |
| Soluble – P (mg kg$^{-1}$) | 163 | 171 | 178 | 215 |
| Soluble – K (mg kg$^{-1}$) | 210 | 218 | 200 | |

study confirmed the earlier observation (Hernandez 1991) that the application of aerobic sludge resulted in the reduction of heavy metal content in the growing plant, and is therefore to be preferred. Incinerator sludge brought about the largest decrease in the heavy metal content of the plants, possibly because of the formation of oxides or because of the alkalinity of the soil treated with incinerator sludge.

## REFERENCES

Al-A'Ame, M.S. & G.F. Nakhle 1995. Wastewater reuse in Jubail, Saudi Arabia. *Water Res.* 29(6): 1579-1584.

American Waterworks Association (AWWA) 1985. *Future of Water Reuse*, Proc. Symp. on Water reuse, Aug. 26-31, 1984, San Diego, Calif., USA, v. III.

Ayers, R.S. 1975. Quality of water for irrigation. *Proc. Irrig. and Drain. Div., Speciality Conf., Amer. Soc. Civil Engineers*, Logan, Utah, 13-15 Aug., 24-56.

Bouwer, H. 1993. From sewage farm to zero discharge. *Europ. Water Pollu. Control.* 3(1): 9-16. Amsterdam: Elsevier

Bouwer, H. 1994. Role of geopurification in future water management. In *Soil and Water Science: Key to Our Understanding Our Global Environment.* Soil. Soc. Amer. Sp. Publ., 41: 73-81.

Dean, R.B. & E. Lund 1981. *Water Reuse – Problems and Solutions.* London: Acad. Press.

Gunnerson, C.G., Shuval, H.I. & Arlosoroff, S. 1985. *Future of Water Reuse,* Proc. Symp. on Water Reuse, Aug. 26-31, 1984, San Diego, Calif., USA, III, 1576-1605.

Hang Xu et al. 1991. Experimental studies on the purification of wastewater in the water peanut ponds. In B.Z. Wang et al. (ed.), *Low-cost and energy-saving wastewater treatment technologies, Water Sci. & Tech.* 24(5): 97-109.

Hernandez, T., J.I. Moreno & F. Costa 1991. Influence of sewage sludge application on crop yields and heavy metal availability. *Soil Sci. Plant Nutr.,* 37(2): 201-210.

Idelovitch, E. & Michail, M. 1985. Groundwater recharge for wastewater reuse in the Dan region project: summary of five-year experience, 1977-1981. In Asano, T. (ed.), *Artificial Recharge of Groundwater,* 481-507. Boston: Butterworth Publishers.

Jones, P.H. 1982. Wastewater treatment technology. In Laconte, P. & Y.Y. Haimes (eds), *Water Resources and Landuse Planning: A Systems Approach,* p. 93-132. The Hague: Martinus Nijhoff Publishers.

Kramer, R.E. 1985. Regulations for the reuse of wastewater in Arizona. *Future of Water Reuse,* Proc. Symp. on Water Reuse, Aug. 26-31, 1984, San Diego, Calif., USA, III, 1666-1672.

Laconte, P. & Y.Y. Haimes (eds) 1982. *Water Resources and Landuse Planning: A Systems Approach.* The Hague: Martinus Nijhoff Publishers.

Matters, M.F. 1981. Arizona rules for irrigating with sewage effluent. In *Proc. Sewage Irrigation Symp.,* Phoenix, Arizona. US Water Conservancy Lab., 6-12.

National Academy of Sciences and National Academy of Engineering. 1972. Report no. 5501-00520. Supt. of Documents.

O'Neill, P. 1985. *Environmental Chemistry.* London: George Allen & Unwin.

Richards, D., T. Asano & G. Tchobanoglous 1992. *The cost of wastewater reclamation in California.* Rep. from Dep. of Civil and Environ. Engg., Univ. of California, Davis.

Sepp, E. 1976. *Use of sewage for irrigation – A literature review.* Bureau of Sanitary Engineering, California, USA

Shi, S. & X.Wang. 1991. The purifying efficiency and mechanism of the aquatic plants in ponds. In B.Z. Wang et al. (ed.), *Low-cost and energy-saving wastewater treatment technologies. Water Sci. & Tech.* 24(5): 63-73.

Shiklomanov, I.A. 1998. World Water Resources: A new appraisal and assessment for the 21st century. Paris: UNESCO

Switzenbaum, M.S. (ed.) 1991. Anaerobic treatment technology for municipal and industrial wastewater. *Water Sci. & Tech..* 24(8).

Wang, B. 1991. Generating and resources recoverable technology for water pollution control in China. In B.Z. Wang et al. (ed.), *Low-cost and energy-saving wastewater treatment technologies, Water Sci. & Tech.* 24(5): 9-19.

Wilson, L. 1985. Potable reuse criteria established for El Paso, Texas. *Future of Water Reuse,* Proc. Symp. on Water Reuse, Aug. 26-31, 1984, San Diego, Calif., USA, III, 1639-1665.

World Health Organization (WHO). 1973. *Reuse of effluents: Methods of wastewater treatment and health safeguards.* Tech. Report 517. Geneva: WHO.

World Health Organization (WHO). 1989. *Health guidelines for the use of wastewater in agriculture and aquaculture.* Tech. Bull. Ser. 77, Geneva: WHO.

CHAPTER 6

# Use of water in agriculture

## 6.1 IRRIGATION

Irrigation has been practiced for the millennia. The increasing demand for food by the growing populations necessitated the development of intensive irrigation on a large scale, as irrigated agriculture is known to be far more productive than rain-fed agriculture. Thus, though only 15% of the cultivated lands in the world are irrigated, they account for almost half of the total crop production in the world in terms of value.

Aswathanarayana (1999, Chapter 6, Soils in relation to Irrigation) gave a comprehensive account of different aspects of use of water in irrigation.

At the beginning of the Twentieth Century, irrigation accounted for 90% of all water use. It was 80% in 1960, and is currently around 70%. Early in the 21st century, it is expected to go down to 60% (Biswas 1993). *This implies that more food for the growing populations will have to be grown with less water.* This would mean a more intensive agriculture with greater use of fertilizers and pesticides, and the consequent greater possibility of groundwater pollution with agricultural chemicals. That would be a disastrous development, considering that the quantity of fresh groundwater within drillable depth is about 67 times more than the water in all rivers and lakes in the world (Bouwer 1978). The values for specific water withdrawal in agriculture vary greatly depending upon the climate, soil and hydrological conditions, etc. As is to be expected, the water withdrawal rates are minimum in northern Europe, and maximum in Africa. The large variation in the withdrawal rates in the case of Asia and South America is a consequence of the great variability in crop composition and watering techniques practiced in those regions.

The specific water withdrawals and returnable water for various parts of the world are as set out in Table 6.1 (Shiklomanov 1998, p. 20).

### 6.1.1 *Water quality criteria for irrigation*

The irrigation water quality is judged in terms of electrical conductivity (Ece, in $\mu S$ cm$^{-1}$ or mS m$^{-1}$ at 25°C), sodium absorption ratio (SAR), the concentration of individual ions, the clay mineral composition of the soil and the crops to be grown

Table 6.1. Specific water withdrawals and returnable water for various parts of the world.

| Region | Rate of specific withdrawals (m³ ha⁻¹) | Returnable water (as % of water intake) |
|---|---|---|
| Northern Europe | 300- 5000 | 20-30 |
| Southern and Eastern Europe | 7000-11,000 | 20-30 |
| USA | 8000-10,000 | 40-50 |
| Asia, Central and South America | 5000- 6000 to 15,000-17,000 | |
| Africa | 20,000-25,000 | |

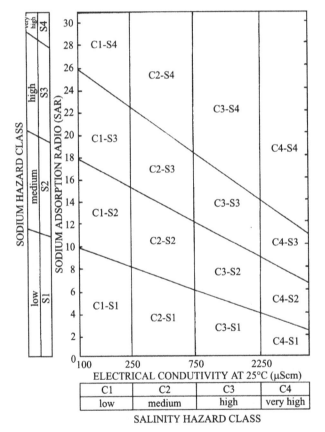

Figure 6.1. Quality of irrigation water in terms of salinity and sodicity (source: US Department of Agriculture; as quoted by Chadwick et al. 1987).

(Chadwick et al. 1987). Particular care should be taken in using water for irrigation in the arid regions, as high evaporation rates tend to concentrate salts.

The classification of the US Department of Agriculture (as quoted by Chadwick et al. 1987) regarding the quality of irrigation water in terms of salinity and sodicity depicted in Figure 6.1. Table 6.2 (US Dept. of Interior 1968), Table 6.3 (USDA), Table 6.4 (Ayers 1975) and Table 6.5 (Nat. Acad. Sci. 1972) give the

Table 6.2. Water quality criteria for agricultural use (Source: US Dept. of Interior 1968).

| Determinand | Livestock | Irrigation continuous | Irrigation iIntermittent |
|---|---|---|---|
| 1. *Elements* (mg $l^{-1}$) | | | |
| Al | | 1.0 | 20.0 |
| As | 0.05 | 1.0 | 10.0 |
| B | | 0.75 | 2.0 |
| Be | | 0.5 | 1.0 |
| Cd | 0.01 | 0.005 | 0.05 |
| Cl | Variable | | |
| Co | | 0.2 | 10.0 |
| Cr | 0.05 | 5.0 | 20.0 |
| Cu | | 0.2 | 5.0 |
| F | 2.4 | | |
| Li | | 5.0 | 5.0 |
| Mn | | 2.0 | 20.0 |
| Mo | | 0.005 | 0.05 |
| Ni | | 0.5 | 2.0 |
| Pb | 0.05 | 5.0 | 20.0 |
| Se | 0.01 | 0.05 | 0.05 |
| Va | | 10.0 | 10.0 |
| Zn | | 5.0 | 10.0 |

2. Total dissolved solids < 10,000 (mg $l^{-1}$)

3. *Herbicides* (mg $l^{-1}$) – range depending upon the crop

| | |
|---|---|
| Acrolein | 15-80 |
| Xylene | 800-3000 |
| Amitrole – τ | 3.5 |
| Dalapon | 0.35-7.0 |
| Diquat | 5-125 |
| Na and K salts of Endothall | 1-10 |
| Dimethylamines | 25 |
| 2,4 – D | 0.7-10 |
| Dichlobenil | 1-10 |
| Fenac | 0.1-10 |
| Picloram | 0.1-10 |

the various aspects of water quality criteria for agricultural use (livestock, and continuous/intermittent irrigation).

Generally, conductivities of less than 100 mS m$^{-1}$ (at 25°C) and SAR of less than 15 are acceptable.

Quality parameter and suggested level of stream standard for irrigation are as follows:

– Boron: Not more than 1.25 mg l$^{-1}$ where there are sensitive crops, and not more than four mg l$^{-1}$ where there are tolerant crops,

Table 6.3. Salinity and sodicity of water usable for irrigation (source: US Dept. of Agriculture, as quoted by Chadwick et al. 1987).

| Salinity/sodicity class | Description and use |
|---|---|
| C1 | Low salinity water which can be used for irrigation with most crops on most soils. |
| C2 | Medium salinity water which can be used for plants with moderate salt tolerance |
| C3 | High salinity water which cannot be used on soil with restricted drainage. |
| C4 | Very high salinity water which is unsuitable for irrigation, except where the soils are highly permeable and high salt-tolerant crops are to be grown. |
| S1 | Low sodium water can be used for irrigation of most soils. Sodium-sensitive crops may still accumulate injurious levels of sodium. |
| S2 | Medium sodium water, presents hazard for fine-textured soils; may be used on coarse-textured or organic soils with good permeability |
| S3 | High sodium water unsuitable, except in the case of gypsiferous soils. |
| S4 | Very high sodium water generally unsatisfactory, except where gypsum amendment is made. |

Table 6.4. Guidelines regarding water quality criteria for irrigation (source: Ayers 1975).

| Problems and quality parameters | No problems | Increasing problems | Severe problems |
|---|---|---|---|
| Salinity effects on crop yields | | | |
| Total Dissolved Solids(TDS) conc. (mg $l^{-1}$) | <480 | 480-1920 | >1920 |
| Deflocculation of clay and reduction in K and infiltration rate | | | |
| Total Dissolved Solids(TDS) conc. (mg $l^{-1}$) | >320 | <320 | < 128 |
| Adjusted Sodium Absorption Ratio (SAR) | < 6 | 6-9 | > 9 |
| Specific ion toxicity | | | |
| Boron (mg $l^{-1}$) | < 0.5 | 0.5-2.0 | 2-10 |
| Sodium (as adjusted SAR), if water is absorbed by roots only | < 3 | 3-9 | > 9 |
| Sodium (mg $l^{-1}$) if water is also absorbed by leaves | < 69 | > 69 | |
| Chloride (mg $l^{-1}$) if water is absorbed by roots only | <142 | 142-355 | > 355 |
| Chloride (mg $l^{-1}$) if water is also absorbed by leaves | <106 | >106 | |
| Quality effects | | | |
| Nitrogen in mg $l^{-1}$ (excess N may delay harvest time and adversely affect yield and quality of sugar beets, grapes, citrus, avocados, apricots, etc.) | < 5 | 5-30 | > 30 |
| Bicarbonate as HCO$_3$ in mg $l^{-1}$ (when water is applied with sprinklers, bicarbonate may cause white carbon deposits on fruits and leaves) | < 90 | 90-520 | > 520 |

- Dissolved oxygen: Greater than two mg $l^{-1}$. A level of two mg $l^{-1}$ should not occur for more than eight hours out of any 24 hour period,
- Faecal coliform density: Not more than 1000 per 100 ml if the water is to be used for unrestricted irrigation. This standard may be relaxed if the crop is not intended for human consumption.

Table 6.5. Recommended maximum limits in mg $l^{-1}$ for trace elements in irrigation water (source: Nat. Acad. Sci. & Nat. Acad. Engg 1972).

|  | Permanent irrigation of all soils | Upto 20-yr irrigation of fine-textured, neutral to alkaline soils (pH 6-8.5) |
|---|---|---|
| Aluminum | 5 | 20 |
| Arsenic | 0.1 | 2 |
| Beryllium | 0.1 | 0.5 |
| Boron-sensitive crops | 0.75 | 2 |
| Boron-semitolerant crops | 1 | |
| Boron-tolerant crops | 2 | |
| Cadmium | 0.01 | 0.05 |
| Chromium | 0.1 | 1 |
| Cobalt | 0.05 | 5 |
| Copper | 0.2 | 5 |
| Fluoride | 1 | 15 |
| Iron | 5 | 20 |
| Lead | 5 | 10 |
| Lithium (citrus) | 0.075 | 0.072 |
| Lithium (other crops) | 2.5 | 2.5 |
| Manganese | 0.2 | 10 |
| Molybdenum | 0.01 | 0.05* |
| Nickel | 0.2 | 2 |
| Selenium | 0.02 | 0.02 |
| Vanadium | 0.1 | 1 |
| Zinc | 2 | 10 |

* For acid soils only.

## 6.2  RESERVOIR MANAGEMENT

Large surface reservoirs are constructed to increase the availability of water during the periods of low flow or dry years. Though the construction of water reservoirs has been going on since ancient times, all the large reservoirs with a total volume of more than 50 $km^3$ have been constructed during the second half of this century. According to Shiklomanov (1998), the total volume of the reservoirs of the world is about 6000 $km^3$, and their total surface area is upto 500,000 $km^2$. The volume of water in the reservoirs corresponds to about 5% of the annual rainfall on land or 15% of the annual runoff. They trap 2-5 Gt $yr^{-1}$ of sediments, which is about a quarter of the total sediment yield of the land. Thus, the reservoirs have a profound impact on the circulation of water on the land surface, discharge of sediments and nutrients to the sea. The evaporation from the surfaces of the reservoirs is considerable, and thus the reservoirs are one of the greatest users of freshwater. Figure 6.2 (source: Vörösmarty et al. 1997) shows the location of the major reservoirs in the world.

Figure 6.2. Location of the major reservoirs in the world (source: Vörösmarty et al. 1997).

### 6.2.1 *Can high dams be avoided?*

High dams, such as the Sardar Sarovar dam on the Narmada River in western India, and the Three Gorges Project on the Yangtze river in China, have drawn much criticism and agitations particularly on account of the hardship caused to the people who have been or will be displaced, on account of the submergence.

Life often faces us with difficult, even cruel choices. A community may have to choose between the interests of a community as a whole and that of a family. A swimmer, when attacked by a killer shark, may have to choose between foregoing his leg rather than his hand, as the only way to save himself (he can use his hand to fire the gun). The construction of a reservoir which could provide drinking water to (say) a million people, may entail the uprooting of (say) 10,000 people from their ancestral homes and avocations, and causing them immense hardship. A choice cannot be avoided. There is nothing like free lunch, and it is not possible to make an omelet without breaking the egg!

The following is a heart-rending story about choices. It is a real story narrated to the author by a Polish Jewish emigrant in Sweden. During the Second World War, he was serving as an Army officer. One day he came to know from a highly reliable source that next day, there was going to be a *pogrom* (massacre of the Jews) by the Nazis, in their village. Immediately, he sent a trusted confidant to his village to tell his wife that she should leave the village immediately the same night, with their two sons, aged seven and two years, without informing any body in the village. The elder son was ill, and was not in a position to walk. It was just impossible for her to carry both the children. She was faced with the cruel choice of losing both the children, and saving the younger child. She abandoned her elder son, knowing full well that the Nazis would surely kill him. She left the house in the dead of the night carrying the younger son.

The choice involved between one high dam and a number of small dams can be illustrated with a simple analogy. Suppose a quantity of bricks need to be transported. There is an option – the job can be done by, say, either 50 donkeys or by one elephant. But a huge log of wood can be transported only by an elephant – any number of donkeys cannot do this. In other words, there is no option.

The following discussion about the high dams takes into account the lucid exposition about high dams by P.V. Indiresan in an article in 'The Hindu' (Madras, India) of Dec. 1, 1999:

– Other things being equal, doubling the height of the dam, increases the water stored *eight* times and the power potential *sixteen* times. Also, for a given amount of storage, the higher the dam the smaller the area that will be submerged. *It therefore follows that the construction of the highest dam that is technically feasible is the best way to minimize submergence and hence the number of people to be displaced.* If, instead of a high dam, it is proposed to have a number of low dams, the aggregate of the areas submerged by them will be considerably more, and the amount of hydroelectricity that could be generated will be minimal. The amount of water available for drinking and irrigation may be unaffected, however.

- In countries with monsoon climate (e.g. India), the rainfall is restricted to a few months of the year. It is also erratic. Instances are known where the annual precipitation in an area occurred in a matter of few days of intense downpour. Consequently, flash floods are common, and very few rivers are perennial. Under these circumstances, harvesting and storage of surface runoff in reservoirs is the only way to provide irrigation and drinking water round the year. The runoff in a watershed can be harvested both through a high dam on a river *plus* a series of small check dams over small streams, and the recharge of groundwater. This is not a case of *either/or* but *and*. However, there may be situations where an area is so arid (e.g. Sinai desert in Egypt or Thar desert in India) that check dams are of no avail. Thus, Egypt is irrigating some areas in the Sinai desert by the transfer of Nile water. But for the water from Bhakra high dam in Punjab, the greening of desert areas in Jaisalmer, Rajasthan, India, would not have been possible.
- In tropical countries, there is intense evaporation. In India, it is about 1.2 m. In other words, irrespective of the depth of the water body, the top 1.2 m layer of water in a reservoir will be lost due to evaporation. If the reservoir is (say) 100 m deep (because of a high dam), the loss due to evaporation would be only 1%. On the other hand, if the reservoir is shallow, (say, about 10 m), the loss due to evaporation would be 10% of the total storage.

Thus, in the ultimate analysis, a country or a community may have to make the choice, one way or the other. Indiresan is critical of the agitators whose sole aim is to stop the construction of high dams. This is a negative approach. Such agitators rarely, if ever, come up with techno-socio-economically viable alternatives which can lead to benefits similar to the high dam. Indiresan suggests some practical steps to ease the problems arising from high dams:

- Engineers should confine themselves only to building dams, and agronomists (not engineers) should manage water allocation, as is the practice in USA,
- The resettlement process should be handled by specialists,
- There should be complete transparency about all aspects of the dam, reservoir, submergence, etc.,
- The displaced persons should be offered a package of benefits (such as retraining for the new jobs that will be created), and not just monetary compensation.

### 6.2.2 *Three Gorges Project (China): A case study*

The advantages and disadvantages of the high dams are sharply manifest in the case of the Three Gorges project on the Yangtze River, which when completed, would be the largest hydroproject in the world. It is a multi-purpose project, involving flood control, power generation, navigation, aquaculture, tourism, ecological protection, transfer of water from north to south, water supply for drinking and irrigation, etc. (Yangbo 1998).

That the dimensions of the Three Gorges project are mind-boggling should be evident from the following particulars: Total storage capacity: 39.3 billion (B) m$^3$; Normal pool level: 175 m; surface area: 1084 km$^2$. Power generation potential of the system: 105 B kwh; Deep navigational channel for 660 km. between Chongquin and Yichang allowing 10,000 t ships to sail – raising the shipping capacity to 50 million tons, and cutting the shipping costs by 37%.

The most important purpose of the Three Gorges Project is flood control. The middle and lower reaches of the Yangtze River are well developed in agriculture and industry. These areas are highly vulnerable to flooding, because the flood level can be 6-17 m higher than the ground level of the adjacent plain. According to statistics, there have been 214 occurrences of disastrous floods during the last 2000 years, with a recurrence interval of about 10 years. The Three Gorges Project can protect the lives and properties of more than 15 million people, and 1.53 million hectares of farmland in the downstream plains, against the biggest flood in 100 years, and even the possible biggest flooding in 1000 years.

The shortage of water in North China is hindering its development. The Three Gorges Project will alleviate this shortage by facilitating the transfer of 70 B m$^3$ of water from south to north.

The Three Gorges Project was criticized on the ground that it is very expensive, that a large number of people will be displaced, and that valuable historical and sacred sites will be permanently lost due to submergence, that it will get silted up, etc. All these criticisms are undoubtedly valid. It should be said to the credit of the Chinese Government that they took great pains to minimize the sufferings of the displaced people. Every village and town have been told long in advance where they will be relocated, and how they will earn their livelihood in the new setting. Habitations and service facilities (such as, schools and hospitals) have been built in advance in the new locations, and the government helped the families to shift their personal effects to the new locations. Many persons have been retrained to take advantage of the new kind of jobs generated by the project. Key historical sites were excavated and shifted to higher positions. There have of course been cases of beauracratic bungling and corruption, and some people were extremely unhappy to leave their ancestral homes.

There is little doubt that on balance, the benefits from the Three Gorges Project far outweigh the disadvantages and the hardship caused to people. *The most important point to bear in mind is that the level of flood control, power generation and navigation achievable by the high dam of the Three Gorges Project cannot just be realized through a number of low dams.*

### 6.2.3 *Allocation of water from reservoirs*

The compendium, 'Irrigation and Water Allocation' (IAHS Pub. no. 169, 1987) is devoted to two issues: the interaction of irrigation and hydrological processes, and allocation of irrigation water. Holý & Kos (1987) aver that the long-term planning of irrigation has to reflect the stochastic character of the irrigation water demands. They developed methods of calculating mean monthly time series of irrigation water demands based on precipitation, temperature, humidity, sunshine and wind velocity, etc. The water resources in the irrigated areas consist of rainfall, snowfall, surface water, soil water and groundwater. Figures 6.3 and 6.4 (source: Gui 1987) show the linkages and interactions between different classes of water. Because of these linkages, the damage to one part of the system will have impact on other parts of the system. Rubin (1987) developed a hierarchical method of designing least cost water distribution networks.

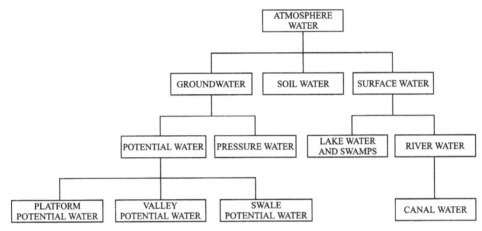

Figure 6.3. Linkages between different kinds of water (source: Gui 1987).

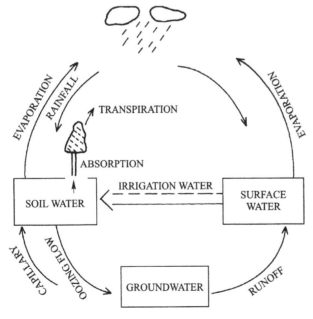

Figure 6.4. Place of irrigation water in the hydrological processes (source: Gui 1987).

Because of the great complexity and socio-economic importance of the irrigation system, a systems approach is used. This involves the application of techniques such as, linear programming, dynamic programming, simulation, hierarchical or multi-level optimization.

Billib et al. (1995) described an interactive management model using the Incremental Dynamic Solving Technique, for the conjunctive use of the San Juan river system. The latter consists of a surface reservoir with a hydropower plant, a groundwater reservoir, artificial recharge areas and pumping fields and a complex channel distribution system for irrigation.

The management model is aimed at optimizing the following objectives:
– Maximum hydropower production by the hydropower plant,
– Minimum costs for groundwater pumping,
– Maximum irrigation areas,
– Maximum/minimum artificial groundwater recharge for dry/wet years,
– Minimum/maximum system outflow for dry/wet years.

The characteristics of the system are as follows:
– Area under irrigation: 67,000 ha,
– Climate: Potential evaporation (1230 mm $yr^{-1}$); Temperature (17.8°C),
– Water requirement (in $hm^3 yr^{-1}$): Irrigation (1060), Population (60), Industry (12),
– Water supply capacities (in $hm^3 yr^{-1}$): Reservoir inflow (2020), Groundwater extraction (600) and Artificial groundwater recharge (170); Surface reservoir (390 $hm^3$).

The parameters of the model are given in Figure 6.5 and the flow network of the system is given in Figure 6.6 (source: Billib et al. 1995). A Sequential Multi-Objective Problem Solving (SEMOPS) technique was employed. The flow-chart of the procedure is summarized below:

*Step 1: Long-term analysis of the hydrological system*
– time series analysis
– groundwater prediction
↓

*Step 2: Short-term analysis (hydrological year)*
D   →  –   quality restriction
E   →  –   selection of objective functions
C   →  –   change of restrictions
I   →  –   IDP for reservoir/groundwater system
S   →  –   SEMOPS for allocations of:
I        irrigation, municipal, hydropower purposes
O   →  –   simulation of groundwater flow and solute transport
N

↑        ↓

*Step 3: Long-term analysis of groundwater quality*

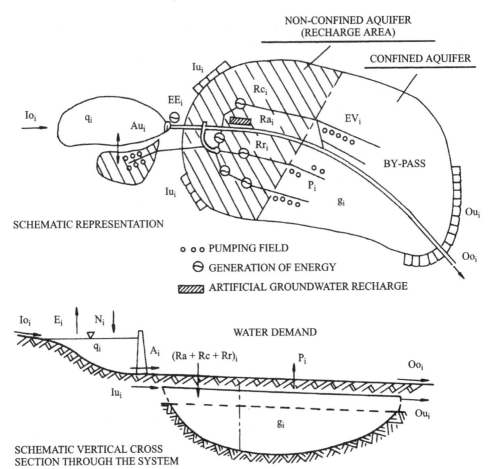

Figure 6.5. Parameters of the model for the conjunctive use of water (source: Billib et al. 1995). Notation: A: release from the surface reservoir, Au: subsuperficial inflow (outflow) to (from) the surface reservoir from (to) the unconfined aquifer, E: evaporation from the surface reservoir, EE: energy generated in the hydropower plants, EV: energy consumed in the pumping fields, g: volume stored in the aquifer system, Io: inflow to the surface reservoir, Iu: subsuperficial inflow to the groundwater reservoir, N: precipitation over the surface reservoir, Oo: superficial outflow from the system, Ou: subsuperficial outflow from the system, q: volume stored in the surface reservoir, P: groundwater extraction, Ra: volume of artificial groundwater recharge, Rc: volume of groundwater recharge from the irrigation areas over the unconfined aquifer, Rr: volume of deep percolation in the river bed, (Ra+Rc+Rr): groundwater recharge, i: index for simulation steps.

Water allocation for various solutions are figured out by SEMOPS, to facilitate informed decision-making. Long-term analysis of groundwater quality (in terms of issues such as nitrate concentration) is made to enable DM to take appropriate corrective steps.

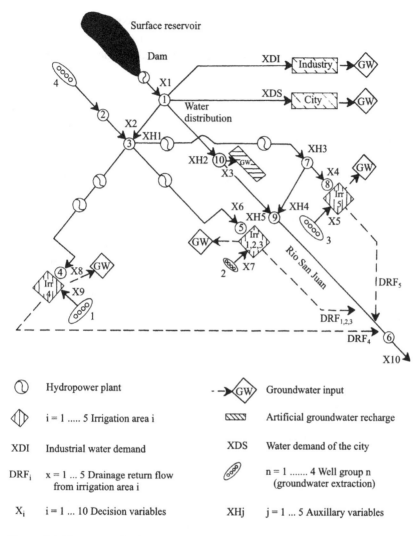

Figure 6.6. Flow network of the reservoir system (source: Billib et al. 1995).

### 6.2.4 *Ecologically sound operation of the reservoirs*

Depending upon their capacity and catchment area, reservoirs have ecological impact not only within the storage area itself but also in the lower reaches. While the ecological changes within the storage area start soon after the construction of the reservoir, the changes in the downstream side may take several years to become manifest. Bed shear stress is an important indicator of the diversity and density of aquatic species. Evidently, under conditions of increased stress, some sensitive organisms will go down in numbers.

The question then arises whether ecologically meaningful hydraulic parameters should be treated as decision variables or constraints which should not be violated. If ecological systems are badly damaged, they may take a long time to re-cover, or the damage may be irreversible. Ostrowski & Lohr (1995) describe a reservoir information system which would facilitate the ecologically-sustainable operation of a reservoir (Fig. 6.7).

### 6.2.5  *Reservoir operation as an aid in sustainable water management*

A reservoir is an inflexible element in a water resource system. When once a res-ervoir is built, it is very difficult to change its storage capacity, and it is impossi-ble to shift it. When a reservoir is built some decades ago, it would have been planned in the context of the then existing situation. The efficiency of the techni-cal structure of the management (reservoirs, water works, conveyance systems, canals, etc.) very much depends upon the software component, such as, operation rules, management models, decision support system for operation, etc. It is this software part that is amenable to adjustment in the context of changes in supply and demand conditions.

Three factors may necessitate the updating of the water management planning in general and reservoir management in particular:

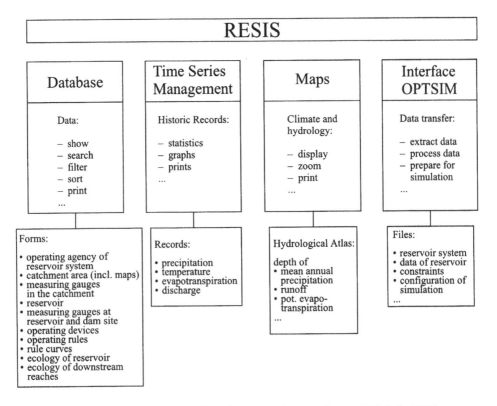

Figure 6.7. Structure of the reservoir information system (source: Ostrowski & Lohr 1995).

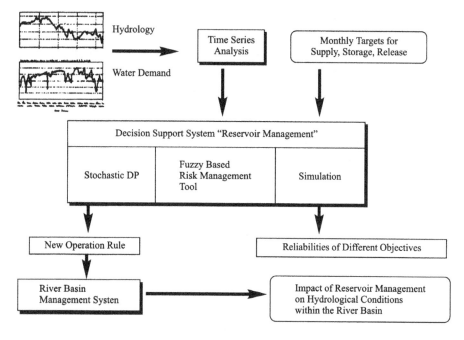

Figure 6.8. Methodology for the adaptation of reservoir management (source: Schumann 1995).

1. Availability of hydrological data for a long period could be used to reduce the uncertainties in the database, and thereby facilitate better decision making about (say) quantum and timing of releases,
2. The loads of harmful substances in the surface water and groundwater might have increased, thereby affecting the quality of water supplied to the users. The ecological impact in the lower reaches of the river should also be considered,
3. The water demand is affected by socio-economic changes. The consumption of water for domestic purposes is not inelastic. If water is subsidized, people will tend to use more water than when the water is higher priced. Also, with increasing industrialization, the demand for water may go up.

The operation of the reservoirs has to be capable of being adapted to changing socio-economic needs at comparatively short intervals of some years. The structure of the methodology for adaptation of reservoir management to changing conditions is schematically shown in Figure 6.8 (source: Schumann 1995).

## 6.3 NATURAL WATER SYSTEM IN AN AGRICULTURAL REGION

Paces (1976) gave a box model to represent the natural water systems (NWS) in an agricultural region (Fig. 6.9; source: Paces 1976) The solid-line boxes represent the moving medium, such as natural water (e.g. soil solution) carrying dissolved and suspended matter. The dashed-line boxes represent the environments (e.g. soil) through which the water moves and with which it interacts. The slanted boxes represent the inputs and outputs of the system by processes other than water transport (e.g. fertilizers). This is followed up by a box model for soil solutions (Fig. 6.10; source: Paces 1976). The composition of the soil solution is determined by the rates of interaction with the various components of the environment. The composition of the soil solution may reach chemical equilibrium under conditions of slow water flow and rapid reaction with the environment.

The contribution of soil environment to water composition may be given by the following simplified equation (Paces 1976):

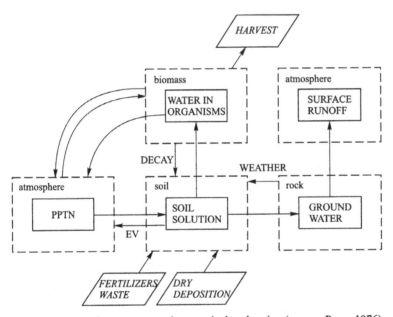

Figure 6.9. Natural water systems in an agricultural region (source: Paces 1976).

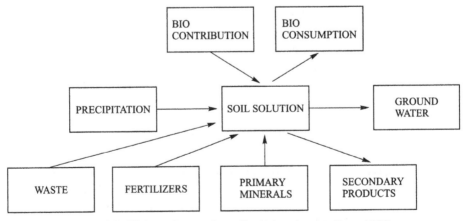

Figure 6.10. Box model of linkages in regard to soil solutions (source: Paces 1976).

$$m - m_o = \frac{t_{\text{mean}}}{fu}(\mathrm{d}g/\mathrm{d}t) \tag{6.1}$$

where $m$ is the concentration in water, $m_o$ is the initial concentration before the water entered the soil, $f$ is the effective porosity of the soil, $u$ is the degree of saturation, $(\mathrm{d}q/\mathrm{d}t)$ is the rate of removal of the component from unit volume of the soil, and $t_{\text{mean}}$ is the mean residence time of the water in the environment.

Chemical and isotopic measurements are made in different watersheds to obtain quantitative figures for the parameters to be used in the above equation.


## 6.4 SOIL MOISTURE

All soils have voids or interstitial spaces, their extent depending upon the composition of the soil. After precipitation, flooding or irrigation, these interstitial spaces get filled with water, temporarily or permanently. When this happens, the soil is said to be *saturated*. The *saturation capacity* of a soil is defined as the maximum amount of water a soil can hold when saturated. It should, however, be borne in mind that it is physically not possible for all the voids to be completely filled, for the simple reason that some air is inevitably trapped.

*Soil moisture* refers to the amount of water present in the soil above the watertable (Chris Barrow 1987, p. 84). Water is held in the soil with varying degrees of tension, ranging from water which is free to flow, to water that is held firmly on the surfaces of the soil particles. On the basis of the process of physical movement involved, soil moisture may be classified into three types: gravity, capillary, and hygroscopic. Gravity (or vadose) water is that water which is present in the soil above the watertable and which drains away under the influence of gravity. Sandy soils take less than a day to drain while this process may take about four days in the case of clay soils. Complete drainage is not possible and considerable amounts of moisture is invariably left behind.

Capillary water moves by capillary action – it 'creeps' above the watertable to a height determined by the texture and composition of the soil. The capillary fringe refers to this zone. Where the watertable is shallow, it may happen that the capillary fringe may extend right up to the surface of the soil. The capillary water which thus moves upward evaporates at the surface, leaving behind the dissolved salts that the water may contain or may have leached from the soil in the process of upward movement.

Hygroscopic water refers to the film of water that adheres to the soil particles by molecular attraction. This moisture is virtually inaccessible to plants because of the firmness with which it is held by the soil particles. However, some *xerophytes* (plants that can survive in arid conditions) are known to be able to extract such water.

*Field capacity* or Field Moisture Capacity of a soil refers to the maximum amount of water remaining in the soil after the gravity water has drained away.

The term has been defined in various ways. Field Capacity has been defined by Carruthers and Clark as the quantity of water held at a particular suction pressure forty-eight hours after wetting. The parameter was to be expressed in mm. According to FitzPatrick (1993, p. 232), Field Capacity is to be expressed as a percentage of the oven-dry soil. Hillel (1987, p. 35) defines the Field Capacity as the water content of specified volume of soil measurable two days after thorough irrigation, expressed as fractional volume (%). As the draining of soils takes place continuously and unevenly, there is no constancy in the field capacity. The term has become largely obsolete.

The relationship between soil type and available moisture is given in Table 6.6 (source: Stern 1979, p. 85).

Plants growing on the soil will extract moisture as long as it is available. When moisture is no longer available, the plants will wilt and die (wilting point). FitzPatrick (1993, p. 37) explains the relationship between the soil moisture characteristics, soil texture, field capacity and wilting point. It may be noted that the wilting point increases with fineness of texture – low for sand and high for clay. The field capacity increases from sand to silt loam but remains unchanged after that.

Water moves in the soil in three ways:
- *Saturated flow*: When water is moving through a soil in which all the pores are filled with water. Saturated flow may be in any direction, vertically downward or laterally;
- *Unsaturated flow*: When water moves from pore to pore, in a situation wherein only some pores are filled with water. Unsaturated flow takes place both in response to gravity (vertical or lateral movement) as well as to moisture gradient (capillary movement upward); and
- *Vapor transfer*: Movement of water as a vapor within the soil and from soil to atmosphere. The rate of movement is determined by the relative humidity, temperature gradient, porosity and permeability, and the degree of saturation.

There are two demands upon the soil moisture – evaporative demand of the atmosphere and water needs of the crop plant. Whereas the evaporative demand is continuous, the supply of water by natural precipitation is sporadic. Under the circumstances, the crop has to rely most of the time on the limited reserves of water present in the soil pores. The water economy of some plants is so delicate that if the available soil moisture is inadequate, they become less productive or may even die.

Table 6.6. Soil texture and available water (after Stern 1979, p. 85).

| Soil type | Available water,% | Available water, mm/m |
|-----------|-------------------|-----------------------|
| Fine sand | 2- 3 | 30- 50 |
| Sandy loam | 3- 6 | 40-150 |
| Silt loam | 6- 8 | 60-120 |
| Clay loam | 8-14 | 90-210 |
| Clay | 13-20 | 190-300 |

### 6.4.1  *Chemistry of soil solutions*

The aqueous liquid phase in the soil constitutes the soil solution. Since soil is an open system, the composition of the soil solution is influenced by flows of matter and energy between the soil, atmosphere, biosphere, and hydrosphere.

*Complexation* is a process whereby an ion acting as a central group attracts other ions and forms associations with them. The associated ions are called *ligands.* The complex may have a positive or negative charge, or it may be neutral. Examples are: $Si^{4+}$, $Al^{3+}$ and $(CO_3)^{2-}$ acting as a central group to form aqueous complexes, Si $(OH)^0_4$, Al $(OH)^{2+}$ and $HCO_3^-$ – respectively, with $OH^-$ or $H^+$ serving as a ligand. The term ligand may also be applied to cations coordinated to an anion, as in the complex, $H_2PO_4^-$. A complex is called a *chelate*, if two or more functional groups of a single ligand are coordinated to a metal cation – for instance, the complex between the metal cation $Al^{3+}$ and citric acid, involving two $COO^-$ groups and one COH group, $[Al\ (COO)^-_2 - COH\ (CH)_2\ COOH]^+$, is a chelate.

A normal soil solution may contain 100-200 different soluble complexes. Table 6.7 (source: Sposito 1989, p. 69) lists the principal chemical species in the soil solutions in acid soils and alkaline soils. It may be noted that for a given element, acid soils tend to contain free metal cations and protonated anions, whereas alkaline soils contain carbonate or hydroxyl complexes.

Table 6.7. Speciation of some chemical elements in acid and alkaline soils.

| Cation | Speciation in acid soils | Speciation in alkaline soils |
|---|---|---|
| $Na^+$ | $Na^+$ | $Na^+$, $(NaHCO_3)^0$, $(NaSO_4)^-$ |
| $Mg^{2+}$ | $Mg^{2+}$, $(MgSO_4)^0$, Org.* | $Mg^{2+}$, $(MgSO_4)^0$, $(MgCO_3)^0$ |
| $Al^{3+}$ | Org., $AlF^{2+}$, $AlOH^{2+}$ | Al $(OH)^-_4$ Org. |
| $Si^{4+}$ | Si $(OH)^0_4$ | Si $(OH)^0_4$ |
| $K^+$ | $K^+$ | $K^+$, $KSO^-_4$ |
| $Ca^{2+}$ | $Ca^{2+}$, $(CaSO_4)^0$, Org | $Ca^{2+}$, $(CaSO_4)^0$, $(CaHCO_3)^+$ |
| $Cr^{3+}$ | $CrOH^{2+}$ | Cr $(OH)^-_4$ |
| $Cr^{6+}$ | $(CrO_4)^{2-}$ | $(CrO_4)^{2-}$ |
| $Mn^{2+}$ | $Mn^{2+}$ $(MnSO_4)^0$, Org. | $Mn^{2+}$, $(MnSO_4)^0$, $(MnHCO_3)^+$, $(MnBOH)^+_4$ |
| $Fe^{2+}$ | $Fe^{2+}$, $(FeSO_4)^0$, $(FeH_2PO_4)^+$ | $(FeCO_3)^0$, $Fe^{2+}$, $(FeHCO_3)^+$, $(FeSO_4)^0$ |
| $Fe^{3+}$ | $(FeOH)^{2+}$, Fe $(OH)^0_3$, org | Fe $(OH)^0_3$, org |
| $Ni^{2+}$ | $Ni^{2+}$, $(NiSO_4)^0$, $(NiHCO_3)^+$, Org. | $(NiCO_3)^0$, $(NiHCO_3)^+$, $Ni^{2+}$, NiB $(OH)^+_4$ |
| $Cu^{2+}$ | Org., $Cu^{2+}$ | $(CuCO_3)^0$, Org. CuB $(OH)^+_4$, Cu $[B(OH)_4]^0_4$ |
| $Zn^{2+}$ | $Zn^{2+}$, $(ZnSO_4)^0$, Org | $(ZnHCO_3)^+$, $(ZnCO_3)^0$, Org, $Zn^{2+}$, $(ZnSO_4)^0$, ZnB $(OH)^+_4$ |
| $Mo^{3+}$ | $(H_2MoO_4)^0$, $(HMoO_4)^-$ | $(HMoO_4)^-$, $(MoO_4)^{2-}$ |
| $Cd^{2+}$ | $(CdSO_4)^0$, $CdCl^+$ | $Cd^{2+}$, $CdCl^+$, $(CdSO_4)^0$, $(CdHCO_3)^+$ |
| $Pb^{2+}$ | $Pb^{2+}$, Org, $(PbSO_4)^0$, $(PbHCO_3)^+$ | $(PbCO_3)^0$, $(PbHCO_3)^+$, Org., Pb $(CO_3)^{2-}_2$, $PbOH^+$ |

Org. refers to organic complexes, such as fulvic acid complexes.

## 6.5 WHY CROP PLANTS NEED WATER

As already stated, modern agriculture accounts for about 80% of the use of water. In order to have enough water in the long run, we must develop techniques to use water more efficiently in agriculture, and keeping it cleaner as we do so.

While the world's population has doubled in the past half century, the consumption of meat has quadrupled. This has important implications for grain harvests needed. About two kg. of grain are needed to produce a weight gain of one kg. of chicken. For pork, it is three kg., and for beef, eight kg. For the annual global production of 200 million tons of meat, livestock are now fed about 40% of all grain harvested. At a given level of nutrition (say, minimum of 2200 daily calories), the grain and soybean needed to feed a pig to produce pork, will feed a man ten times the number of days, if consumed directly instead of as pork (source: *National Geographic*, Oct. 98).

### 6.5.1 *Role of water in the physiology of crop plants*

The growth and well-being of a crop plant are critically dependent upon the availability of water. Water provides the hydrogen atoms for reducing carbon dioxide in the process of photosynthesis. Nutrients are transported into, within and out of the plants through the medium of water. Water constitutes the basic structural component of plant – more than 90% of plant mass is composed of water.

Water has to be delivered to the plant in tune with the rooting depth of the crop, and available water capacity of the soil (150-200 mm per meter-depth of the soil for clayey soils, 100-150 mm for loamy soils, and 50-100 mm for sandy soils).

The rate of water uptake by the plants depends upon
– Characteristics of the plant (rooting density, rooting depth, rate of root extension, etc.),
– Properties of the soil (water retention and conductivity), and
– Weather conditions (which determine the quantum of transpiration) (Hillel 1987, p. 31).

It was earlier believed that the plant functions are not much affected until the permanent wilting point is reached. Recent studies, however, show that plant suffers water stress and reduction in growth long before the permanent wilting point is reached. This has to be borne in mind in the application of irrigation.

Crop yields can be maximized only by keeping water readily available to the plant throughout the growing season.

When plants are stressed and transpiration decreases even temporarily, this is manifested in the form of rise in temperature of the crop canopy. Thus water stress can be remotely sensed through infrared radiation thermometers by the measurement of crop canopy temperatures. Since a satellite takes about 90 minutes to go round the world, frequent and repetitive monitoring of the water stress is feasible. In advanced countries, irrigation decisions (how much, where, when and how) are now-a-days taken on the basis of such remotely sensed satellite data.

### 6.5.2 *Crop evapotranspiration and water requirements*

Since the water requirement of a crop is conditioned by the rate of evapotranspiration in a particular situation, a knowledge of the rate of evapotranspiration (mm $d^{-1}$) is necessary to plan irrigation.

The rate of evapotranspiration depends both upon meteorological factors (radiation, atmospheric humidity, temperature, wind, etc.) and field factors (wetness and surface properties). Potential Evapotranspiration (PET) corresponds to the 'maximal evaporation rate which the atmosphere is capable of extracting from a well-watered field under given climatic conditions' (Hillel 1987, p. 14). The equations developed by Penman of England in 1948 for determining PET are still used, though with some refinements.

The typical PET values (mm $d^{-1}$) for different agro-climatic zones are summarized below (Table 6.8 quoted from Hillel 1987, p. 16):

MET (Maximal Evapotranspiration) is a useful parameter. It corresponds to maximal seasonal evapotranspiration from a well-watered crop stand of optimal density (Hillel 1987, p. 22). It is usually 0.6 to 0.9 of total seasonal PET. MET is a direct measure of the water requirements of a crop.

The crop water requirements (CWR) are computed from the following equation:

$$CWR = KC \times PET \tag{6.2}$$

where KC is an empirical crop coefficient, and PET is Potential Evapotranspiration.

The KC values depend upon climate (higher values for hot, windy and dry climates, and lower values for cool, calm and humid climates) and the crop (stage of growth, crop reflectivity, crop height and roughness, degree of ground cover, canopy resistance to transpiration, etc.). For most crops, the KC value for the total growing season varies between 0.6 to 0.9.

Hillel (1987, p. 22) gave the relationship between crop coefficient and the stages of growth of an annual crop.

It is important to bear in mind that soil water availability does not depend on the soil alone, but on the soil-crop-climate system. For instance, a crop with ex-

Table 6.8. PET values in various climatic zones.

| Climatic zones | Mean daily temp. cool (below 20°C) | Mean daily temp. warm ( above 20°C) |
|---|---|---|
| Tropics & sub-tropics | | |
|   Humid & subhumid | 3-5 | 6-8 |
|   Arid & semiarid | 5-7 | 8-10 |
| Temperate zones | | |
|   Humid & subhumid | 2-4 | 5-7 |
|   Arid & semiarid | 3-5 | 6-9 |

tensive and dense roots (such as those of small grains) can utilize the soil moisture more effectively than plants with sparse and shallow roots (such as those of
potato). A crop growing in arid climates (with consequent high evaporative demand) tends to experience water stress more frequently than the same crop growing in moderate or humid climates.

## 6.6   WATER USE – YIELD RELATIONSHIPS

Transpiration is a necessary and productive part of plant activity, but evaporation
should be deemed to be a loss since the plant does not benefit by it. It therefore
follows that the evaporation component of evapotranspiration needs to be minimized. This consideration has a bearing on the efficiency of application of an irrigation system. For instance, in the sprinkle irrigation, that part of the water which
is intercepted by the foliage evaporates rapidly without entering the transpiration
system. Similarly, when the whole surface is repeatedly wetted through surface irrigation, there will be avoidable loss of water through evaporation. On the other
hand, inadequate irrigation will lead to soil desiccation and water stress. Thus, a
balance has to be maintained between the various parameters. Evaporational
losses can be minimized through some kinds of irrigation (say, drip irrigation)
which wet only a small fraction of the soil surface but do not wet the foliage. This
consideration holds good even if the irrigation is frequent.

One would think intuitively that, other things being equal, the more the application of water to the crop, the more would be its yield. The actual relationship between the two parameters is not a simple one. When the water supply is not a limiting factor throughout the growing season, a crop plant may be able to transpire
at MET rate and attain full potential yield. If water is applied beyond MET needs,
it becomes counter-productive and the yield gets reduced.

Doorenbos & Kassam (1979) gave the following relationship between yield
and applied water:

$$1 - (Y/Ym) = f \, [1 - (AET/PET)] \tag{6.3}$$

where $Y$ is the actual yield, $Ym$ is the maximum attainable yield when full water
requirements are met, $f$ is an empirical yield response factor, AET is the actual
evapotranspiration and PET is the potential evapotranspiration. The values of $f$ for
different crops and agro-climatic settings have been reported in the literature. In
the above equation, yield refers to dry matter yield. As the grain yield is directly
related to dry matter yield, it is safe to state that grain yield depends upon the
quantum of applied water, assuming that there are no complications, such as pests
and nutrient deficiencies.

Irrigation is of course unnecessary when it rains, and the rain is free. But the
catch in it is the uncertainty involved. We can make a projection of the water
needs of the crops in time and space, but we cannot be sure as to what proportion

of this could be taken care of by rain. Statistical models have been developed to address this issue.

## 6.7 SPAC (SOIL – PLANT- ATMOSPHERE – CONTINUUM)

It is now well recognized that the processes taking place in the soil – plant – water system are inter-linked. For instance, the availability of moisture in the soil does not depend upon the characteristics of the soil alone, but is regulated by the inter-actions between the plant, soil and climate. It is therefore necessary to study the system in a dynamic and holistic way. Such a system is called SPAC (Soil – plant – Atmosphere – Continuum).

As Hillel (1987, p. 19) put it elegantly, a plant has to live simultaneously in two very different realms, the atmosphere and the soil. The conditions in the two realms vary constantly, but not necessarily in the same manner. A plant therefore has to respond to the sometimes conflicting demands from the two realms. Water is a case in point. A plant has to keep its stomates open in order to absorb carbon dioxide from the atmosphere for the purpose of photosynthesis. The plant loses water, i.e. transpires, through the same stomates. Plants therefore have to extract water from the soil through the root systems, and store the water. However, if the soil has no moisture for the roots to extract, the plant suffers moisture stress. The plant reacts to moisture stress by partially closing the stomates and reducing the loss of water through stomates. Transpiration of water cannot be stopped entirely by the closure of stomates, because the leaves continue to lose some water through the cuticular surfaces. This process, however, has an adverse effect on the productivity of the plant. The closure of stomates leads to decreased absorption of carbon dioxide, decreased photosynthetic activity and ultimately decreased yield of the crop plant. The plant suffers dehydration and wilting when it is faced with sustained transpiration without sufficient replenishment of moisture.

Pedro (1984), as quoted by Darnley et al. (1995), illustrated comprehensively the relationship between the soils, types of duricrust, thickness of the weathering zone, mean annual temperature, mean annual rainfall, Eh, pH, electrical conduc-tivity and activity of some major cationic and anionic species. It may be noted that prolonged and deep weathering in tropical regions has a profound effect on bedrock chemistry.

The dynamics of the movement of moisture in the soil- plant-atmosphere sys-tem, has been schematically shown by Barrow (1987, p. 82)

## 6.8 MOISTURE CONSERVATION IN DROUGHT-PRONE AREAS

There is much variation in the capacity of the soils to provide water and nutrients to crops, and the needs of the crops for water and nutrients. For instance, one hectare of sugarcane needs 300 ha cm of water. The same amount of water could irrigate about eight ha of wheat or 30 ha of pearl millet, known as bajri in India

(Falkenmark et al. 1990). Both the supply and demand for water are determined by climate (precipitation, evaporation, transpiration, temperature and radiation, etc.).

### 6.8.1  *Collection and storage of precipitation*

The issues have been covered earlier under Sections 4.1 and 4.2.

Ancient civilizations in parts of Asia and South America have developed agriculture under a variety of adverse conditions – high slopes, rugged terrain, low rainfall, lack of suitable supplies of groundwater, too remote for bringing water through canals, and so on. They could do so because they developed simple but efficient techniques of collecting and concentrating precipitation.

For instance, two cm. of rain falling over a catchment of one ha has a volume of 200 m$^3$. Even if 50% of this precipitation (i.e. 100 m$^3$) is lost due to infiltration and evaporation, it is still possible to save 100 m$^3$ of water if appropriate techniques are used. The water thus stored can be used for irrigation. Any technique of conserving water almost invariably ameliorates soil erosion.

Chris Barrow (1987, p. 172) suggested the application of the following methods of making use of runoff, while at the same time preventing soil erosion:
1. Simple tied ridge and furrow (box ridging), with ties being made with hoe or disc-plough,
2. Natural catchment and ephemeral stream, with contour bunding to reduce rate of runoff and to guide water to the storage tank(s) – rainfed crops can be grown in the catchment area, and irrigated crops can be grown with water from tank(s),
3. ICRISAT system of broad-beds and grassed furrows/drains – water stored in the tanks can be made use of as supplemental irrigation, or through sprinkler irrigation through bullock-drawn water carts,
4. Cross-section of 3.

### 6.8.2  *Rain and storm water harvesting*

All rainfall does not produce runoff. There is a threshold level of precipitation which is the minimum rainfall needed to produce useful runoff. An area may experience frequent showers, but still there may be no runoff because the precipitation may simply evaporate or soak away.

Techniques are available to improve the runoff efficiency, which is defined as the yield expressed as a percent of the total precipitation on the catchment. Runoff efficiencies greater than 80% can be achieved by covering the catchment with sprayed silicone compounds, concrete, aluminum foil, butyl rubber sheet, gravel-covered plastic sheet, paraffin wax, asphalt, polyvinylfluoride, etc. Apart from the problem of expense, there is an ever present risk of the materials being stolen. Besides, plastics are susceptible to damage by wind, sunlight, livestock or pests.

A labor-intensive, and inexpensive treatment is the method of soil grading and compaction. This does not require any purchased material, and it cannot be stolen. This can last for 5-10 years and costs only USD 0.04-0.06 cents/m$^2$. The cost per 1000 l of yield is USD 0.07-0.19 cents.

Model experiments made by ICRISAT, Hyderabad, India, show that two installments of supplementary irrigation of 20 mm each, will greatly enhance crop productivity. This can be accomplished if one ha of maize field has a water pond of 800 m$^3$ capacity. Harvested rainwater is recommended to be stored in ponds with limited surface area (to reduce evaporation), and lined with suitable impervious material such as bentonite clay (to reduce loss due to seepage). In the case of deep Alfisols, impervious mixtures of clay, silt and sodium carbonate, or soil, sand and cement in the ratio of 4:4:1, could reduce the seepage by 47 to 97%. Spraying of long-chain alkanols and monomolecular films on the surface of water reduces evaporation by 10-35%. It has been reported that emplacement of foamed wax blocks on the water surface is capable of reducing evaporation by 33 to 87%. These techniques are not recommended to be used in Africa, as the materials are not readily available locally.

### 6.8.3 *Strategies for improving the availability and use of moisture*

In the case of drought-prone areas (as in, say, Sub-Saharan Africa), it is crucially important to improve the availability and use of soil moisture. Four groups of techniques are available for the purpose:
1. Improving soil moisture intake,
2. Reduction of evaporation losses,
3. Reduction of evapotranspiration losses, and
4. Reduction of percolation losses.

Though some techniques are essentially agronomic, they are still included for purposes of comprehensive coverage (Table 6.9; source; Barrow 1987, p. 151-154).

The following agronomic methods are available to optimize the use of soil moisture:
– Sowing of seeds before the break of the main rains,
– Maintaining seeding density in inverse proportion to aridity,
– Fallowing: in some African countries, nitrogen-fixing trees (e.g. *Acacia)* or shrubs (e.g. *Stylosanthes)* are grown during the fallow period,
– Planting of crops in such a manner that they mature one after another,
– Planting of a crop to go through the previous crop, and
– Growing of two or more crops or two or more varieties of a given crop to spread the risk (so that there will not be total failure of the crop) – for instance, the practice of intercropping of groundnuts, maize and pumpkin in rainfed agriculture in Mozambique, is the most sensible one under the circumstances.

Table 6.9. Strategies for the conservation of moisture (after Barrow 1987, pp. 151-154).

| Description of the technique | Comments |
| --- | --- |
| (a) *Improving soil moisture intake* | |
| 1. Planting of cover crop(s) – they provide ground cover and reduce runoff. | Inexpensive and easy. Possible benefits are: addition of organic matter to the soil and fixation of nitrogen. Also, the crop may be grazed. |
| 2. Leaving crop stubble/debris (this slows the runoff). | Inexpensive and easy. Reduces erosion. Could add organic matter and nitrogen to the soil. Has grazing potential. |
| 3. Strip cropping – strips of crops, perennials, pasture, etc. arranged roughly along contour. | Inexpensive and easy. Reduces erosion. May have beneficial symbiotic effects, by way of fixing nitrogen, deterring pests, etc. Strongly recommended. |
| 4. Mulches – organic or inorganic material spread on land to slow runoff and reduce evaporation and weed growth. | Mulches appear to be more valuable in increasing infiltration (by slowing the surface flow) than in moisture conservation. Can reduce mud-splash of crop, and weed growth. |
| 5. Tillage – for improving infiltration into poorly permeable soils. | By facilitating infiltration when rains come, tillage speeds planting. But in other situations, can cause erosion and oxidation of organic matter. |
| 6. Chemical treatment – application of 'wetting agents' which speed up infiltration. | Easy. Possible pollution risks and costs. Not fully tested. |
| 7. Runoff bunds and terraces – physical structures which retard runoff and promote infiltration. | Inexpensive, involving essentially the labor of the farmer. |
| (b) *Reduction of evaporation losses* | |
| 8. Mulches – peat, plant debris, straw, vermiculite, ash, sand, dust, gravel, sawdust, etc. | Reportedly capable of reducing evaporation, but its effectiveness for the purpose is debatable. Reduces rain-splashing and erosion. May deter pests and retard weed growth.Usually applied after the infiltration of moisture. Has to be replaced for each new cropping cycle, as the mulch is ploughed into soil. |
| 9. Antievaporation compounds – these are sprayable emulsions of water and oil or wax, hexadecanol, bitumen, asphalt, latex, etc. | Easy to apply. Binds sandy soils. Reduces erosion due to wind and water. Can be used to build shelter-belts. May be costly or may have pollution risks. |
| 10. Removal of weeds, by the use of herbicides, pulling or burning | Can reduce interception losses. May have pollution risks. |
| 11. Planting of hedges or shelter-belts, in such a manner that they do not take away moisture from shallow-rooted crop plants. | Though these take time to establish, and may have to be protected initially from animals and people, they bring several benefits: reduction of evaporation and wind damage to crops; may increase dew precipitation downwind. May fix nitrogen in the soil; may provide compost for soil improvement; may provide fodder, fuel, and fruits. |

Table 6.9. Continued.

| Description of the technique | Comments |
| --- | --- |
| 12. Wind breaks (structures, built of brushwood, palm fronds or stone barriers). | Reduce evaporation. Protect crops from wind damage. |

*(c) Reduction of evapotranspiration losses*

| | |
| --- | --- |
| 13. Cultivating crops with low evapotranspiration. | Such crops have to be identified for a given agro-climatic situation. |
| 14. Removal of deep-rooting weeds from ground crops. | Reduces evapotranspiration. However, may reduce infiltration and increase soil erosion. |
| 15. Spray crops with a suitable compound (such as, kaolinite and water) to reduce the albedo. | Such a spraying reduces solar heating and pest damage. Possible adverse consequences are reduction in yields because of reduction in photosynthesis or respiration of crop. Care should be taken to ensure that the compound sprayed is not toxic, and does not leave a bad taste on the crop material. |
| 16. Spraying of crops with compound(s) such as wax emulsion, silicone, latex emulsion, or apply chemicals which cause crop stomata to close, to retard transpiration. | Same considerations as above |

*(d) Reduction of percolation losses*

| | |
| --- | --- |
| 17. Addition of hydrophytic material, such as organic matter or the new 'Agrosoak' compounds. | Can be applied easily and fast. May be costly. |
| 18. Insertion of an impermeable film, made of plastic, rubber bitumen, below the crop roots. | The film prevents percolation losses. Also, the nutrients are retained, thereby reducing the need for fertilizer application. On the other hand, the insertion may need much labor or special mechanical equipment, and may cause soil drainage problems and salinization. |

### 6.8.4 *Concurrent production and conservation*

There are various ways to increase the production of food, fuel, fodder and fiber. The Green Revolution, which is based on the use of high-yielding varieties of seeds, and related technical and chemical inputs, did increase the yields of several cereals. The well-off farmer who practiced it derived much benefit from it. But the subsistence farmer working poor soils in arid areas, did not share in this prosperity. An alternative strategy has been developed in Maharashtra province, India, for this kind of situation (Paranjpe et al. 1987, Rao 1989, Gadgil 1989, quoted by Falkenmark et al. 1990). The strategy has the following features:
– Controlling soil erosion, maintaining organic matter, moisture and physical properties of the soil, and promoting efficient nutrient cycling,

- Efficient use of photosynthesis and total stock of water and nutrients in the ecosystem,
- Combining perennials with seasonal crops, through integrated tree crop – agriculture and intercropping. The annual precipitation in the area is 500 to 700 mm, with considerable variation between the years. Rain-water is harvested and stored, and the *assured* water-supply is used to grow *low water-need food crops* only. In years when there is better than average rainfall, the additional amount of water available is used to grow perennials and cash crops. Preference is given to low-water need crops, such as maize, sorghum and millet, and special effort is made to preserve soil organic matter. The project has been able to achieve 20 t ha$^{-1}$ (dry weight) in the forestry part, and six t ha$^{-1}$ of seasonal crops (sorghum).

## 6.9   INFILTRATION UNDER VARIOUS IRRIGATION REGIMES

### 6.9.1   *What is infiltration?*

Infiltration (or surface intake) refers to the rate of entry of water into the soil under the action of gravity. It is greatest when the land is dry, and would take place either at the start of precipitation or the earliest irrigation application. The rate of infiltration decreases as the top soil gets saturated, and finally gets stabilized at a particular rate depending upon the particle size/texture of the soil.

The infiltration rate is a measure of how much water a soil can soak up in a given period of time. As should be expected, the more permeable a soil is, the more will be its infiltration rate. Hence, clay soils are characterized by low infiltration rates, and sandy soils have high infiltration rates. The typical infiltration rates (mm h$^{-1}$) for various types of soils are as follows (Stern 1979, p. 88): Clay: 1-5; clay loam: 5-10; silt loam: 10-20; sand loam: 20-30; sand: 30-100.

The infiltration rate will determine how long it will take for the rains or irrigation to water a unit area of land having a particular soil. For instance, a 60 mm precipitation or irrigation will take about 12 hours to water a unit area of clay soil, but only two hours to water the same areal extent if it is composed of sandy loam. In the case of the clay soils which develop mud-cracks during the dry season, infiltration takes place rapidly to start with, but as the soil cracks get sealed, and the soil swells, the infiltration slows down, before getting stabilized. The infiltration characteristics of a soil is an important consideration in irrigation. When irrigation is applied in excess of the infiltration rate, water may be wasted (this is often the case, where the water cess is charged per hectare, but not on the basis of water used). Besides, the flowing water may cause erosion or it may form puddles, from which it evaporates faster.

### 6.9.2   *Irrigation methods*

Figure 6.11 (source: Hillel 1987, p. 61) shows the pattern of infiltration under alternative irrigation methods.

The principal methods of irrigation, and their advantages and disadvantages are described as follows:

– *Surface irrigation or gravity irrigation*, whereby water introduced at the head of the field spreads and infiltrates throughout the field through forces of gravity and hydrostatic pressure. In this method, the soil is the medium through which water is *conveyed, distributed and infiltrated*. For this reason, the physical and chemical properties of the soil, and way the land has to be prepared for irrigation enter the picture prominently. This is the most ancient of all irrigation technologies, and it has been estimated that this method serves about 95% of the irrigated land worldwide. This method is mechanically simple, has low energy requirements (as it is based on gravity), and can be easily adapted to small holdings. Its serious deficiencies are low application efficiency, high conveyance losses and wastage of water, which lead to adverse consequences such as water-logging and salinization.

– *Sprinkle irrigation*, whereby sprayed water falls on plants like rain. This system does not depend upon the soil surface for conveyance and distribution of water. It is hence not necessary to level the land (as required in the case of surface irrigation) and conveyance losses are minimal. Water can be applied at a rate less than that of the soil infiltrability. This allows soil aeration. The disadvantages are the high capital costs, maintenance requirements and energy costs (for maintaining high pressure). Also, sprinkle irrigation is strongly affected by wind.

– *Drip irrigation*, whereby water is applied directly to the root zone through a set of polyethylene tubes laid along the ground or buried at a depth of 15-30 cm. Water is delivered drop by drop through perforations or emitters in the pipes. The trickling rate is maintained at less than the rate of infiltrability in the soil. The operating pressures are 15 psi (~104 kPa) to 45 psi (~310 kPa). Under this system, it is not necessary to level the land, and it is not affected by wind. By

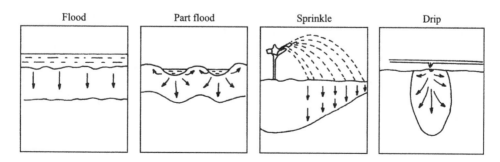

Figure 6.11. Alternative irrigation methods (source: Hillel 1987, p. 61).

providing the rooting volume with the requisite amounts of water (and nutrients, which can be added to water), the soil can be kept continuously moist and aerated. As only a portion of the soil surface is wetted, the amount of direct evaporation is reduced. Slightly brackish water (upto 1000 mg $l^{-1}$) can be used under the method for some crops which are not too sensitive. As the water does not come into direct contact with foliage, it will not cause saline scorching. As the chosen volume of the soil is kept constantly wet, the salts in the brackish water do not have a chance of concentrating and affecting the crop.

Drip irrigation can be used any where – in rugged terrains, in sandy soils of low moisture storage capacity, and arid climates of high evaporativity.

The capital costs of drip irrigation system is high. The method needs very rigorous adherence to scheduling and maintenance. In an effort to cope with the extreme scarcity of water, Israelis developed the drip irrigation into a fine art. In the context of rising cost and scarcity of water, and lower costs of plastic tubing, this method is likely to be widely adopted.

A microprocessor-based, drip irrigation system suitable for use in Developing countries, is now commercially available. This device uses low cost ceramic sensors, and operates on a solar cell-charged battery. The device continuously monitors the soil moisture and controls the drip rate so as to maintain the moisture within the desired limits.

– *Micro-sprayer:* Micro-sprayers have several of the advantages of the drip systems in that the water is applied only to a fraction of the ground surface, high frequency irrigation is possible, and fertilizers can be injected, if desired. Besides, they have the following advantages over the drip systems: 1. As micro-sprayers have larger nozzle orifices, clogging of emitters is not a serious problem, and it is not necessary to filter the irrigation water, 2. Micro-sprayers are operated at pressures of the order of two atmospheres, which are much lower than for the drip systems, and 3. Micro-sprayers can be scaled down for use in small farms in the developing countries.

The micro-sprayers have the following minor disadvantages: 1. As wetting of leaves will be involved, brackish water cannot be used in irrigation, and 2. Since the area wetted is larger area than in drip systems, there are some evaporational losses.

– *Low-head, bubbler irrigation:* A closed-conduit irrigation system has the great advantage in that it avoids conveyance losses and maintains uniformity in application. But any such system needs energy to pressurize water for distribution. Low-head, bubbler irrigation has the advantages of a piped system, but it does not require any pumps or nozzles. Even the low head available from a surface ditch may be adequate. In this arrangement, water is simply allowed to bubble out from open, vertical standpipes, 1-3 cm diameter which rise from buried lateral irrigation tubes. Bubbler systems are particularly suitable for widely spaced crops like fruit trees and grape vines. Small circular basins can be constructed around the trees for the water to bubble into. Low-volume, high-frequency, partial area irrigation is possible with the system. The initial cost of the system (about USD 1600/ha) is comparable or even lower than other systems. As the system works by gravity, there are no energy costs. It also has a longer life, because it is buried. Rawlins (1977) gave a techno-economic analysis of the system.

## 6.10  IRRIGATION SCHEDULING

In the course of irrigation, water is introduced into that part of the soil profile that serves as the root zone of a crop plant. The soil moisture thus produced serves as a kind of a bank providing water to the roots as needed. An ideal irrigation system is the one whereby the spatial and temporal distribution of water in the soil is just what the crop needs to achieve the highest productivity.

The timing and quantity of water application are decided upon on the basis of the monitoring of the soil, the plant and the micro-climate.

As the days pass after rains or irrigation, the moisture reserve in the root zone steadily gets reduced due to evaporation and root extraction. It is critically important to ensure that the soil moisture never goes down to the level of permanent wilting point, lest it adversely affect the yield of the plant. Hence it would be prudent to keep the allowable depletion midway between the field capacity (PC) and the permanent wilting point (PWP). If the field capacity is 20%, and the wilting point is 8%, the irrigator should apply water when the field capacity has dropped to 14%.

The monitoring of the soil needs information on the following aspects:
- Rooting depth of the crop – it may be shallow (0.3-0.7 m for vegetables), medium (0.9-1.5 m for wheat) or deep (1.0-2.0 m for maize, sorghum and sugar cane),
- Potential and actual water contents of the root zone, and field capacity (expressed as fractional volume), on the basis of the determination of soil moisture content, moisture tension and moisture release characteristics of the soil (Hillel 1987, p. 35).

As soil moisture diminishes, the tensiometers record increased tension. This information can be used to estimate when the plants will suffer stress, and need application of water. Because of the complexity of the variables involved, laboratory measurement of soil moisture may give misleading results. Soil moisture is best determined *in situ* with a portable neutron moisture meter. Though expensive, it has several positive features: measurement is made *in situ*, it can be made conveniently and instantly and it gives meaningful results. Moisture tension is determined with a tensiometer (Hillel 1987, p. 37 and 38).

An experienced farmer or agronomist can assess the water status of the crop just by direct visual inspection of the foliage. Irrigation scheduling has to take into account the meteorologically imposed evapotranspirational demand. Time-domain reflectometry (TDR) is being increasingly used to determine irrigation scheduling (Topp & Davis 1985).

## 6.11   IRRIGATION EFFICIENCY

The World Bank estimates that irrigation efficiency (i.e. net amount of water added to the root zone divided by the amount of water drawn from some source) is almost always below 50% (it may be as low as 30%). It is demonstrably possible to achieve an irrigation efficiency of 85 to 90% by proper management practices, particularly conveyance systems. It is therefore cost-effective to spend money on improving the irrigation efficiency rather than on constructing new irrigation reservoirs.

The efficiency of any process is usually taken as the measure of the output obtainable from a given input. In the case of irrigation, this could be defined in financial, physiological, socialogical, etc. terms. Thus, irrigation efficiency may be considered as the financial return for a particular amount of investment in water supply. This can vary tremendously from year to year and from place to place. Besides, it is not always possible to quantify the long term sociological benefits of irrigation. In drought-prone situations, where the incidence of drought is not predictable, even a modest contribution to food security arising from assured irrigation can have very profound consequences on the quality of life of the people.

The conceptual basis for different kinds of irrigation efficiencies has been explained by Barrow (1987, p. 108).

The water supply efficiency is estimated in terms of the flow at four points:
– A: Flow at source,
– B: Flow at turnout,
– C: Flow into the field, and
– D: Return flow, i.e. excess, 'used' or waste water from the scheme (some water may leave the system through subsurface flow).
Project efficiency deals with the relative proportion between flow at source and the amount of water that is applied to crops. Farm efficiency deals with the proportion of water at the turn out to the amount of water that is applied to the crops. Field efficiency deals with the proportion of water flowing into the field to the amount of water that is applied to crops.

## 6.12   DRAINAGE AND SALINITY CONTROL

### 6.12.1   *Why drainage?*

In the context of agriculture, drainage refers to artificial removal of excess water from cropped fields. Drainage may be of two kinds, surface drainage and groundwater drainage. Excess water accumulating on the surface is drained by shaping the land so as to facilitate overland flow (surface drainage). Excess water within the soil or subsoil is drained by lowering the water-table or preventing its rise (groundwater drainage).

While soil saturation for short periods is not harmful by itself, prolonged saturation (water logging) can adversely affect the plant growth. Plants need to respire constantly. Excess water in the soil blocks the soil pores, impedes the movement of oxygen from the atmosphere, and thereby severely limits the respiration activity of the plants. When the soil thus becomes anaerobic, reducing conditions may prevail, leading to the development of toxic concentrations of ferrous, and manganous ions and sulfides and the formation of methane. Nitrification is prevented, and denitrification may occur. Besides, wet soils allow the growth of harmful root pathogens, such as fungi (Hillel 1987, p. 81).

Decrease in the solubility of oxygen in water and increase in the respiration rate of both plants and microorganisms, result in temperature rise. Thus, water-logging has more severe consequences in warm climates than in cold climates.

The application of too much water and lack of drainage would raise the water table, and the leached salts would come to the surface by capillary action, thus rendering the soil saline. In the medium-textured soils, the upward flow of salt-bearing water during the dry season may be one to two meters.

Thus, it follows that *all irrigated lands must have drainage*. Irrigation without drainage can have disastrous consequences. Ancient civilizations based on irrigated agriculture in river valleys (e.g. Mesopotamia) collapsed, as they did not provide for drainage. The land became infertile due to the steady rise of water-table, water-logging and salinization. The salinization problem persists in many river valleys in the world, such as Indus (India and Pakistan), Nile (Egypt), Murray (Australia), Colorado (USA), etc. Groundwater drainage is undoubtedly expensive, but it should always be planned for right from the beginning. Reclamation of saline land is prohibitively expensive, and in some cases, even impossible.

### 6.12.2 *Methods of subsoil drainage*

Excess water from the soil or subsoil is drained through ditches, perforated pipes, or machine-formed 'mole' channels, by gravity flow or pumping. The water thus drained is let into a stream or a lake or an evaporation pond or the sea. As the drainage water invariably contains undesirable concentrations of harmful salts, fertilizer nutrients or pesticide residues, it should never be let into a river whose water will be used for drinking purposes by people in the downstream side. The drainage water is often recycled or reused for agricultural, industrial or recreational purposes. For instance, the drainage water from the fertile Imperial Valley of California, USA, is let into a lake (Salton Sea) where it is used for recreational purposes (e.g. boating and water-skiing).

The rate of flow from the soil to the drains depends upon the following factors:
- Permeability of the soil,
- Depth of the water-table,
- Depth of the drain,
- Horizontal spacing between the drains,
- Configuration of the drains (open ditches or tubes), their diameter and slope,

- Nature of the drain-enveloping material (such as gravel) used to increase the seepage surface, and
- Rate of recharge of groundwater (i.e. excess of infiltration over evapotranspiration) (Hillel 1987, p. 83).

As the field conditions tend to be highly complex and variable, recourse has to be taken to the use of empirical equations to estimate the desirable depths and spacing of drain pipes under particular soil and groundwater conditions.

The classical equation of Hooghoudt can be used for the purpose:

$$H = QX (S - X)/2 \, KD \tag{6.4}$$

where $H$ = Height above the drain, $X$ = Horizontal distance to the nearest drain, $Q$ = Percolation flux (the excess of infiltration over evapotranspiration), $S$ = Spacing between adjacent drains, $K$ = Hydraulic conductivity of the saturated soil, $D$ = Height of the drains above the impervious layer that is presumed to exist at some depth in the subsoil.

It can be seen from the equation that the height of rise of watertable between the drains is directly related to the recharging flux and to the square of the distance between the drains, and inversely related to the hydraulic conductivity of the soil (Hillel 1987, p. 84).

In Holland, a country which developed drainage into a fine art, drainage is aimed at removing about seven mm/day and preventing the watertable rise above 50 cm from the soil surface. In arid regions, the watertable must be kept considerably deeper, because of greater evaporation and consequent faster rate of increase in salinity.

Hillel (1987, p. 85) described the groundwater drainage under steady and nonsteady flow conditions. He gave the drainage layout for a 25 ha farm unit.

### 6.12.3  *Irrigation water quality and salinization*

Chemically, irrigation waters are electrolyte solutions. According to Shainberg & Oster (1978), the quality of irrigation water is determined on the basis of the following criteria:
- Total salinity (total concentration of all salts in the water),
- Sodicity (concentration of sodium relative to other cations),
- Anion composition, particularly the content of carbonate and bicarbonate ions, and
- Concentration of toxic elements, chiefly boron.

Irrigation waters drawn from surface or groundwater sources typically contain 200 to 2000 ppm of salts (or 200-2000 g of Total Dissolved Solids per $m^3$). Thus, the application in a single season of 1000 mm of medium quality water (containing, say, 500 ppm of salts), would add five tonnes of salt to a hectare of land. If salt is allowed to accumulate at this rate, the land will become saline and infertile in a matter of a few seasons. *Hence, the imperative need for leaching and drainage to remove the salt, particularly if the irrigation water is even slightly brackish.*

As the demand for irrigation water of good quality is in excess of supply of such water, it is inevitable that water of inferior quality (in terms of dissolved salts, and contamination) has to be used. This would involve either preprocessing of water before application, or growing crops which can tolerate the inferior quality water. Shuval et al. (1986) gave an account of using waste water for irrigation. Use of brackish water for irrigation has been found to be entirely feasible under certain conditions (growing tolerant crops and high frequency irrigation) (Hillel 1987, p. 89). There are a number of salinity tolerant plants which can be grown in the saline coastal wastelands and yield valuable non-edible oils and products usable in industries. Jojoba (*Simmondsia chinensis*) provides seed oil (50-55%) similar to sperm whale oil. *Salvadora persica* yields 35-40% seed oil rich in lauric acid which can replace coconut oil in soap and detergent industries. *Atriplex nummularia* is a potential biological desalinator. *Juncus rigidus* is a source of paper pulp. Some commercially important plant species are being biogenetically altered to be able to have high salinity tolerance (vide brochure of CSMCRI, Bhavnagar, India).

The chemical composition of the irrigation waters would depend upon the source, and post-withdrawal treatment. Before applying the water for irrigation, it is necessary to project as to how the chemistry of irrigation waters would affect the kind of compounds and weathering processes that exist in the soil. The impact of irrigation water quality on irrigated agriculture is studied in terms of salinity hazard, sodicity hazard and toxicity hazard.

Soils may have primary salinization arising from the content of residual salts, or secondary salinization may be caused by natural or human causes. Soils may be rendered saline either due to insufficient downward movement, application of poor quality irrigation water, or the groundwater being too close to the surface. Kovda found that even if good quality irrigation water is used, salinization can still take place if the water-table were to rise above the critical depth for a given soil.

The critical depth (l, in cm.) can be computed from the following formula of Kovda (1980, p. 205):

$$l = (17 \times 8\ t) + 15 \tag{6.5}$$

where $t$ is the average annual temperature in °C.

Crops can be broadly divided into categories, depending upon their tolerance of salinity: Sensitive crops require the application of irrigation water with electrolytic conductivity ($EC_w$) of less than 0.7 dS m$^{-1}$ (equivalent to ionic strength < 10 mol m$^{-3}$). Relatively tolerant crops can withstand upto 3 dS m$^{-1}$ (44 mol m$^{-3}$). Water with $EC_w$ > 3 dS m$^{-1}$ (> 44 mol m$^{-3}$) should not ordinarily be used for irrigation, except with appropriate amendment.

It should be borne in mind that soil salinity is defined in terms of $EC_e$, and not in terms of $EC_w$ of applied water. The relationship between $EC_e$ and $EC_w$ in the root zones is complex because it depends on many factors. The usual practice therefore is to use empirical 'rules of the thumb'.

The steady-state value of $EC_e$ resulting from the application of water with conductivity $EC_w$ can be estimated from a knowledge of the leaching fraction ($LF$).

$$LF = \frac{\text{Volume of water leached below root zone}}{\text{Volume of water applied}} \qquad (6.6)$$

Typically, $LF$ is in the range of 0.15-0.20. This would mean that 15-20% of the water applied leaches below the root zone, whereas evapotranspiration accounts for 80-85% of the applied water.

$EC_e$ in the root zone can be computed from $EC_w$ of applied water, from the following equation:

$$EC_e = X(LF). EC_w \qquad (6.7)$$

where $X(LF)$ is an empirically estimated parameter based on experience with irrigated, cropped soils. Ayers & Wescot (1985) gave $X(LF)$ factors for different values of $LF$. For instance, if the $EC_w$ is 1.5 dS m$^{-1}$, and $LF$ is 0.20 (with the corresponding $X(LF)$ factor of 1.3), then $EC_e$ would be 1.95 dS m$^{-1}$. For $LF$ values greater than 0.3, the $X(LF)$ value will be less than 1.0, and $EC_e$ becomes less than $EC_w$. Application of water with $EC_w > 2$ dS m$^{-1}$, under conditions of $LF < 0.1$, will render the soil saline.

The US Salinity Laboratory developed the concept of 'leaching requirement' as a monitoring mechanism to ensure that no excessive build up of salt takes place in the root zone of the plant. Leaching requirement has been defined as the fraction of the irrigation water that must be leached out of the bottom of the root zone in order to prevent average soil salinity from rising above some permissible limit (Richards 1954). Thus, the leaching requirement is determined by the salt content of the irrigation water, rate of evapotranspiration and the specific salt tolerance of the crop concerned.

## 6.13 ENVIRONMENTAL IMPACT OF IRRIGATION

Table 6.10 (source: Holy 1982) summarizes the impact of irrigation on environment.

The principal purpose of the irrigation schemes is to ensure that the soil-water-plant-air system works optimally, such that the crop plants do not suffer from periodical or regular moisture deficiency. Irrigation facilitates maximum efficiency and economic effectiveness of agrotechnological measures such as improved seeds, reclamation, application of fertilizers and pesticides.

Depending upon the geographic, climatic and technical conditions, irrigation systems may have complex social and biophysical impacts. When the Volta dam was built in Ghana, large areas were flooded. People who were farmers for generations, have had to make their living as fishermen. Boats replaced carts as means of trans-

Table 6.10. Environmental impacts of irrigation (source: Holy 1982, p. 79).

| Factor | Positive impact | Negative impact |
| --- | --- | --- |
| Engineering | Improvement of the water regimes of the irrigated soils. Improvement of micro-climate. Possibility for use and disposal of waste water; Retention of water in the reservoirs and possibility of multipurpose use thereof | Rise in groundwater table, danger of waterlogging and salinization of soils; Changes in properties of the reservoir water; Removal of vegetation in the area to be irrigated, and resultant changes in the water regime in the area; Abrasion of the banks of the reservoir |
| Health | Greater crop and fish production improves the nutritional status and health of the people. Recreation facilities in the irrigation canals and reservoirs | Possible spread of diseases such as schistosomiasis; Danger of pollution of water by return flow from the irrigation; possible infections from waste water irrigation |
| Social and cultural | Tourism promotion around the water reservoirs | Displacement of population from the area flooded by the water of the reservoir; protection of cultural monuments in the irrigated area |
| Aesthetic | New artificial lakes improve the aesthetic setting of the area | Project's architecture does not fit in with the area |

port. Reference has already made to salinization of soil due to the rise of the water table (as has happened in India and Pakistan, for instance) and the rise in the incidence of schistosomiasis (as has happened in the Nile Delta in Egypt). The chemicalization of agriculture led to high nitrogen and phosphorus contents in the runoff into surface and groundwaters, with consequent eutrophication of surface water bodies.

## 6.14 METHODOLOGY OF OPTIMAL IRRIGATION MANAGEMENT

The problems in the case of irrigation arise from a fallacy in human thinking – if some thing is good, more of it should be better. If a certain amount of irrigation raises crop productivity, which it surely does, more irrigation should produce more crops. As Hillel (1987) put it perceptively in his monograph, 'The Efficient use of Water for Irrigation', *Just Enough is Best – no less and certainly no more.*

For too long, irrigation has been thought of as simply a water delivery system, and this had disastrous consequences. None would disagree with the dictum, *Just Enough is Best.* But to determine what is enough is an enormously complicated job.

The following scenario is envisaged for optimizing an irrigation system:
1. Determine the water requirements of the crop to be grown in a particular agro-climatic setting, on the basis of the Potential Evapotranspiration (PET) and empirical crop coefficient (KC) data,
2. Withdrawal of water from the source (river, reservoir or aquifer) in tune with the water requirements, on time-variable and space-variable basis,

3. Delivery of water in tune with the rooting depth of the crop, and available water capacity of the soil,
4. Planning the drainage system right at the outset as an integral part of the of the irrigation system. This is absolutely essential in the case of river valleys prone to high water-table conditions. Soil salinity should be monitored continuously in order to alert the farmer about the accumulation of injurious levels of salinity in the root zone of the plant.

The two facets of the traditional irrigation procedures are:

1. Timing of application: irrigation is applied when the soil moisture is almost exhausted,
2. Quantum of application: irrigation is applied until the soil root zone is refilled to field capacity. The new methodology has different answers to the same questions:
   – Timing of application: irrigate as frequently as possible, even daily, and
   – Quantum of application: irrigation is applied in quantities enough to meet the current evaporative demand and to prevent salinization of the root zone (Hillel 1987, p. 33).

In some countries, irrigation water is released on fixed dates in specified quantities. The farmer has to take it or leave it. Almost invariably, he takes it, just to be on the safe side. This often results in over-irrigation in some areas, with consequential problems of disposal of return flow, waterlogging, elevation of water-table, salinization, etc. The farmer has no incentive to conserve water. He pays water-cess at a flat rate in proportion to the area of land. So a farmer near the top end of the irrigation scheme uses water excessively, whereas the tailenders complain that their lands do not receive enough irrigation. It is not uncommon for irrigation water to be highly subsidized.

There are two ways to ensure that irrigation water is not wasted:

1. Irrigation water should be available on demand. For instance, nobody would buy postage stamps needed for six months, for the simple reason that nothing is gained by doing so – one could buy any quantity of stamps at any time without any hassles,
2. Irrigation water should be properly priced in proportion to the quantity used per unit area of land. Pricing could have a built-in incentive to consume less water – *if a farmer uses lesser amount of water per unit area of land, he should get the water at a cheaper rate.*

The recommended methodology of irrigation management involves the optimization of quantity and increasing the frequency of irrigation, with small daily rather than massive weekly or monthly application of irrigation water. The high frequency-low volume irrigation technique has profound consequences:

1. Since the added water is small, it will only wet the few cms or few tens of cms of the surface. The flow below this point is essentially steady. The rate of through-flow through the soil, leaching of soil salts and the salinization conse-

quences can be controlled. It becomes possible to maintain a wet root zone, while at the same time minimizing the drainage rate.

2. Infiltration, rather than extraction, would in consequence become the dominant process. Since water is supplied to the crop at nearly the precise rate needed by the plant, the ability of soil to store water and supply it to plant when there is no irrigation or rain, becomes a property of minor importance. Thus, the 'water holding capacity' or 'field capacity' of a soil need no longer be the constraining factor about the irrigability of a given soil. Sandy or gravelly soils have very little moisture-holding capacity. Hence, even a short interruption in providing irrigation may result in deprivation, and severe stress. It is perfectly possible to irrigate coarse sands and gravels, even on slopes, by using drip or micro-sprayer irrigation.

### 6.14.1 *Directions of future development*

Several regions in the world have rivers and water bodies that cross or form international boundaries. For instance, Africa has ten river basins (including Lake Chad) which have a total drainage area of more than 350,000 km$^2$, and affect 33 Sub-Saharan countries and Egypt. Sharma et al. (1996) find that few of the transboundary basins are effectively jointly managed. National interests often override regional objectives. Either the countries concerned learn to work together, or they all stand to lose. Kaczmarek et al. (1996) developed a number of models of the use of water resources for agriculture and other purposes in the context of hydrological and climatic uncertainties.

Irrigation units in developing countries have an enormous range – from several thousand hectares in the case of Government-sponsored commercial farms, to less than a hectare in the case of small family units. For instance, high-pressure sprinkle systems are cost-effective in the case of large farms, whereas micro-sprayer or bubbler systems may turn out to be more suitable for family units.

It is not possible to design a universally applicable system of efficient use of water in irrigation, because of the complexity of the variables in regard to soil, water, climate, crop and people, and because of the need to be compatible with other inputs such seed varieties, fertilizers, tillage, pest control, etc.

Advances in information technology have made it possible to optimize the various system variables. Efficiency of water delivery needs to be optimized in terms of conveyance of water with minimal losses (say, in closed conduits), capability to provide measured amounts of water calibrated to meet the needs of crops in time and space, while preventing wastage, salinity and rise in the water-table. Efficiency of water utilization is to be optimized to low-volume, low-pressure, high-frequency, partial-area irrigation to achieve high crop yields (Hillel 1987, p. 99). Farmers in countries like USA are already using high-tech systems whereby the computers ensure the delivery of the precise requirements of irrigation and nutrients to crop plants, as sensed by satellites in a pixel area (of about 25 m$^2$) in the context of a particular weather setting. Thanks to the availability of inexpensive PC's, groups of small farmers in developing countries also should be able to avail of this approach, where the irrigation source is decentralized and compact, and is directly under their control. Irrigation is not a just biophysical – economic process. It is a human activity, carried out in a particular sociological setting. There is little doubt that modern irrigation is going to be knowledge-intensive, and the principal input, which would also be the cheapest input, would be the education of the farmer.

## 6.15  IRRIGATION IN NILE DELTA – A CASE STUDY

Irrigation has been practiced in Nile Delta since about 5000 BC. Egypt has three M ha (million hectares) of arable land, 80% of which is agriculturally productive because of irrigation with Nile waters. The Aswan High Dam provides Egypt with 55.5 billion cubic metres (B m$^3$) of waters annually.Out of this 30 to 33 B m$^3$ is used for crop production, and 12.5 to 13 B m$^3$ are discharged as drainage waters into the sea or into the saline lakes near the Mediterranean. About 6-7 B m$^3$ of the drainage waters is recycled in the Nile Delta by mixing it with fresh water at pumping stations constructed by the Government or by individual farmers. Extensive studies that have been made about the effectiveness and consequences of the use of drainage water for crop cultivation, and its impact on both the chemical and physical properties of the irrigated soils, led to the following conclusions:
1. Rcycled water can be used in irrigation so long as that water's salinity is lower than that of the soil, and
2. It is critically important to continuously monitor the salinity of the recycled water and soils, so as to ensure that they are within the limits of tolerance of the crops that are being grown.

The population of Egypt which is growing at the rate of 1.9%, is expected to reach 68 million in 2000. Egypt uses 97% (about the highest in the world) of its internal renewable water resources (estimated at 56.4 km$^3$). The per capita per annum use of water is 1202 m$^3$ (84 domestic, and 1118 industrial and agricultural). Since Egypt has very little new water to be developed, and in order to provide for burgeoning population, the emphasis has shifted towards improving irrigation efficiency and developing systems of use of drainage water. The Ministry of Public Works and Water Resources has embarked upon an ambitious Irrigation Improvement Project. Water losses are sought to be minimized through improved distribution canals, pumping stations, automatic hydraulic control structures, and new concrete-lined canals. Farmers are advised as to what kinds of recycled drainage water can be used in what kinds of soils to grow what kind of crops, and the costs of on-farm improvements are recovered from the farmers (sources: *'Berger World'* 1997; World Development Report 1992, of the World Bank).

It has been reported that a study made under the auspices of the Ministry of Public Works and Water Resources of Egypt, has recommended the following steps, among others, for the conservation of water in the West Nile Delta:

1. Reduction in the rice area, and the substitution of rice by maize,
2. Reuse of drainage water: assuming that the salinity of the canal water and drain water is 500 ppm and 1500 ppm respectively, a blend of three parts of canal water and one part of drain water, could yield water with salinity of 750 ppm, which is acceptable for the irrigation of some salt-tolerant crops.

## REFERENCES

\* Suggested reading

\*Aswathanarayana, U. 1999. *Soil Resources and the Environment*. Enfield, N.H., USA: Science Publishers.

Ayers, R.S. 1975. Quality of water for irrigation. *Proc. Irrig. and Drain. Div., Speciality Conf., Amer. Soc. Civil Engineers*, Logan, Utah 13-15 Aug.: 24-56.

Ayers, R.S. & D.W. Wescot 1985. Water quality for agriculture. FAO Irrig. and Drain., Paper 29, Rome: Food and Agricultural Organization.

\*Barrow, C. 1987. *Water Resources and the Agricultural Development in the Tropics*. England: Longman's.

Billib, M.H.A., P.W. Boochs, A. Matheja & B. Rusterberg 1995. In S.P. Simnovic et al. (ed.), *Modelling and Management of Sustainable Basin-scale Water Resource Systems*. IAHS Pub. no. 231: 273-280.

Biswas, A.K. 1993. Water for agricultural development: Opportunities and constraints. *Int. J. Water Resour. Dev.* 9(1): 3-12.

Bouwer, H. 1978. *Groundwater Hydrology*. New York: McGraw-Hill.

Chadwick, M.J., N.H. Highton & J.P. Palmer (eds) 1987. *Mining Projects in Developing Countries*. Stockholm: Beijer Institute.

Darnley, A.G. et al. 1995. *A Global Geochemical Database – for environmental and resource management*. Paris: Unesco.

Doorenbos, J. & A.H. Kassam. 1979. *Yield response to water*. Irrig. Drain. Paper 33. Rome: FAO.

Falkenmark, M., J. Lundquist & C. Widstrand 1990. *Water scarcity – An Ultimate Constraint in Third World Development*. Tema V, Report 14. Linköping, Sweden: Univ. of Linköping.

FitzPatrick, E.A. 1993. *An Introduction to Soil Science*. ELBS 2nd. ed. England: Longman's.

Gui, C.X. 1987. Effects of irrigation on the chemical balance in the ecological environment of water and soil. IAHS Publication no. 169, p. 111-124

\*Hillel, D. 1987. *The Efficient Use of Water in Irrigation*. World Bank Tech. Paper no. 64. Washington, D.C.: World Bank.

Holy, M. & Kos, Z. 1987. Irrigation systems management related to meteorological factors and water resources. IAHS Publ. no. 169: 5-14.

\*Holy, M. 1982. Environmental aspects of water management. In P. Laconte & Y.Y.Haimes (eds), *Water Resources and Land-use Planning: A Systems Approach*. The Hague, Netherlands: Martinus Nijhoff Publishers, p. 69-91.

Kaczmarek, Z. & J. Napiorkowski 1996. Water Resources Adaptation Stategy in an Uncertain Environment. In J. Smith et al. (ed.), *Adapting to Climate Change: Assessment and Issues*. Berlin: Springer-Verlag, p. 211-224.

Kovda, V.A. 1980. *Land Aridization and Drought Control*. Boulder, Colo., USA: Westview Press.

Ostrowski, M.W. & H. Lohr 1995. Development of planning tools for the ecologically sound operation of the reservoir. IAHS Publ. no. 231, p. 181-187.

Paces, T. 1976. Kinetics of natural water systems. In *Interpretation of Environmental Isotope and hydrochemical data in groundwater hydrology*. Vienna: IAEA, p. 85-108.

Pedro, G. 1984. La gense des argiles pedologiques, ses implications mineralogiques, physico-chimiques et hydriques. *Sci. Geol. Bull.* 37(4): 333-347.

Rawlins, S.L. 1977. Uniform irrigation with low-head bubbler system. *Agriculture and Water management.* 1: 167-178.

Richards, L.A. (ed.) 1954. *Diagnosis and Improvement of Saline and Alkaline Soils*. (USDA Handbook 60). Riverside, Calif., USA: US Soil Salinity Lab.

Rubin, Y. 1987. A hierarchical method for the design of water allocation and water distribution networks based on graph theory. *IAHS Pub.* 169: 207-220.

Shainberg, I. & J.D. Oster 1978. *Quality of Irrigation Water*. Bet Dagan, Israel: Int. Irrig. Info. Ctr.

Sharma, N. et al. 1996. *African Water Resources: Challenges and Opportunities for Sustainable Development*. World Bank Tech. Paper no. 331. Washington, DC: World Bank

Shiklomanov, I.A. 1998. *World Water Resources – A new appraisal and assessment for the 21st. century*. Paris: Unesco.

Schumann, A.H. 1995. Flexibility and adjustability of reservoir operation as an aid for sustainable water management. IAHS Publ. 231: 291-297.

Shuval, H.I. et al. 1986. *Waste Water Irrigation in Developing Countries*. Tech. paper no. 51. Washington, DC: World Bank.

Sposito, G. 1989. *The Chemistry of Soils*. New York: Oxford Univ. Press.

Stern, P.H. 1979. *Small Scale Irrigation: A Manual for low-cost water technology*. Nottingham, UK: Intermediate Technology Publications & Russell Press.

Topp, G.C. & J.L. Davis 1985. Time-domain reflectometry (TDR) and its application to irrigation scheduling. In D. Hillel (ed.), *Advances in Irrigation*, v.3. Orlando, Florida, USA: Academic Press.

Vörösmarty, C.J. et al. 1997. The storage and aging of continental runoff in large reservoir system of the world. *Ambio*, 26: 210-219.

Yangbo, C. 1998. Three Gorges Project: The key project for transferring water from south to north. In H. Zebidi (ed.), *Water: A Looming Crisis?*. Paris: Unesco, p. 309-313.

CHAPTER 7

# Industrial and municipal uses of water

## 7.1 INTRODUCTION

The issue of uses of water may be looked at from two angles:
1. User angle: specifications of water quality that are required by a particular user (e.g. petroleum refining), and
2. Use angle: if a water has certain characteristics naturally, the uses to which it can be put.

The sectoral water withdrawals, by country income group are summarized in Table 7.1 (source: World Resources Institute 1990).

The data in Table 7.1 allows us to draw the following conclusions:
– The total withdrawal of water by the low-income groups is less than one-third that of the high-income groups,
– The low-income groups use most of their water (91%) in agriculture (evidently inefficiently, as they are unable to grow adequate amount of food for themselves),
– The use of water in the industry by the high-income groups is about 30 times more than that by the low-income groups, and
– The quantity of water used for domestic purposes by the low-income groups (about 40 l d$^{-1}$) is less than one-tenth (about 450 l d$^{-1}$) of that used by the high-income groups.

The very limited quantity of water that the low-income groups use for domestic purposes, cannot but have profound health consequences. More than one billion people in the Developing countries are still without access to safe water and about 1.7 billion people are without access to adequate sanitation facilities (p. 113, World Development Report 1992, of the World Bank). This has disastrous health consequences. About 900 million cases of diarrheal diseases, resulting in the death more than three million children, are attributed to the lack of safe water and proper sanitation. At any time, 200 million people are suffering from schistosomiasis or bilhazaria, and 900 million from hook worm. Other water-related diseases like cholera, and typhoid are rampant. During 1997-1998, the cholera epidemic caused by El Niño -induced flooding in East Africa killed an estimated 5000 people. Apart from human suffering involved, these illnesses cause enormous economic losses.

Table 7.1. Withdrawals by sector for different income groups.

| Income group | Annual withdrawals per capita (in m³) | Withdrawals by sector (% and m³) | | |
| --- | --- | --- | --- | --- |
| | | Domestic | Industry | Agriculture |
| Low-income | 386 | 4% ( 15 m³) | 5% ( 19 m³) | 91% (351 m³) |
| Middle-income | 453 | 13% ( 59 m³) | 18% ( 82 m³) | 69% (313 m³) |
| High-income | 1167 | 14% (163 m³) | 47% (548 m³) | 39% (455 m³) |

$(1 \text{ m}^3 \text{ yr}^{-1} = 2.74 \text{ l d}^{-1})$

Table 7.2 Patterns of water use in the world (km³ yr⁻¹) (source: Unesco 1978).

| User | 1970: Full water use (km³ yr⁻¹)/ irecoverable use (km³ yr⁻¹), and percentage of water irrecoverably consumed | 2000 (projected): Full water use (km³ yr⁻¹)/irrecoverable use (km³ yr⁻¹),and percentage of water irrecoverably consumed |
| --- | --- | --- |
| Public utilities | 120/  20 ( 17%) | 440/  65 ( 14%) |
| Industry | 510/  20 (  4%) | 1900/  70 (  4%) |
| Agriculture | 1900/1500 ( 79%) | 3400/2600 ( 77%) |
| Water-storages | 70/  70 (100%) | 240/ 240 (100%) |
| Total (rounded) | 2600/1600 ( 62%) | 6000/3000 ( 50%) |

The sharp differences between various sectors in respect of consumptive use of water should be evident from Table 7.2 (source: ('World Water Balance and Water Resources of the Earth', Unesco 1978).

## 7.2   USE OF WATER IN INDUSTRY

Thermal and atomic power generation constitute the principal users of water in the industry. Other users are chemical and petroleum plants, ferrous and non-ferrous metallurgy, wood pulp and paper industry, and machinery manufacture. The volumes of water withdrawn and consumed vary greatly among the industries, depending upon the manufacturing process and climate (for instance, more cooling water is needed in the tropical areas than in cold areas, for the same unit of production). The extent of industrial water consumption is usually only a fraction of the actual intake – it may be as low as 0.5 to 3% in the case of thermal power. But in some industries where water enters the composition of the finished product (e.g. food and beverage industries), the industrial water consumption can be 30-40% of the water intake (Shiklomanov 1998).

Physico-chemical water quality criteria for use in various industries, such as, steam generation under low pressure and high pressure, textile industry, paper industry, chemical industry, petrochemical industry, iron and steel industry, cement industry, and leather industry, are given in Table 7.3 (source: US Dept. of the Interior 1968). Attention is drawn to some features of the criteria:

Table 7.3. Water quality criteria for industrial use (Source: US Dept. of the Interior 1968).

| Determinand | 1 | 2 | 3 | 4 | 5 | 6 | 7 | 8 | 9 |
|---|---|---|---|---|---|---|---|---|---|
| *1. Elements* (mg $l^{-1}$) | | | | | | | | | |
| Al | 5.0 | 0.01 | | | | | | | |
| Ca | | | | 20.0 | 68.0 | 75.0 | | | 60.0 |
| Cl | | | | 200 | 500 | 300 | | 250 | 250 |
| Cu | 0.5 | 0.01 | 0.01 | | | | | | |
| Fe | 1.5 | 0.01 | 0.1 | 0.1 | 0.1 | 1.0 | | 25 | 0.1 |
| Mg | | | | 12.0 | 19.0 | 30.0 | | | |
| Mn | 0.3 | | 0.01 | 0.05 | 0.1 | | | 0.5 | 0.01 |
| *2. Inorganic compounds* | | | | | | | | | |
| $NH_3 - N$ | 0.07 | 0.07 | | | | | | | |
| $NO_3 - N$ | | | | | 5.0 | | | | |
| $PO_4$ | | | | | 100 | | | 250 | 250 |
| $SiO_2$ | 30.0 | 0.01 | | 50.0 | | | | 35.0 | |
| *3. Quality parameters* | | | | | | | | | |
| Hardness* | 20 | | 25 | 100 | 250 | 350 | 100 | | 150 |
| Alkalinity * | 140 | | | | 125 | | | 400 | |
| $HCO_3$ | 17.0 | | | | 128 | | | | |
| Total Dissolved Solids (TDS) | 700 | 0.5 | 100 | | 1000 | 1000 | | 600 | |
| Suspended Solids (SS) | 10 | 0 | 5 | 10.0 | 30 | 10 | | 500 | |
| Dissolved Oxygen | 2.5 | 0.007 | | | | | | | |
| pH units | | | | 6-10 | 6-8 | 6-9 | 5-9 | 6.5-8.5 | 6-8 |

*expressed as $CaCO_3$. 1. Steam generation – low pressure, 2. Steam generation – high pressure, 3. Textile industry, 4. Paper industry, 5. Chemical industry, 6. Petrochemical industry, 7. Iron & Steel industry; 8. Cement industry, 9. Leather industry

- There is enormous range in the acceptability of water for use in particular industry – for instance, the acceptable Fe content of water ranges from 0.01 mg $l^{-1}$ for high pressure steam generation to 25 mg $l^{-1}$ in cement industry, and
- There is a sharp difference of two orders of magnitude between the water quality criteria for low pressure and high pressure steam generation – high pressure steam generation is far more demanding. The biological criteria involved in the case of food industries, and electrical conductivity characteristics of waters used in electronics industry, etc. came into the picture later.

Industrial energy is expressed in terms of *Toe* – tonne of oil equivalent = 41.7 GJ (Giga Joules or $1 \times 10^9$ J) = 1.44 t of bituminous coal. Fuel value of one tonne of coal is 28.9 GJ, and that of 1 $m^3$ of fuelwood is 9.4 GJ.

The requirements of water for the energy industry are summarized as follows (source: 'Water and energy: demand and effects', UNESCO 1985):

*Production:*
Coal mining – 100 l/toe; Oil production – 1100 l/toe;

Gas processing and transmission – 240 l/toe

*Oil shale processing:*
Tar-sand processing/coal conversion process – 3000 l/toe

*Refining:*
Oil refining – 1500 l/toe; Nuclear fuel cycle – 500 l/toe

*Steam – electric power generation:*
Fossil fueled plants – 2.2 l/kwh
Nuclear plants – 3.0 l/kwh
Geothermal plants – 15.0 l/kwh

Two contrasting developments are at work in the case of the use of water for industries. On one hand, expanding industrialization all over the world would demand more water. On the other hand, intensive research is going on to bring down the consumption of water needed per unit of production.

### 7.2.1 *Water for cooling*

Previously, about 3 l of water is evaporated in cooling turbogenerators per kwh of power generated. A simple calculation would show that a 1000 MW thermal power station would need about 25 M m$^3$ of water per year or about 0.8 m$^3$ s$^{-1}$. This is a lot of water – it is almost like a small river. If, as projected by L'vovich (1979), the thermal power generation in the world reaches $5 \times 10^{13}$ kwh by 2000, it would need about 100 km$^3$ of water per year, which is obviously untenable. Though the water used for cooling does not get contaminated in the process, it will get hot and will also get depleted in oxygen. When such hot water is let into a stream, it will kill the organic life in the stream. It also will have to travel a long way in the stream channel before it recovers the oxygen that has been lost. Hence the effluents from thermal power plants should not be discharged into rivers.

Several steps can be taken to minimize the water requirement for industrial cooling:
– It is now possible to bring down the water requirement to less than two l per kwh (through techniques such as high pressure circulation),
– Water circulation can be done in a closed loop,
– Instead of fresh water, municipal wastewater can be used after removing the undesirable constituents, and
– seawater or brackish water could be used. In the case of nuclear power plants located along the coast, seawater is generally used for secondary cooling.

### 7.2.2 *Water use in process industries*

Water consumption in various industries is summarized below (L'vovich 1979). Industries are either using less and less water per unit of production or switching over to dry technologies, for the simple reason that the less the quantity of water used in the processing, the less would be the effluents that need to be disposed off. For instance, in the chemical, and paper and pulp industries, the industrial effluents tends to be highly polluted, and it is extremely expensive to purify them.

*Textile industry:*
Factory cotton fabric: 250 $m^3$ $t^{-1}$; Synthetic fibre: 2500-5000 $m^3$ $t^{-1}$

*Chemical industries:*
Ammonia production: 1000 $m^3$ $t^{-1}$; Synthetic rubber: 2000 $m^3$ $t^{-1}$

*Non-ferrous metallurgy:*
Nickel: 4000 $m^3$ $t^{-1}$

*Ferrous metallurgy:*
Pig iron: 180-200 $m^3$ $t^{-1}$ (which is just 5% of that for nickel)

*Petroleum refining:*
Initially, about 35 $m^3$ $t^{-1}$ were used. Research has brought this down steadily. Modern refineries use less than 0.12 $m^3$ $t^{-1.}$

## 7.3  MUNICIPAL USE OF WATER

Municipal use of water refers to the quantity of water withdrawn by populations in cities and towns for domestic use in residences, and housing estates, and in public services and enterprises (such as hospitals, schools, transport systems). Such water has to be of the highest quality. Municipal water use also encompasses good quality water used in some industries (bakeries, breweries, food processing, beverages, etc.) directly serving the urban populations, as also the water used for watering gardens and domestic vegetable plots.

Globally, there is enormous variation in the withdrawal of water for municipal purposes. In many large cities in the Industrialized countries, the present water withdrawal is 300-600 l per day per person. This figure may go up to 500-1000 l per day per capita. On the other hand, in the Developing countries of Asia, Africa and Latin America, public water withdrawal is about 50 l per capita per day. In some localities in the Developing countries, the domestic water consumption may be as extremely low as 10 l per capita per day. This deplorable situation may not always be due to scarcity of water arising from meager rainfall. In some towns in Mozambique (say, Pemba), a municipal domestic water connection is a veritable gold mine. The poor have no option except to buy water at exorbitant prices, from householders having water taps, and consequently their water consumption is exceedingly low.

Some amount of water is irrevocably consumed in the process of watering the gardens and parks, washing the streets, and evaporation from leaking pipes, etc. To some extent, the climatic conditions determine the extent of water lost in this process. Evidently, the consumptive loss would be greater in hot, dry regions than in colder areas. Theoretically, most of the water withdrawn for municipal use is supposed to be returned to the hydrologic system as waste water. This would indeed be true if there is efficient sewerage system whereby all the wastewater is collected. In the modern cities in the Industrialized countries with efficient system of water supply and sewerage, only 5-10% of the water is actually consumed out of the specific water withdrawal of (say) 400-600 l d$^{-1}$. In small towns where all the consuming centres are not connected to central sewerage, the water consumption may reach 40-60% out of the withdrawal of 100-150 l d$^{-1}$, depending upon the climate. When a person uses 10 l d$^{-1}$ in a tropical country, the consumption would be almost total.

Whether a person is poor or well-to-do, a person needs on an average 2.5 to 3 l of water per capita per day for drinking and preparation of food. This works out to about one m$^3$ per capita per annum. Thus, the world population of six billion would need six B m$^3$ of six km$^3$ of potable water for physiological purposes. The difference between the rich and the poor lies in regard to the amount of water used for non-potable uses. Very broadly, for populations in the Developed countries which use (say) 300 l d$^{-1}$, or 100 m$^3$ yr$^{-1}$, drinking water component is just 1% of the total water withdrawal. For populations in some LDCs which use (say) 30 l d$^{-1}$, or 10 m$^3$ yr$^{-1}$, the drinking water component constitutes 10% of the total water withdrawal.

Some cities, like Paris, and Frankfurt-am-Main, have tried to solve the problem by setting up of two separate water systems – one for potable use, and another for non-potable use (bathing, washing, sanitation, gardening, etc.). In many urban areas round the world, where the quality of tap water is unreliable, many families simply buy drinking water.

Many municipalities in the Developing countries which do not have sewerage facilities, discharge the wastewaters either inadequately treated or untreated, into stream or into the sea. This has disastrous consequences. Every m$^3$ of contaminated water discharged into water bodies or water courses spoils upto 8-10 m$^3$ of good water (Shiklomanov 1998). Many water bodies in the world have already suffered serious qualitative degradation because of this practice.

The quantity of water for sanitation can be reduced in a number of ways:
– Use of water-efficient commodes (as is done in California, USA), or
– Use of stainless-steel, water-seal, pour-flush commodes which use just 2.5 l of water per flush (as is done in India), or
– Use of Zimbabwe-type, dry ventilated pit latrines. In future, the kind of vacuum-suction toilets that are used in airplanes may come into more common use in homes. Also, bathrooms can be so designed that the wastewater after bathing could be used for flushing the toilets.

### 7.3.1 *Water quality standards of potability*

The US Environmental Protection Agency, 401 M St., SW, Washington, DC 20460, USA, has brought out a computerized Environmental Monitoring Methods Index (EMMI) – 1994 (obtainable from National Technical Information Services, Tel: + 1-703-487 4650). It contains information on the recommended methods of analysis of various elements and compounds in different media, ways of reporting data, 'Permissible Exposure Limits' to toxic substances, etc.

Water quality for drinking purposes has been dealt with earlier in Section 3.3.2.

## 7.4 MISCELLANEOUS USES OF WATER

### 7.4.1 *Fisheries*

As explained in Chapter 3, the quality of water needed for fisheries purposes is almost on par with that for drinking water, i.e. the highest quality. There are of course fish varieties which can grow in wastewater (see Section 5.3), in the shallow water of the paddy fields and in the brackish water, etc. in the terrestrial environment, but they have to be carefully chosen for their ability to survive and grow in these environments.

The most important quality requirement in the case of fisheries use is the level of dissolved oxygen in the water. This level is reduced when an effluent exerts an oxygen demand on the receiving water. Tropical fish communities can tolerate lower levels of dissolved oxygen.

Quality parameters and suggested level of stream standard for fishing:

$CO_2$: < 12 mg $l^{-1}$; pH: 6.5-8.5; $NH_3$: < 1 mg $l^{-1}$

Heavy metals: < 1 mg $l^{-1}$;

Cu – < 0.02 mg $l^{-1}$; As: < 1 mg $l^{-1}$; Pb: < 0.1 mg $l^{-1}$; Se: < 0.1 mg $l^{-1}$

Cyanides: < 0.012 mg $l^{-1}$; Phenols: < 0.02 mg $l^{-1}$; Dissolved solids: < 1000 mg $l^{-1}$; Detergents: < 0.2 mg $l^{-1}$; Dissolved oxygen: > 2 mg $l^{-1}$; Pesticides: DDT: < 0.002 mg $l^{-1}$; Endrin: < 0.004 mg $l^{-1}$; BHC: < 0.21 mg $l^{-1}$; Methyl parathion: < 0.10 mg $l^{-1}$; Malathion: < 0.16 mg $l^{-1}$.

### 7.4.2 *Water quality criteria for bathing*

Water quality for recreational or religious bathing is not usually a serious problem in most countries. In some developing countries, bathing may involve the risk of schistosomiasis infection. In areas of schistosomiasis transmission, any freshwater body receiving urine or faecal pollution is likely to contain schistosome cercaria. Children who bathe or splash in such ditches or ponds, are at risk. Fastflowing streams, and deep water far out from the shore of the lake, may be safe.

## 7.5 MEDICINAL, INDUSTRIAL AND THERMAL WATERS

Some groundwaters have special properties naturally, which make them useful for specific purposes: medicinal mineral waters (for treatment of diseases), industrial waters (for the extraction of chemicals), and thermal waters (for heat supplies and power production). The following account is drawn from Klimentov (1983, Chapter XIV).

### 7.5.1 *Medicinal (mineral) waters*

Balneology is the science of mineral (medicinal) waters. Mineral (medicinal) waters are those waters that have a beneficial effect on human body, because of their salt, ion or gas contents, or the presence of micro-constituents or radioactive elements of therapeutic value, alkalinity or acidity, or higher temperatures. There are many health resorts (e.g. Karlovy Vary in Czechoslovakia, Borzhomi in Georgia, etc.) where medicinal waters are used for treatment of cardiovascular, rheumatic, nervous, skin, etc. disorders.

Waters whose salt content ranges from 8-12 g $l^{-1}$ are regarded as mineral (medicinal) drinking water. Depending upon the composition and therapeutic value, the waters are classified into the following types: A – waters without any 'specific' constituents or properties, B – Carbonated waters, C – Hydrogen sulfide (sulfide) waters, D – Ferruginous waters and those with high concentrations of As, Mn, Cu, Al, etc. E – Bromine, iodine waters with high organic contents, F – radonic (radioactive) waters, and G – Siliceous thermal waters.

The maximum permissible concentrations (mg $l^{-1}$) allowable for mineral waters are as follows: ammonium (2.0), nitrite (2.0), nitrate (50.0), vanadium (0.4), mercury (0.02), lead (0.3), selenium (0.05), chromium (0.5), phenols (0.001), uranium (0.5), radium (5. $10^{-10}$). For some constituents, there is a difference in the permissible levels of concentration between medicinal waters and medicinal table waters, as follows:

| Constituent | Medicinal waters (mg $l^{-1}$) | Medicinal, table waters (mg $l^{-1}$) |
| --- | --- | --- |
| As (metal) | 3.0 | 1.5 |
| F | 8.0 | 5.0 |
| Organic matter | 30.0 | 10.0 |

*Carbonated waters*:
The therapeutic effect of carbonated waters arises from their contents of dissolved carbon dioxide which may constitute 80-100% of the total gases. Of all the mineral waters, the carbonated waters contain the largest amount of dissolved gases. The released gas yield versus water yield (in $m^3/m^3$) is usually 1.5 to 4.6, but could be as high as 18. It is believed that the carbon dioxide dissolved in water is of mantle origin, and could also have been generated due to biochemical reac-

tions. Five different kinds of carbonated waters have been recognized depending upon the salinities (4 to 50 g $l^{-1}$), carbon dioxide contents (2.5-3.5 g $l^{-1}$), and temperatures (20 to 37°C).

Carbonated waters typically occur in the regions of young Alpine folding (e.g. Carpathian mountains), and active volcanic regions (e.g. Kamachatka). When they emerge at the surface, carbonated waters lose a portion of their carbon dioxide, thus giving rise to the precipitation of tufa or travertine. In the Mbeya region of Tanzania, carbon dioxide of magmatic origin issues along the fissures. The gas is collected and used to make carbonated soft drinks. There are extensive deposits of travertine which is soft and takes good polish – souvenirs are fabricated from it.

*Hydrogen sulfide (sulfide) mineral waters*:
The therapeutic value of these waters arises from the presence of free hydrogen sulfide and hydrosulfide ions. These waters are used for the hydropathic (i.e. baths) treatment of skin, muscular and nervous disorders. The requirement for being designated as hydrogen sulfide water is that its total hydrogen sulfide content should be at least 10 mg $l^{-1}$. Highly concentrated waters may contain upto 250 mg $l^{-1}$ of hydrogen sulfide contents.

Carbonated hydrogen – sulfide waters are related to volcanic activity and recent magmatism. Some fumarolic waters may contain upto 3000 mg $l^{-1}$ of hydrogen sulfide. On the other hand, the methane – hydrogen sulfide waters are produced in reducing environments, and are genetically related to the accumulations of petroleum and bitumen.

*Radioactive mineral waters*:
$^{222}$Radon is the gaseous daughter of $^{226}$Radium. One ppm of uranium corresponds to 12.3 Bq $kg^{-1}$ of $^{226}$Ra, and could result in radon concentration in soil gas of 4000 Bq $m^{-3}$ (gross estimate). Radon generated by the radioactive decay of radium diffuses into the flowing water. Waters with the following concentrations of radioactive elements are designated as radioactive waters: Ra >> $1.10^{-11}$ g $l^{-1}$, U > $3.10^{-5}$ g $l^{-1}$, Rn > 1.85 Bq $l^{-1}$. Some waters are highly enriched in radon, upto $7.4 \times 10^3$ Bq $l^{-1}$. The most widely used waters are the radonic – radium waters – these waters contain excess radon (i.e.) more radon than is in secular equilibrium with radium. Such radon-excess waters form when radium-rich waters penetrate reservoirs whose waters already contain radon.

Radioactive waters are used in hydropathic (baths) treatment of the peripheral nervous system, gynaecological and skin diseases.

### 7.5.2 *Industrial waters*

Industrial groundwaters constitute the feedstock for extracting iodine, bromine, and common salt. Some groundwaters also yield compounds of boron, lithium, rubidium, germanium, uranium, tungsten, etc. The water types which bear the

Table 7.4. Concentrations of economically valuable constituents in industrial waters.

| Water type | Constituent | Minimum concentration (mg g$^{-1}$) | Range of concentrations |
|---|---|---|---|
| Specific bromine-bearing | Bromine | 25 | $\times$ mg l$^{-1}$ – 10 g l$^{-1}$ |
| Industrial bromine-bearing | Bromine | 250 | |
| Specific iodine-bearing | Iodine | 1 | Traces to > 80 mg l$^{-1}$ |
| Industrial iodine-bearing | Iodine | 18 | |
| Specific iodine & bromine-bearing | Iodine | 1 | |
| | Bromine | 25 | |
| Industrial iodine & bromine-bearing | Iodine | 10 | |
| | Bromine | 200 | |
| Specific boron-bearing | Boron | 10 | |
| Industrial iodine & boron-bearing | Iodine | 65 | |
| | Boron | 162.5 | |

economic constituents, minimum concentrations needed for economic use, and maximum concentrations that have been recorded, are given in Table 7.4 (source: Klimentov 1983, p. 224).

The industrial waters tend to occur in the deeper portions of large artesian systems. The depths at which they occur vary from several tens of metres to 4-5 km. The normal range, however, is 1000 to 3000 m. The yields range from a few m$^3$ d$^{-1}$ to 400-500 m$^3$ d$^{-1}$. Some wells yield as much as 3000 m$^3$ d$^{-1}$. The common host rocks for these waters are clastics (sands, sandstones, gravels, conglomerates, etc.), carbonates (limestones and dolomites), and anhydrites interbedded with carbonates, and occasionally, volcanics. They range in age upto Proterozoic.

The total salinities of the industrial waters range from 20-250 g l$^{-1}$. The content of economic constituents, such as bromine and strontium, increase with increasing salinity, calcium chloride concentrations, depth and age of the host rock. The presence of halide deposits is a favorable indication for industrial waters.

## 7.5.3  *Thermal waters*

Customarily, waters having temperatures higher than that of human body (37°C) are called thermal waters: thermal (37-42°C), superthermal (42-100°C) and superheated (more than 100°C). As should be expected, the economic use for a given thermal water is strongly temperature-dependent:

| Temperature (°C) | Economic purpose of the thermal water |
|---|---|
| Up to 20°C | Water supplies |
| 20- 50°C | Balneology; production of iodine and bromine |
| 50- 70°C | Warming of greenhouses and hot beds, and central heating of farms, and hot-beds; Balneology |
| 75-100°C | Power and heat supply to towns, resorts, greenhouse complexes |
| > 100°C | Power production |

Total salinities of the thermal waters are highly variable, from 1 to 650 g $l^{-1}$. Thermal waters are characteristic of regions with geothermal gradients of more than 3°C/100 m. They occur in the volcanic regions, folded regions with positive geothermal anomalies, cratonic structural units, foredeeps, and intermontane troughs filled with Mesozoic and Cenozoic formations (Klimentov 1983, p. 230). Geysers are common in active volcanic areas, such as, the Yellowstone National Park in USA, Kamachatka in Russia, New Zealand, etc. They are big tourist attractions. Some countries, such as, Iceland, New Zealand, Italy, Kenya, etc. have succeeded in utilizing the thermal waters for hot water supplies, space heating and power generation. Experience has shown that not only is geothermal power is cost-effective, it is also environmentally less polluting.

## REFERENCES

*Suggested reading

Klimentov, P. 1983. *General Hydrogeology*. Moscow: Mir Publishers.
L'vovich, M.I. 1979. *World Water Resources and their Future*. Washington, D.C.: Amer. Geophy. Uni.
*Shiklomanov, I.A. 1998. *World Water Resources – A new appraisal and Assessment for the 21st century*. Paris: Unesco.
Unesco 1978. *World Water Balance and Water Resources of the Earth*. Paris: Unesco.
Unesco 1985. *Water and Energy: Demand and Effects*. Paris: Unesco.

CHAPTER 8

# Water pollution

## 8.1 PATHWAYS OF POLLUTION

Geoenvironment comprises rocks, soils, fluids, gases and organisms. It is linked to and is influenced by, climate, terrain and vegetal cover. Human activities affect the geological, physical, chemical, and biochemical processes taking place in rocks, soils, hydrological systems and associated media. Surface water and groundwater are by far the most dynamic components of the geoenvironment and are most affected by the activities of man. Protection of surface water and groundwater from depletion and pollution (chemical, organic, thermal, mechanical, etc. contamination) constitute the most important environmental task of a community. Microorganisms affect the hydrochemical processes and water quality. For instance, thionine bacteria oxidize hydrogen sulfide to sulfuric acid, and render the water more corrosive. Atmospheric pollution can penetrate the hydrosphere, biosphere and lithosphere, and can bring about undesirable changes in climate, soils, vegetation and water quality.

This chapter draws extensively from an excellent summary (paper 1.1, Groundwater contamination, in 'Water Resources and Land use planning', Laconte & Haimes 1982). The issues of water pollution are discussed under different headings: sources (contaminants generated by activities, such as, agriculture, mining, industrial and households, etc.), and types (inorganic and organic, biological and radioactive contaminants).

Pollutants may enter the surface water directly, and the groundwater through fractures/pores in rocks and soils. Dissolved pollutants in the surface water move much faster than the dissolved constituents in the groundwater. The hydraulic connection between the groundwater and surface water has a profound implication – for instance, steps taken to protect groundwater quality by reducing infiltration into the groundwater may lead to greater surface runoff, and consequent effect on river flow and quality.

### 8.1.1 *Activities that can cause groundwater pollution*

The type of activities that can cause groundwater contamination at various levels are summarized as follows (source: US Environmental Protection Agency's Citizen's Guide to Groundwater Protection, April 1990):

*Surface of the ground*: Infiltration of polluted surface water, Land disposal of wastes, Stockpiles, Dumps, Sewage sludge disposal, Use of salt in de-icing of roads (in cold countries), Animal feedlots (and manure heaps), Fertilizers and pesticides. Accidental spills. Airborne source particulates. Human and animal faeces, etc.

*In the ground, above the water-table*: Septic tanks, cesspools and privies, Holding ponds and lagoons, Sanitary landfills, Waste disposal in excavations, Leaks from underground storage tanks, Leaks from underground pipelines, Artificial recharge, Sumps and dry wells, Graveyards, etc.

*In the ground, below the water-table*: Waste disposal in wells, Drainage wells and canals, Underground storage, Mines, Exploratory wells, Abandoned wells, water-supply wells, groundwater withdrawal, etc.

In the public mind, chemical pollution is almost always associated with industrial effluents. It is not adequately realized that household activities (particularly, in high income homes) could introduce potentially harmful compounds into the groundwater. The toxic and hazardous components contained in some of the common household products (such as, motor oils, paints, refrigerants, toilet cleaners, swimming pool chemicals, etc.) include hydrocarbons, heavy metals, toluene, xylene, benzene, trichloroethylene, acetone, 1,2,2 trifluoroethane, chlorinated phenols, sulfinates, pentachlorophenols, sodium hypochlorite, etc. Thus, households can play a significant role in limiting groundwater contamination by conserving water, maintaining the septic system in good condition, using fertilizers and pesticides for the home gardens to the minimum extent possible, and taking care of the disposal of the toxic and hazardous chemicals, etc.

## 8.2   LEACHATES FROM SOLID WASTES, SOURCE – WISE

Contamination of groundwater can occur when the solid wastes are leached. The leachate may reach the groundwater through either natural filtration or the deliberate disposal of liquids in association with solids (paper 1.1, Groundwater contamination, in 'Water Resources and Land use planning' 1982).

### 8.2.1   *Solid wastes from agriculture*

While solid and liquid agricultural wastes do sometimes occur in concentrated form, the more common situation is the low-level groundwater contamination in rural areas as a consequence of leaching of excess nutrients from the inorganic and organic fertilizers applied to both arable and pastoral land, and the residues of pesticides and herbicides that may be leached from the soil.

Nitrate which is the principal contaminant arising from agricultural activities, arises from two sources:

- Use of fertilizers: The consumption of commercial fertilizers has been increasing rapidly in all parts of the world – for instance, the fertilizer consumption in 1981/1982 was 10 times more than it was in 1950's. It ranges from about 120 kg ha$^{-1}$ in the Developed countries to less than 40 kg ha$^{-1}$ in the Developing countries. The application of ammonium sulfate and potassium chloride fertilizers manifest themselves in the form of increased concentrations of sulfate and chloride in the drainage water,
- Mineralization: This is the process whereby the bacteria transform the organically-bound nitrogen in the soil to inorganic forms, following the ploughing of the established or temporary grasslands. Studies show that the potential quantities of nitrate released by ploughing may exceed the total, annual quantities of nitrogen normally applied.

Where the livestock is free ranging, the faeces and urine of the animals get deposited in a disseminated manner. In such a situation, the soil bacteria can readily degrade the animal wastes, and there is little risk of contamination of groundwater due to faecal bacteria. But serious problems can arise in the case of stock yards and dairy farms, where large number of animals are kept. In USA, livestock produces 130 times more waste than humans. It has been reported that one large hog farm in Utah produces more sewage than the City of Los Angeles (California, USA)!

The solid wastes produced in agriculture and animal husbandry activities which could contaminate water resources through leaching and effluent production, are listed in Table 8.1 (source: Laconte & Haimes 1982).

### 8.2.2 *Solid wastes from mining*

Mining industry produces more solid wastes than any other industry. Presently, there are more than 40,000 mines in the world, which produce an aggregate volume of $33 \times 10^9$ m$^3$ yr$^{-1}$ of rock (Vartanyan 1989). Certain kinds of solid wastes, such as, waste products of quarrying for building stone, lime for cement and agricultural use, filters (e.g. gypsum, barytes), and roadstone, are generally inert. Consequently, water percolating through them does not undergo any significant chemical changes. If the waste contains crushed material, the Total Suspended Solids (TSS) content of the percolating water could increase.

When the soil or sediment cover is removed in the process of quarrying, their filtering and attenuation capabilities would have been lost, thus exposing the groundwater to greater risks of pollution. Groundwater could also be contaminated due to some ancillary activities associated with quarrying, such as, accidental spillages of fuel oil, leakages from storage tanks or toilets for workers, or draining of water from the surrounding areas into the quarry, etc.

Solid wastes arising from the mining of coal, lignite, metallic sulfides, uranium, etc. tend to contain pyrite (Fe S$_2$). Under oxidizing conditions, and in the

Table 8.1. Solid waste production from agriculture and animal husbandry (source: 'Water Resources and Land Use Planning').

| Source | Potential characteristics of leachate/effluent | Rate of effluent or solid waste production |
|---|---|---|
| Arable crop, fertilizers, pesticides | Increased nitrate, ammonia, sulfate, chloride and phosphate from fertilizers. Fecal bacterial contamination from organic fertilizers. Organochlorine compounds from pesticides | Rate of leachate production dependent on local climate /irrigation regime. Fertilizer applications range from 10 to $10^3$ kg ha$^{-1}$yr$^{-1}$. Partial uptake of fertilizer components by crops. |
| Livestock, faeces and urine | *Cattle*<br>Suspended solids  – 90,000 mg l$^{-1}$<br>BOD  – 10,000 mg l$^{-1}$<br>Total N  – 4000 mg l$^{-1}$<br>Fecal coliforms  – $2 \times 10^5$ g$^{-1}$ | 15 m$^3$ per year per animal |
| | *Pigs*<br>Suspended solids  – 100,000 mg l$^{-1}$<br>BOD  – 30,000 mg l$^{-1}$<br>Total N  – 5000 mg l$^{-1}$<br>Faecal coliforms  – $3 \times 10^6$ g$^{-1}$<br>Fecal streptococci  – $8 \times 10^7$ g$^{-1}$<br>Heavy metals  – variable | 3 m$^3$ per year per animal |
| | *Sheep*<br>Similar SS, BOD, and N for cattle<br>Fecal coliforms  – $1\text{-}2 \times 10^7$ g$^{-1}$<br>Fecal streptococci  – $4 \times 10^7$ g$^{-1}$ | 2 m$^3$ per year per animal |
| | *Poultry*<br>Suspended solids  – 36,000 mg l$^{-1}$<br>BOD  – 36,000 mg l$^{-1}$<br>Total N  – 12,000 mg l$^{-1}$<br>Fecal coliforms  – $10^5 - 10^7$ g$^{-1}$<br>Fecal streptococci – $10^6 - 10^7$ g$^{-1}$ | 50 m$^3$ per year per 1000 birds |

presence of catalytic bacteria, such as *Thiobacillus ferrooxidans*, pyrite gets oxidized into sulfuric acid and iron sulfate. Thus, surface runoff and groundwater seepages associated with waste piles tend to be highly acidic, and corrosive, and contain high concentrations of iron, aluminum, manganese, copper, lead, nickel and zinc, etc. in solution and suspension. The discharge of such waters (known as Acid Mine Drainage or AMD) into streams destroys the aquatic life, and the stream water is rendered non-potable. The ubiquitous gangue minerals, such as calcite and quartz, are less soluble and reactive (ways and means of ameliorating AMD have been discussed in detail under Section 4.5).

The solid wastes produced in mining activities which could contaminate water resources through leaching and effluent production, are listed in Table 8.2 (source: Laconte & Haimes 1982).

Table 8.2. Solid wastes from mining (source: 'Water Resources and Land Use Planning').

| Source | Potential characteristics of leachate/effluent | Rate of effluent or solid waste production |
|---|---|---|
| Coal-mine drainage | High total dissolved solids. Suspended solids. Iron. Often acid. May contain high chlorides from connate water | $10^5$-$10^7$ m$^3$ yr$^{-1}$ |
| Colliery waste | Leachate similar to mine drainage waters | $10^5$-$10^7$ t yr$^{-1}$ of wastes per colliery. Quantity of leachate depends on climate |
| Metals | High total suspended solids. Possibly low pH. High sulfates from oxidation of sulfides. Dissolved and particulate metals. Washing and mineral dressing waters may contain organic flocculents | $10^5$-$10^7$ t yr$^{-1}$ of wastes per mine. Quantity of leachate depends on climate |

Figure 8.1 (source: Laconte & Haimes 1982, p. 4) is a schematic depiction of how water resources can be contaminated by agricultural and mining activities.

### 8.2.3  *Solid wastes from household, commercial and industrial sources*

Household solid wastes are largely composed of biodegradable, putrescible matter. The process of biodegradation is accompanied by a rise in the initial temperature within the waste mass, and the generation of carbon dioxide and methane gases (methane spontaneously catches fire, and billowing smoke is a common sight in the municipal waste piles; instances are known of people dying due to methane poisoning, when they enter chambers where organic matter is undergoing putrefaction). The content of Total Organic Carbon (TOC) tends to be very high in the leachates from the domestic wastes. More than 80% of TOC is in the form of volatile fatty acids (acetic, butyric, etc.) which are responsible for the characteristic odor of municipal garbage dumps. In due course, the organic carbon content will change over into higher molecular weight substances, such as carbohydrates. The rate of change depends upon temperature and moisture content. The change may take place in a matter of 5-10 years, in the case of humid, temperate climates, and probably in a shorter period in the case of humid, tropical regions. In arid regions, bacterial degradation may be impeded due to lack of moisture.

The generally high sulfate, chloride and ammonia concentrations in the leachates from the household wastes, reach their peak concentrations within a year or two of disposal of wastes, and then steadily decrease over a long period of time (of the order of decades). There may be seasonal fluctuations in the concentrations of these components as a consequence of the changes in ambient temperature and infiltration rates.

Heavy metals (such as, lead, cadmium, mercury and arsenic) may be immobilized, either by getting adsorbed on the organic matter within the fill, or precipitated as sulfides. Municipal garbage invariably contains plenty of cellulose derived from

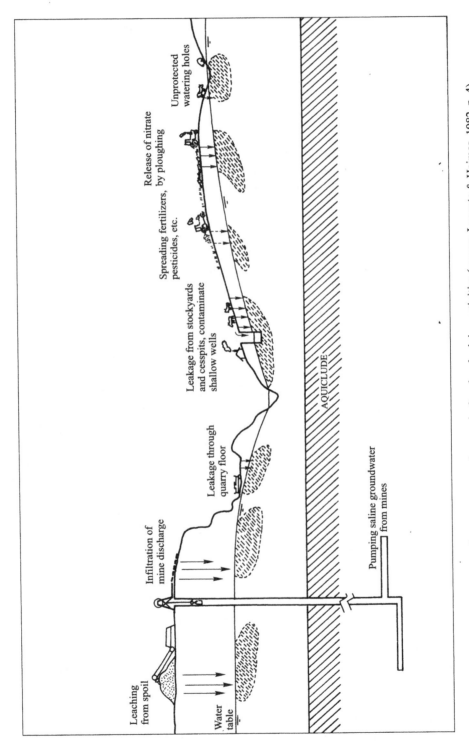

Figure 8.1. Sources of contamination of water resources from agricultural and mining activities (source: Laconte & Haimes, 1982, p. 4).

waste paper. The halogenated hydrocarbons may get adsorbed on cellulose and then evaporated or degraded. Most municipalities co-dispose domestic wastes, with commercial wastes. The two kinds of wastes may not differ greatly in content, except that the commercial wastes may contains oils, phenols and hydrocarbon solvents which may not be biodegradable.

By virtue of their dissolved constituents and high BOD, the leachates from domestic and commercial wastes could degrade the aquifers by producing anoxic conditions.

In February 2000, leachates from the cyanide wastes of an Australian-operated gold mining company in Romania entered the Danube river through the Tisza (a tributary), and caused an ecological disaster. For several tens of km of the stretch of the Danube river in Hungary, and Yugoslavia, hundreds of tonnes of dead fish were found floating in the waters, and the birds which ate the fish also died.

Fly ash is reactive because of its high surface area: volume ratio. Leaching of the fly ash may produce effluents containing toxic elements, such as, Mo, F, Se, B and As. Low pH leachates from fly ash may give rise to problems of iron floc formation in surface waters. When the flue gases are scrubbed, the resulting sludge will typically contain cyanide and heavy metals. Its pH will be low, unless neutralized by lime. It has been reported that mixtures of sludge, lime and fly ash will set rapidly to a load-bearing, low-permeability, solid which is not easily leachable. Two benefits accrue from this process – on one hand we will have a useful construction material, and on the other, we would have minimized the pollution risk.

An environmentally-sound, and technoeconomically viable approach to minimize the contamination potential of the wastes, such as fly ash and red mud, is to put them to some useful purpose soon after they are produced. Fly ash – clay bricks can be prepared by the incorporation of 10-40% of fly ash to the clay, and subsequent firing in conventional continuous type Bull's kiln or intermittent type kilns at a temperature of 950°C and 1050°C. With the use of fly ash, 40% more bricks can be produced from the same quantity of soil. The clay – fly ash bricks are lighter, have better thermal insulation and lower dead load than clay bricks. Incorporation of siliceous materials would make the red mud suitable for making building bricks. Its alkali content (4-5%) provides a fluxing action, resulting in good plasticity and bonding of the bricks. Red mud bricks develop a pleasing pale brown, orange or golden yellow color, depending upon the composition of the red mud and firing temperature. They have therefore good architectural value as facing bricks (source: *CSIR Rural Technologies*, New Delhi, India 1995, p. 82-83).

Practitioners of most of the religions in the world bury their dead (with the exception of the Hindus who cremate their dead and immerse the ashes in flowing waters – by an extraordinary coincidence, the aboriginal Yanomami Indians in northeastern Brazil cremate their dead, and promptly disperse the ashes, though not necessarily in waters!). Drainage water from the graveyards may contain organic residues and pathogens, and could contaminate the surface water and groundwater with which they may come into contact. The realization of such a risk led to legislation in mid-nineteenth century England, prohibiting the construction of shallow wells within 200 m of the graveyards. It can happen that due to rapid growth of a town, a cemetery which was initially located in the outskirts of the town, may now be located right in the heart of the town, and cannot be shifted due to religious sensibilities. It is critically important that no shallow borewells be allowed to be constructed within 500 m of a cemetery. If such wells already exist and their use is unavoidable, the water should be regularly be monitored for pathogens and organic residues, before being allowed to be used.

The solid wastes produced in household, commercial and industrial activities which could contaminate water resources through leaching and effluent production, are listed in Table 8.3 (source: Laconte & Haimes 1982).

## 8.3 POLLUTION FROM LIQUID WASTES, SOURCE-WISE

### 8.3.1 *Agriculture and animal husbandry*

The large volumes of liquid and semi-solid faecal matter produced by the animals could readily infiltrate to the water-table, and seriously impair the quality of

Table 8.3. Solid waste production from household, commercial and industrial sectors (source: 'Water Resources and Land Use Planning' 1982).

| Source | Potential characteristics of leachate | Rate of solid waste production |
|---|---|---|
| Household wastes | High sulfate, chloride, ammonia, BOD, TOC, and suspended solids from fresh wastes. Bacterial impurities. In humid climates leachate composition changes with time. Initial TOC mainly volatile, fatty acids (acetic, butyric, proprionic), subsequently changing to high molecular weight organics (humic substances, carbohydrates). Period of change, ~5 to 10 yrs after deposition of wastes in humid, temperate regions. | $0.2\text{-}0.4 \text{ t yr}^{-1} \text{ person}^{-1}$. Characteristic landfill size: $10^4\text{-}10^8$ $m^3$. Rate of leachate production dependent on climate. |
| Commercial wastes | Similar to domestic wastes. may also include phenols, mineral oil, and hydrocarbon solvent wastes. | Co-disposal with domestic wastes. |
| Industrial wastes | Variable. May contain toxic substances – heavy metals, oils, phenols, solvents, pesticide /herbicide residues | $0.3 \text{ t yr}^{-1} \text{ person}^{-1}$ in industrialized societies; Industrial landfills: $10^4\text{-}10^6 m^3$. May be co-deposited with domestic, commercial wastes. |
| Power generation (thermal) | Pulverized fuel ash. Upto 2% by weight of soluble constituents, sulfate. May contain concentrations of Germanium and Selenium. Fly ash and flue gas scrubber sludges. Finely particulate, containing disseminated heavy metals. Sludges of low pH unless neutralized by lime addition. | $10^4\text{-}10^6 \text{ t yr}^{-1}$. |

groundwater. Where domestic water supplies are drawn from shallow wells, or boreholes located near the animal farms, the contaminated water could cause serious health problems.

Because of their extremely high BOD (Biological Oxygen Demand) values and the presence of organic contaminants arising from fermentation, the farmyard slurries and silage effluents are capable of causing intense groundwater contamination, including the onset of anoxic conditions. Consequently, special precautions need to be taken to protect the water resources from contamination, such as:
− Covering the wells to protect the ingress of contaminants from above, and
− Provision of casings and linings, or sills and impermeable aprons around the wellhead. If these steps are found to be inadequate, recourse has to be taken to either collect and treat the wastes, or to confine the wastes hermetically through the use of water-impermeable geomembranes.

Table 8.4 (source: Laconte & Haimes 1982) describes the contaminant effluents from agriculture and animal husbandry activities.

### 8.3.2 *Liquid wastes from mining*

Drainage waters from coal collieries tend to have high suspended and dissolved solids of iron and sulfate (derived from the oxidation of sulfates), and chlorides (derived from connate water trapped within the sedimentary rocks). Discharge of such waters on the surface and their subsequent percolation could contaminate the groundwater seriously (in the early part of the last century, the discharge of mine drainage water severely contaminated about 13 km$^2$ of Chalk aquifer in southern England). The drainage waters from metallic mines tend to be acidic, and have higher concentrations of dissolved metals. The drainage waters may also contain organic flocculents used in the screening and dressing of metallic ores.

Oil deposits are often associated with hot brines carrying traces of hydrocarbons. In the early phases of development of the hydrocarbons, it is not uncommon

Table 8.4. Contaminants resulting from effluents related to agriculture and animal husbandry (source: 'Water Resources and Land Use Planning').

| Source | Potential characteristics of leachate/effluent | Rate of effluent or solid waste production |
|---|---|---|
| Intensive units, livestock yards | Effluent formed by washing down, diluting faeces and urine, 3 to 10 times. SS, BOD and N reduced. Chloride – 200-400 mg l$^{-1}$ | Cattle units: $10^3$–$10^5$ m$^3$ yr$^{-1}$ Pig units: $10^3$–$10^4$ m$^3$ yr$^{-1}$ Poultry units: $10^4$–$10^5$ m$^3$ yr$^{-1}$ |
| Washing/drainage from farm-buildings and yards | High suspended solids, high organic content, high BOD, Mineral oil from machinery. Fecal bacteria | Variable quality ($10^3$ m$^3$ yr$^{-1}$) washing wastes/head of livestock |
| Silage | High suspended solids, BOD 1-6 × $10^4$ mg l$^{-1}$. Organic components – carbohydrates, phenols | 0.3 m$^3$ yr$^{-1}$ t$^{-1}$ $10^2$-$10^3$ m$^3$ yr$^{-1}$ livestock$^{-1}$ |

for the hydrocarbon/water systems to be under artesian conditions. In such a situation, the contaminated water may spill on the ground and percolate into shallow aquifers, or it could leak upwards into incompletely grouted production well. When hydraulic mining is employed to mine evaporite deposits, care should be taken to ensure that the brines do not contaminate the groundwater through surface spills and pipeline leakages.

The heavy withdrawal of groundwater in the coastal regions and estuaries could lead to the incursion of saline water into the coastal freshwater aquifers, thereby degrading them. There have been several instances of giant tidal waves generated by tropical cyclones, salinising the arable land, surface water and the groundwater (On Oct. 29, 1999, a 10 m high tidal wave generated by a super-cyclone swept a stretch of 150 km. along the coast of Orissa province in eastern India, destroying every thing in its path, and rendering the surface and groundwaters saline).

Mineralized waters may sometimes occur at depth – in the form of connate water trapped below the zone of natural groundwater circulation, or they may arise from the leaching of evaporite beds terminating against an aquifer. Salinization of fresh groundwater can occur in inland areas due to upconing of such mineralized waters.

Overpumping of the groundwater may result in the lowering of the water-table below the stream bed levels. If the river concerned is perennial or seasonally influent, and if the river water is already contaminated, this would inevitably induce undesirable recharge of the aquifer with the contaminated water of the river.

Table 8.5 (source: Laconte & Haimes 1982) describes the contaminant effluents from mining activities.

### 8.3.3 *Liquid wastes from household, commercial and industrial sources*

The potential of the human sewage to contaminate the groundwater arises out of its high BOD, suspended solids, faecal bacteria and viruses, chloride and ammo-

Table 8.5. Contaminants resulting from liquid wastes from mining (source: 'Water Resources and Land Use Planning')

| Source | Potential characteristics of leachate/effluent | Rate of effluent/leachate production |
|---|---|---|
| Oil and gas well brines | High total solids ($10^3$-$10^5$ mg $l^{-1}$), High $Ca^{2+}$ and $Mg^{2+}$ ($10^3$-$10^5$ mg $l^{-1}$), High $Na^+$ and $K^+$ ($\sim 10^4$ mg $l^{-1}$), High $Cl^-$ ($10^4$-$10^5$ mg $l^{-1}$), High $SO_4^{2-}$ ($10$-$10^3$ mg $l^{-1}$), Oil, upto $10^3$ mg $l^{-1}$, Possibly high temp. | $10^3$-$10^4$ m$^3$ d$^{-1}$ per well |
| Saline intrusions, due to overpumping close to coastlines | $Na^+$ ($10^3$-$10^4$ mg $l^{-1}$), $Mg^{2+}$ ($10^2$-$10^3$ mg $l^{-1}$), $Ca^{2+}$ ($10^2$ mg $l^{-1}$), $K^+$ ($10$-$10^2$ mg $l^{-1}$), $Cl^-$ ($10^3$-$10^4$ mg $l^{-1}$), $SO_4^{2-}$ ($10^2$-$10^3$ mg $l^{-1}$), Alkalinity (as $CaCO_3$) $10^2$ mg $l^{-1}$ | Rate of landward movement of saline incursion varies with pumping regime and aquifer type (example: 4 km. in 40 yr along the estuary of River Thames in England) |

nia. Most of these are removed when the sewage is subjected to Primary (mechanical) treatment, and Secondary (biological treatment) (see Chapter 5 for details). But some anions persist in the final effluent (typically in the ranges of 100-200 mg l$^{-1}$ for Cl, and 30-40 mg l$^{-1}$ for NO$_3^-$ – N). Where an aquifer is contaminated by such treated effluents, the movement of the contaminant plume could be readily monitored on the basis of increased chloride concentrations.

In some countries, the mains which carry the sewage water also carry storm water washed from paved surfaces. The storm water generally contains considerable particulate matter of vegetable, animal and mineral origin. In the cold countries, common salt is used for de-icing of roads, and consequently the storm water may contain seasonally high concentrations of sodium chloride. Contamination due to automobile exhausts and accidental oil spills may lead to immiscible layers and emulsions containing hydrocarbons and organo-metallic complexes, such as tetraethyl lead. Bacterial contamination may be high initially, but intense flushing due to large quantities of storm water will quickly dissipate the pollution.

When wastewater is treated in the sewage works, some quantity of sludge is produced. The sludge typically contains 4 to 7% solids, with organic residues constituting more than half of the solids. Sewage sludge is used in agriculture as a fertilizer and soil conditioner. It has been observed that sludge application reduces the surface runoff, and gives some protection against soil erosion.

The composition of sewage sludge varies greatly, depending on the source of sewage (domestic, industrial, etc.). The mean concentration level and range (percent content of dry matter) of the fertilizer content are as follows: N – 2.2 (1.5-3.5), P – 1.7 (1.0-2.5), K – 0.15 (0.05-0.3). The undesirable consequences of sludge application arise from the heavy metal content of the sludge, which may be taken up by the plants.

The metals present in sewage sludge may be divided into three categories on the basis of their availability to plants:

- Unavailable forms, such as insoluble compounds (oxides or sulfides),
- Potentially available forms, such as insoluble complexes, metals linked to ligands, or forms attached to clays and organic matter, and
- Mobile and available forms, such as hydrated ions or soluble complexes (Peter O'Neill 1985, p. 207). A soil which has high pH (i.e. alkaline) and high cation exchange capacity, will be able to immobilize the metals added to it via the sewage sludge. A soil which is presently alkaline may not always remain that way. Thus, if at a later stage the pH drops (i.e. becomes acidic), the metals may get released.

Elements such as iron, zinc and copper are essential elements needed by man. The same elements may become toxic at high concentrations, however. Cadmium is a highly toxic element. Because of its geochemical affinity with zinc, it enters the plants along with zinc.

In a study made in Norway, sludge application at rates of 60-120 t (dry matter) per ha increased the Ni and Zn contents of plants, but had no significant effect on the concentration levels of Cd and Pb. Excessive application of sludge could increase the content of heavy metals in the crops to levels which render them unfit for human consumption.

Liming (1.5-6 t of $CaCO_3$ $ha^{-1}$) reduces the heavy metal uptake by plants *in the short run.* The heavy metals will continue to persist in the top soil, however.

It has been reported that sludge-amended soils almost invariably contain higher levels of Cd than garden soils. Potential increase in the cadmium content of plants due to the application of sewage sludge constitutes the most important health hazard, which can be mitigated by increasing the available content of zinc in the soil. Sludge should not be applied to soils which are used for growing vegetablés.

Toxic, liquid wastes are some times disposed by deep well injection (In 1962, in the Denver Basin, USA, the forceful injection of toxic, fluid wastes into a borehole triggered low-magnitude earthquakes, but that is a different story). The following precautions need to be taken to ensure that the injection of toxic wastes does not contaminate the overlying aquifer:
– Impermeable, confining layers exist above and/or beneath the layer into which the wastes are injected,
– The injection borehole should be so constructed that there is no possibility of leakage of wastes around the casings,
– The injection zone and local aquifers should be regularly monitored,
– Any abandoned oil or gas or water supply or exploration boreholes present in the vicinity of the injection wells, should be sealed, and
– The injection pressures should not be so high as to lead to hydraulic fracturing of the confining beds.
There may be accidental spills when liquid wastes stored in tanks are transported by road or rail. Also, there may be leakages when the wastes are transported by pipeline. Such spills or leakages could contaminate the groundwater, particularly in the case of shallow watertable aquifers. The magnitude of such contamination may have an enormous range – while the leakage of a few cubic metres of oil from a domestic tank may contaminate a water well nearby, an undetected leak of several thousand cubic metres of oil from a pipeline could jeopardize a whole aquifer.

Table 8.6 (source: Laconte & Haimes 1982) describes the contaminant effluents from domestic sources.

Figure 8.2 (source: Laconte & Haimes 1982, p. 8) shows how groundwater abstraction and effluent discharges could contaminate water resources.

## 8.4  CONTAMINANTS, TYPE-WISE

The presence of certain contaminants in water supplies can cause health risks to humans and animals. The effects of contaminants vary greatly depending upon the general state of health, age, diet, body weight, genetic factors, etc. of the individuals. Consequently, it can happen that different members of a community

Table 8.6. Leachates/effluents from domestic sources (source: Paper 1.1 in Laconte & Haimes 1982).

| Source | Potential characteristics of leachate/effluent | | Rate of leachate/effluent production |
|---|---|---|---|
| Raw sewage (human faeces and urine) | Suspended solids | $\sim 6 \times 10^4$ mg/l | 0.4-0.6 m$^3$ /yr/person (0.3-0.5 t /yr/person) |
| | BOD | $\sim 5 \times 10^4$ mg/l | |
| | Total N | $\sim 1 \times 10^4$ mg/l | |
| | Faecal coliforms | $\sim 1 \times 10^7$/g | |
| | Faecal streptococci | $\sim 1 \times 10^6$/g | |
| Septic tanks | Suspended solids | 100-300 mg/l | 40-60 m$^3$ /yr/ person as- |
| | BOD | 50-400 mg/l | suming connection to |
| | TOC | 100-300 mg/l | water-flush toilet system |
| | Total solids | 300-600 mg/l | |
| | Ammonia | 20-40 mg /l | |
| | Chloride | 100-200 mg/l | |
| | High fecal coliforms, and streptococci | | |
| | Trace organisms, grease | | |
| Sewage, primary treatment | Suspended solids | 25-150 mg/l | |
| | BOD | 30-250 mg/l | |
| | Grease removed | | |
| Sewage, secondary treatment | Suspended solids | 14-45 mg/l | $10^4$-$10^8$ m$^3$ /yr |
| | BOD | 10-50 mg/l | |
| | Ammonia nitrified | 2.0 mg N/l | |
| | Nitrate | 30-40 mg/l | |
| | Chloride | 100-200 mg/l | |
| | Coliforms | 1000-4000 mg/l | |
| Sewage, tertiary treatment | Suspended solids | 0 | |
| | BOD | 0-10 mg/l | |
| Sewage, quaternary treatment | Virus and bacteria free | | |
| Sewage sludge, digested | Solids (4-7%) and organic matter (2-4%). Heavy metals, if mixed with storm water discharges. High Al, P, K, and N (as NH$_3$). Leachate contains: Ammonia (10-25 mg/l), Potassium (1-10 mg/l), Sodium (1-50 mg/l), Zinc (1000-4000 mg/l). | | $10^4$-$10^5$ m$^3$ /yr |
| Storm water drains – street drainage | High suspended solids ($\sim$ 1000 mg/l). Hydrocarbons, minerals, etc. from roads, service areas, etc. Wide variety of compounds from accidental spillages. Bacterial contamination high, but order of magnitude lower than sewage | | Variable rates. May be treated at sewage works, mixed with foul water drainage |

drinking water from (say) the same contaminated source, may be affected differently. For convenience, the contaminants may be classified type-wise, as chemical, biological and radioactive.

As should be expected, the composition of the industrial effluents depends upon the industrial process involved (Table 8.7; source: Laconte & Haimes 1982). Table 8.8 quantifies the rate of production of contaminant effluent depending upon the size of the industrial unit (source: Laconte & Haimes 1982).

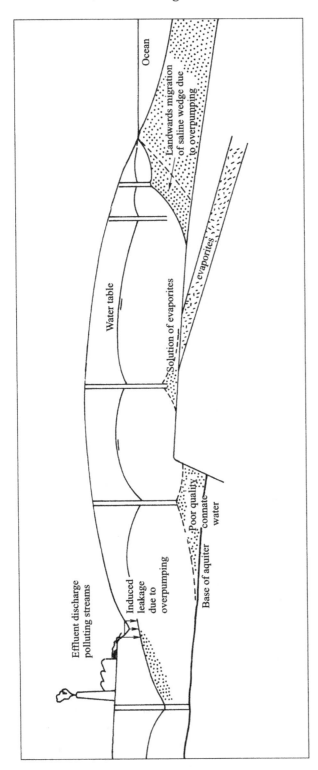

Figure 8.2. Sources of contamination of water resources arising from groundwater abstraction and effluent discharges (source: Laconte & Haimes, 1982, p. 8).

Table 8.7. Characteristics of the waste effluents from different industries.

| Industries | Characteristics of the waste effluents |
|---|---|
| Food and drink | High in BOD and suspended matter; taste and odor problems |
| Textile | Characteristically alkaline |
| Tannery | High concentration of dissolved chlorides, sulfides and chromium |
| Petrochemicals | High BOD, toxic sulfur compounds and phenols |
| Metallurgical & metal finishing | Characteristically acid, with high suspended solids; metal finishing wastes additionally contain heavy metals, phenols and oils |
| Thermal power production (cooling water) | Suspended matter in cooling water will not increase as it does not come into contact with any particulate matter during circulation. It may, however, get reduced in volume due to evaporation, and raised in temperature by about 10°C. Consequently, the concentration of dissolved constituents will increase fourfold, and carbonate minerals may be partly precipitated due to loss of dissolved carbon oxide because of increasing temperature. Loss of dissolved oxygen may also occur. The disposal of such oxygen-depleted, heated water to the ground may have the following consequences:<br>– soil or aquifer material may be leached,<br>– when the warm water cools down as a result of contact with cooler, less mineralized groundwater, some components may be reprecipitated within the rock pores, thus reducing the permeability of the rock, and sequestering the recharged waters within an oxygen depleted zone. If the heated water is recharged via boreholes, microbial growth may take place, and clog the wall screens. |
| Production of gas by coal distillation (by, say, horizontal retort method) | Crude tar oils with significant content of phenolic compounds are produced. When hydrated ferric oxide is used to purify the raw gas, it may be contaminated with sulfides, free sulfur and cyanides. Impurities in the gas itself may dissolve in water. Leaching of these wastes could contaminate the groundwater with phenols and cyanide. |

### 8.4.1 *Chemical contaminants*

In 1984, the World Health Organization (WHO) has prescribed norms for the quality of potable (i.e. drinking) water, and these have been updated in 1993. The Federal Drinking Water Standards, prescribed by the US Environmental Protection Agency (EPA) in 1988 (vide Tables 3.4 and 3.5), are widely used all over the world. US EPA has elaborate procedures for continuously updating the figures for the Maximum Contaminant Levels (MCLs) and their enforceability for various components (inorganic, pesticides, and radioactive contaminants) on the basis of laboratory investigations, etiological and epidemiological studies and consultations with the authorities, stakeholders and interest groups (for instance, US EPA has organized a consultation in San Diego in June, 2000, to examine the need for, and implications of, the downward revision of the MCL for arsenic from its current level of 50 µg $l^{-1}$).

The health and other risks arising from the presence of some contaminants in the groundwater supplies, are summarized in Chapter 10.

Pesticides (e.g. herbicides, insecticides, fungicides, nematocides, etc.) are an important group of chemicals which could contaminate groundwater.

Table 8.8. Quantum of production of contaminants from industrial and miscellaneous sources (source: Paper 1.1 in Laconte & Haimes 1982).

| | | |
|---|---|---|
| Food and drink manufacturing | High BOD. Suspended solids often high. Colloidal and dissolved organic substances, odors | $10^3$-$10^7$ m$^3$/yr |
| Textiles and clothing | High Suspended solids. High BOD. Alkaline effluent | $10^4$-$10^6$ m$^3$/yr |
| Tanneries | High BOD, total solids, hardness, chlorides, sulfides, chromium | $10^3$-$10^6$ m$^3$/yr |
| *Chemicals-* | | $10^5$-$10^9$ m$^3$/yr |
| Acids | Low pH | |
| Detergents | High BOD, saponified residues, low pH, high organic acids | |
| Explosives | Low pH, High organic acids, alcohols, oils | |
| Insecticides/herbicides | High TOC, toxic benzene derivatives, low pH | |
| Synthetic resins and fibers | High BOD | |
| Petroleum and petro-chemical | | |
| refining | High BOD, chloride, phenols, sulfur comp. | $10^6$-$10^8$ m$^3$/yr |
| process | High BOD, Suspended solids, chlorides, variable pH | |
| Thermal power | Increased water temperature. Slight increase in dissolved solids by evaporation of cooling wastes | $10^3$-$10^4$ m$^3$/yr/ megawatt |
| Engineering works | High suspended solids, soluble cutting oils, trace heavy metals, variable BOD, pH | $10^4$-$10^7$ m$^3$/yr |
| Foundries | Low pH, High suspended solids, phenols, oil | $10^7$-$10^9$ m$^3$/yr |
| Plating and metal finishing | Low pH, High content of toxic heavy metals, sometimes as sludge | $10^7$-$10^9$ m$^3$/yr |
| Deep well injection | Various concentrated liquid wastes, often toxic. Brines. Acid and alkaline wastes. Organic wastes | $10^4$-$10^6$ m$^3$/yr |
| Leakage from storage tanks and pipelines | Aqueous solutions, hydrocarbons, petrochemicals, sewage | |
| Accidental spillages | Various liquids in transit, hydrocarbons, petro-chemicals, acids, alkalis, solvents. Liquids may enter surface drains or soakaways | Generally 10 m$^3$ per incident |

Pesticides are widely used in agriculture, public health, and several other purposes. There are more than 10,000 formulations of pesticides, of which about 450 are widely used. The extensive use of pesticides during the last fifty years not only led to their becoming ubiquitous pollutants in soils, crops, groundwater, human and animal tissues, etc. in industrialized countries, but also led to their dispersal to far-off places; pesticides residues have been found in Antarctic penguins. Phreatic or unconfined aquifers (i.e. aquifers extending to the earth's surface) are most vulnerable to pesticide pollution.

The main types of compounds used as pesticides are as follows (Manahan 1991):

*Insecticides (insect killers):*
Organochlorines (e.g. DDT, Lindane, Aldrin, Heptachlor)

Organophosphates (e.g. Parathion, Malathion)
Carbamates (e.g. Carbaryl, Carbofuran)

*Herbicides (weed killers)*:
Phenoxyacetic acids (e.g. 2,4-D; 2,4,5-T; MCPA)
Toluidines (e.g. Trifluralin)
Triazines (e.g. Simazine, Atrazine)
Phenylureas (e.g. Fenuron, Isoproturon)
Bipyridyls (e.g. Diquat, Paraquat)
Glycines (e.g. Glyphosate 'Tumbleweed')
Phenoxypropionates (e.g. 'Mecoprop')
Translocated carbamates (e.g. Barban, Asulam)
Hydroxy nitriles (e.g. Ioxynil, Bromoxydynil)

*Fungicides (fungus killers)*:
Nonsystemic fungicides
   Inorganic and heavy metal compounds (e.g. Bordeaux Mixture – Cu)
   Dithiocarbamates (e.g. Maneb, Zineb, Mancozeb)
   Pthalimides (e.g. Captan, Captafol, Dichofluanid)
Systemic fungicides
   Antibiotics (e.g. Cycloheximide, Blasticidin S, Kasugamycin)
   Benzimidazoles (e.g. Carbendazim, Benomyl, Thiabendazole)
   Pyrimidines (e.g. Ethirimol, Triforine)

As evidence mounted on the toxic effects of pesticides on human beings and animals, some pesticides, such as DDT, Aldrin and Dieldrin were completely banned in the Industrialized countries. Though the ban was effected in the 1980s, these pesticides continue to persist in the environment to varying extents:

| Pesticide | % decay | Time taken (years) |
|-----------|---------|--------------------|
| DDT | 44 | 8 |
| BHC | 50 | 1 |
| Dieldrin | 49-53 | 3 |

In the place of the banned pesticides, substitutes such as pyrethroids and organophosphorus compounds, have been introduced in developed countries. These substitutes are environmentally more acceptable, though more expensive. For instance, pyrethrins (derived from the flower of the plant, *Chrysanthemum cinerariaefolium*), undergo rapid photolytically induced oxidation, and are therefore short-lived. Several synthetic analogues of pyrethrin have been developed. Intensive work is underway to develop a whole range of pesticides, based on the seeds of the Indian tree, *neem* (*Azirdirachta Indica*). Some of the organophosphorus pesticides, such as Chlorpyrifos, are highly effective against pests, but less toxic to mammals. They are applied in relatively small quantities (200-

1200 g ha$^{-1}$). They break down by hydrolysis and their persistence in the soil is in the range of 60-120 days.

Soil processes have great ability to decompose organic matter. The enzymatic activities of the microorganisms are capable of degrading and detoxifying a variety of substances added to the soil. However, the chemical structures of some pesticides are such as to preclude their enzymatic degradation. Several synthetic pesticides (such as 2,4,5-T or 2,4,5 – trichlorophenoxyacetic acid) are heavily substituted with chlorine, bromine, fluorine or nitro or sulfonate groups which are not found in biological tissues. Microorganisms cannot degrade such pesticides, which therefore persist in the soil for long periods (say, about 15 years).

Those pesticides characterized by low vapor pressure and low solubility in water tend to persist in the soil. The extent of persistence depends upon the temperature, soil type, and soil microbiology. For instance, the herbicide, trifluralin, may persist in the soil between 15 to 30 weeks. Herbicides tend to be lost more readily from moist soil than from the same soil when it is dry. The molecules of herbicides that are strongly adsorbed on dry soils get dislodged under damp conditions.

Herbicides interact with soil components in complex ways. Urea derivatives, triazines, carbamates and nitrophenyl ethers, etc. tend to be strongly adsorbed on soil organic matter. Quaternary ammonium compounds are adsorbed on clays and metal hydroxides. The herbicides may also interact with the exudates from soil microorganisms.

The probable range of persistence (in weeks) of some important herbicides has been given by Hassal (1987, p. 253). Some herbicides, such as, Propham, Dalapon and Aminotriazole, have shorter persistence periods of 3-10 weeks, whereas some others, such as, Diuron and Simazine, have much longer persistence periods of 30-120 weeks. The persistence is greatly influenced by the rate of application, rainfall, soil type, temperature and microbial population. Besides, some crops are more tolerant than others for the same herbicide.

The soil type not only influences the period of persistence of a herbicide, but also affects the toxicity of a given dose to the weeds. For instance, carbamate and urea herbicides are more toxic to the weeds in light soils than to the same weeds in heavy organic soils. Consequently, these herbicides persist in organic soils for longer periods.

### 8.4.2 *Biological contaminants*

Waste waters from urban areas contain a variety of faecally-excreted human pathogens, such as helminths, protozoans, bacteria and viruses. The concentration of pathogens in human faeces is high ($10^4$-$10^7$ organisms per gm. of faeces), and they have persistence times ranging from weeks to months (see Section 5.7, and Fig. 5.7).

Bacteria and viruses transmissible by groundwater, may cause a variety of diseases, such as, typhoid/paratyphoid, enteric fever, cholera, bacterial dysentery, amoebic dysentery, non-specific gastroenteritis, leptospiral jaundice (Weil's disease), poliomyelitis, infective hepatitis, etc.

The biological contamination of groundwater may be caused by

– Leaching of deposits of solid faecal material,
– Spray irrigation with sewage water, and
– Migration of domestic sewage from septic tank or pit latrine.

Though aesthetically objectionable, the presence of live algae in water does not cause disease. But water with dead algae is unacceptable, as they release toxins. Bacteria are unicellular organisms, with sizes of the order of 0.5 to 10 um. Thus, they are similar in size to clay particles. The mobility of bacteria in aquifers may be constrained by their ability to move through pore spaces in rocks. Viruses are smaller than bacteria, with sizes of the order of 20 to 200 nm. They are incapable of reproduction outside a host organisms, and they have no metabolic functions. Since viruses tend to be negatively charged, they can only be adsorbed at positively charged sites. The adsorption of viruses reaches its peak at pH 7, which is also the normal pH level of groundwater. At higher pH, the adsorption of viruses decreases because of the lower availability of positively charged sites.

The maximum length of travel of the biological pollutants in aquifer sediments is of the order of 20 to 30 m.

### 8.4.3 *Radioactive contaminants*

The following six nuclides are of particular hydrological concern because of their toxicity, environmental mobility and relatively long half-life: $^3$H, $^{90}$Sr, $^{129}$I, $^{137}$Cs, $^{226}$Ra, and $^{239}$Pu. Uraniferous rocks contain $^{226}$Ra, and leaching of such rocks could contaminate the surface waters and groundwaters. $^{226}$Ra is also released during the processing of the uranium ores. The rest of the nuclides are produced in the course of nuclear power generation, or during nuclear accidents (see Aswathanarayana 1995, p. 162-165, for a summary account of the Chernobyl Reactor Accident of April 26, 1986). They may enter the groundwater systems due to accidental escape or intentional release. A variety of radioactive tracers (both natural and man-made) are used to estimate the transport and attenuation of inorganic contaminants in the groundwater flow systems and age-dating of groundwaters (see Section 2.5).

Groundwater is depleted in some isotopes because of adsorption in the soil. For instance, $^{137}$Cs is strongly retained in the soil, and very little of it goes into the groundwater. In contrast, $^{90}$Sr is not bound in the soil, and is therefore passed on into groundwater. As groundwater is an important source of drinking water in many communities, the extent of infiltration of radioisotopes in the soil assumes great importance. The extent of lime-lag between the time of emission of a radionuclide into the atmosphere, and the time it reaches the hydrosphere, could be illustrated with an example. For instance, $^{131}$I in the hydrosphere due to Chernobyl has reached the peak level of 1.1 k Bq l$^{-1}$ on May 3, 1986 (i.e. one week after the accident).

The organ in which a given radionuclide accumulates preferentially, is called the critical organ. For instance, bone is the critical organ for $^{90}$Sr, because of the geochemical coherence of Sr with Ca in the bone. As $^{3}$H and $^{137}$Cs are transported by the blood stream, the whole body constitutes the critical organ for these isotopes.

## 8.5 ANTHROPOGENIC ACIDIFICATION OF WATERS

### 8.5.1 *Acid rain*

Combustion of fossil fuels (coal, oil, natural gas, etc.) for power generation, transport and other industries, produces the emissions of $NO_x$ and $SO_x$. These gases get oxidized in the atmosphere and precipitate as dilute nitric and sulfuric acid solutions, generally known as Acid Rain.

Acid rain degrades the quality of air, soils, surface and groundwaters, biota, etc. It has emerged as a serious environmental problem in the industrialized countries. For instance, the pH of the rainwater in the Netherlands decreased by about one unit – from 5.4 in 1938 to 4.52 in 1980, due to increase in the concentrations of $SO_4^{2-}$ and $NO_3^-$ (Appelo 1985). As a consequence of evapotranspiration, the acidity of the solution that enters the soil goes down to a pH to about 3. Appelo et al. (1982) reported that the youngest groundwater got acidified down to a pH of about 4 in the carbonate-free sandy aquifer in the Veluwe area of the Netherlands.

The lowering of the pH down to 3, is a consequence of the oxidation of the higher concentration of ammonia in the 1980 rainwater, by the following reaction:

$$NH_4^+ + 2\,O_2 \rightarrow NO_3^- + 2\,H^+ + H_2O \tag{8.1}$$

Ammonia and manure are used as fertilizers in large amounts. When ammonia is oxidized by oxygen, nitrate is produced by the following reaction:

$$NH_3 + 2\,O_2 \rightarrow NO_3^- + H^+ + H_2O \tag{8.2}$$

Plants use the nitrate thus produced, and proton production is balanced by the production of $HCO_3^-$ in the denitrification process, by the following reaction:

$$5\,CH_2O + 4\,NO_3^- \rightarrow 2\,N_2 + 4\,HCO_3^- + CO_2 + 3H_2O \tag{8.3}$$

Another source of acidification of waters is the reduction of pyrites, $FeS_2$. Reducing sediments such as those with organic matter, invariably contain some amount of pyrite. When the watertable in a well goes down due to pumping, pyrite that might be present in the sediments may get oxidized to sulfate by the following reaction:

$$2\,FeS_2 + 15/2\,O_2 + 5\,H_2O \rightarrow 2\,FeOOH + 4\,SO_4^{2-} + 8\,H^+ \tag{8.4}$$

The oxidation of pyrite is one of the most strongly acid-producing reactions in nature.

Pyrite is a ubiquitous mineral in mine dumps of not only coal but also of sulfide ores. Acid Mine Drainage also involves the oxidation of pyrite.

### 8.5.2 *Susceptibility of rocks to acidification*

Susceptibility of rocks to acidification is dependent upon the rate of dissolution of the constituent minerals. Aquifers of limestones, dolomites and other rocks containing carbonate minerals are unlikely to develop acid groundwater, because of the rapid dissolution of the carbonate minerals. Quartzose rocks like granites and granodiorites, clean sandstones and quartzites which contain slowly-dissolving minerals, are most susceptible to acidification. Basic and ultrabasic rocks which contain more quickly dissolving silicates are less susceptible to acidification.

Figure 8.3 (source: Böttcher et al. 1985) indicates the nature and geochemical consequences of the acidification of groundwater in the saturated zone of a sandy aquifer below a coniferous forest in Germany. The $SO_4$ content which is about 50 mg $l^{-1}$ near the surface rises to about 85 mg $l^{-1}$ at a depth of 2 m, and remains more or less unchanged at that level. The youngest groundwater is most acid (pH of about 4), and pH remains unchanged upto about 8 m. It reaches a maximum of 6 at a depth of about 12 m. The concentrations of Al are generally high (6-10 mg $l^{-1}$) under near-surface conditions of acid pH, and become negligible below 10 m. It is evident that high Al dissolution in the shallow environment is a direct consequence of acid pH.

Thus Al dominates the buffering processes in the aquifers. Al is an oxyphile element, and always occurs in nature as oxides and hydroxides, such as gibbsite, Al $(OH)_3$ and clay minerals. When acid waters come into contact with a soil or an aquifer, they interact with gibbsite releasing Al as per the following equation:

$$Al\,(OH)_3 \rightarrow Al^{3+} + 3\,OH^- \qquad\qquad (8.5)$$

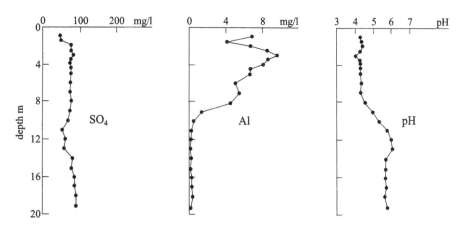

Figure 8.3. Groundwater acidification in the saturated zone of a sandy aquifer below a coniferous forest (source: Böttcher et al. 1985).

## 8.6 WATER POLLUTION ARISING FROM WASTE DISPOSAL

The relevance of waste disposal to water resources management arises from the possibility that leachates from landfill sites could contaminate water and resources.

### 8.6.1 *What is waste disposal?*

Some authorities differentiate between the terms, waste deposit and waste disposal. They hold that 'Disposal' has a wider meaning and greater finality than 'Deposit'. In other words, every deposit on land does not constitute disposal. If the purpose of a deposit is not disposal, the operation will be simply a deposit and not a disposal.

Wastes arise from municipal, industrial, agricultural, mining, demolition, sewage sludge, etc. sources. Hazardous wastes are special wastes which are difficult to handle or cause harm to human health or the environment. They may be inflammable, corrosive, toxic, reactive, carcinogenic, infectious, irritant, harmful, or ecotoxic.

The issues of waste disposal have been comprehensively dealt with by Bagchi (1990) and Petts & Eduljee (1994). The ways in which the treatment and disposal of wastes have impacts on the flows and quality of surface water and groundwater are shown in Tables 8.9 and 8.10 (Petts & Eduljee 1994, p. 167).

Figure 8.4 (source: Petts & Eduljee 1994, p. 229) shows possible pathways of chemicals released from a landfill, and Figure 8.5 (source: Petts & Eduljee 1994,

Table 8.9. Sources and effects of waste treatment and disposal upon surface water (Source: Petts & Eduljee 1994, p. 167).

| Phase | Source of effect on run-off and flows | Source of effect on water quality |
|---|---|---|
| Construction | Soil compaction by heavy vehicles, Provision of temporary drainage, Earth works, Elimination of on-site depression storage, Diversion of surface waters. | Earthworks, Washing effluents from construction vehicles, Leakage or spillage of oils, hydraulic fluids, Accidental release of surface run-off. |
| Operation | Uncontrolled discharge of surface water, Leachate level break-out, Provision of artificial ground surface, Provision of engineered drainage systems, Raised landfill areas, Removal of vegetated areas. | Accidental release of surface run-off, Spillage from transport accidents, Discharge of effluents, Leachate level break-out to surface, Leachate leakage to groundwater, and then to river, Spillage of chemicals, etc. from unloading or filling, Leakage from storage tanks, Uncontrolled discharge from surface water, Deposition of pollutants from air. |
| Restoration/ closure | Provision of artificial ground surface, Engineered low permeability cap, Shape of the final landform – slope profile, Leachate level break-out. | Leachate break-out, Leachate leakage. |

Table 8.10. Sources and effects of waste treatment and disposal upon groundwater (Source: Petts & Eduljee 1994, p. 167).

| Phase | Source of effect on groundwater level | Source of effect on groundwater quality |
|---|---|---|
| Construction | Removal of topsoil and exposure of subsoil of lower permeability, Soil compaction | Leaching of contaminants in soils, Leakage of spillage of oils, hydraulic fluids, etc. |
| Operation | Structures/buildings, Provision of artificial surfaces, Provision of engineered drainage systems | Leachate leakage, Contaminated drainage from paved areas, Spillage of chemicals, etc. Leakage from storage tanks, Leakage from underground sumps or drains, Contaminated drainage from rainfall on unpaved areas |
| Restoration/ closure | Provision of artificial ground surface | Leachate leakage |

p. 166) depicts the effects of waste treatment and disposal facilities on surface water.

A desk study should precede field investigations. The sources of information are maps, archival data about the surveys made earlier, research studies and monitoring information available from relevant statutory authorities. The following account lists the relevant areas of information (Petts and Eduljee, 1994, p. 172):

1. Land-use: current land-uses including identification of possible polluting industries; aquatic habitats that need to be conserved; nature and extent of water use in the catchment area,

2. Topography: relief of the area, based on topographic maps, aerial photographs or space photographs; patterns and flows of surface water drainage; kind and depth of soils,

3. Geological setting: stratigraphic sequence; thickness and lateral extent of layers; geological structures and their geometries; mineral resources, if any, and the extent of their present and projected utilization.

4. Hydrogeology and water supply: Aquifers - their nature, lithology, porosity and permeability, depth of the water-table, direction of flow; sources of recharge; water quality; present level of abstractions; site flood potential; availability of sewage treatment, etc.

Time was when the guiding philosophy of waste disposal was simply, "Out of sight, Out of Mind". A well-known case is that of Love Canal, near Niagara Falls, USA. The company which produced the chemical wastes, simply covered them up with soil, and left. Later, in 1978, land developers built houses on the disposal site. It was found that the community suffered bouts of ill health, and some children were born with birth defects. These were traced to leachates and toxic vapors from the chemical wastes, contaminating the groundwater and indoor air.

Till about 1950s, wastes were deemed to be unwanted, useless material, which had to be got rid of at the lowest cost. By 1990's, the position changed drastically. Waste management, involving the production, handling, storage, transport, pro-

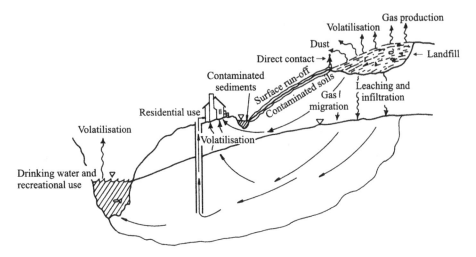

Figure 8.4. Possible pathways of chemicals released from a landfill (source: Petts & Eduljee 1994, p. 229).

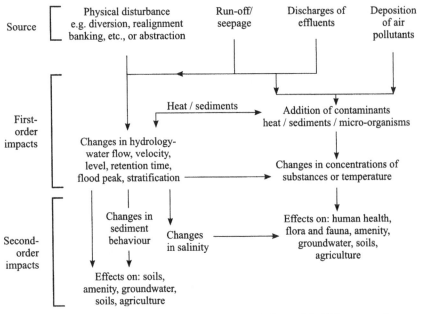

Figure 8.5. Sources and effects of waste treatment and disposal facilities on surface water (source: Petts & Eduljee 1994, p. 166).

cessing, treatment and ultimate disposal of the wastes, has become a big industry (for instance, good part of the multi-billion dollar budget of US EPA is devoted to the clearance of Superfund sites). The objective of the waste disposal is to minimize the environmental impact and maximize the resource recovery.

Presently, the basic paradigm in regard to waste management involves the following approaches: Prevention of waste (closing of cycles), Maximum recycling and reuse of material, Safe disposal of any waste which cannot be reused through combustion as fuel, incineration and landfill. Landfill is the most preferred option for waste disposal. Landfill is described as 'engineered deposit of wastes onto and into land, with deposit usually taking place predominantly below the ground surface in voids, which have often been formed by mineral extraction or quarrying' (Petts & Eduljee 1994, p. 25).

Prior to disposal, wastes are treated to recover materials and/or energy content of the wastes, and to convert the wastes to a form which permit safe and efficient disposal. The treatments are custom-made, depending upon the nature of the waste, on the basis of the Best Practicable Treatment Option (BEPO) (summarized from Salcedo et al 1989):

- *Chemical treatment*: Neutralization, precipitation, ion exchange, oxidation/reduction, solidification,
- *Physical treatment*: Screening, sedimentation, flotation, filtration, centrifugation, reverse osmosis, ultrafiltration, distillation/stream stripping, adsorption,
- *Biological treatment*: Activated sludge, aerated lagoons, composting, anaerobic digestion,
- *Thermal treatment*: Wet air oxidation, incineration, vitrification.

Such treatments benefit waste disposal through steps, such as reduction in volume (e.g. wastes to be landfilled can be reduced by 80-90% through incineration), reduction in toxicity (e.g. conversion of cyanide wastes to non-toxic carbon dioxide and water), change in the physical form (e.g. removal of heavy metals in liquid wastes through precipitation), etc.

### 8.6.2 *Leachate management*

Leachate is produced when water (rainfall, snow, surface or groundwater intrusion, or water in the waste itself) percolates through the landfill material. As should be expected, the composition of the leachate would depend upon the extent of water infiltration, the nature of the waste and the rate of its degradation, the method of operation of the waste site and the procedures of leachate management. The production of leachate can be minimized by good landfill practices, but cannot be avoided altogether. The leachate may be brown to nearly black in color, and its polluting potential can be 10-100 times that of raw sewage.

The leachate may contain the following components:

- Major ions: Ca, Mg, K, Fe, Na, ammonium, bicarbonate, sulfate and chloride,
- Trace metals: Mn, Zn, Cu, Cr, Ni, Pb, Cd,
- Organic compounds: These are usually expressed in terms of TOC, COD, or BOD. Besides, hazardous substances, such as pesticides, benzene, phenols, may be present,
- Microbiological components.

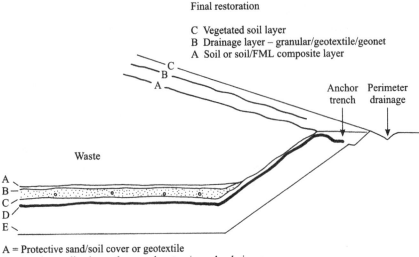

Figure 8.6. Engineered landfill design using composite liners (source: Petts & Eduljee 1994, p. 171).

Farquhar & Rovers (1973) described the decomposition processes that occur at the waste site. The *acetogenic* leachates produced in the early stages are characterized by high organic strength, whereas in the subsequent *methanogenic* phase, landfill gases such as methane, are produced, leaving behind a poorly degradable residue.

The cardinal principle in the disposal of wastes, particularly hazardous wastes, is that they should be effectively isolated from the biosphere for a sufficient length of time, until they no longer present a risk to the biosphere. Two kinds of barriers are envisaged: the *natural* barrier of soil or rock which should prevent or keep within acceptable limits, both the flow of water into the waste or seepage of contaminated water from it, and the *technical* barrier composed of man-made material whose purpose is to seal the waste in order to reduce leaching to a minimum, and to hinder chemical reaction of the waste material with the soil or rock which could adversely affect the capacity of the natural barrier to prevent contact with the biosphere (Archer et al 1987).

Composite impermeable barriers, involving HDPE (High-density Polyethylene) liners, and natural materials, such as bentonite, are fairly effective in preventing leachate migration (Fig. 8.6; source: Petts & Eduljee 1994, p. 171). The manufacturers quote a design life of 25-40 years for the geomembranes, but it should be kept in mind that the period of leachate generation on site could extend to 100 years or more.

US EPA's HELP (Hydraulic Evaluation of Landfill Performance) model can be used to model the landfill water balance.

The operational period for a landfill barrier system cannot be precisely determined in advance. However, assuming that it is possible to eliminate groundwater intrusion and external surface drainage, an order of magnitude figure for the operational period could be estimated from the following equation (Petts & Eduljee 1994, p. 175):

$$Q = I - E - a W \qquad (8.6)$$

where, $Q$ = free leachate generated ($m^3 yr^{-1}$), $I$ = total liquid input including liquid waste ($m^3 yr^{-1}$), $E$ = actual evaporative losses ($m^3 yr^{-1}$), $a$ = absorptive capacity of waste ($m^3 t^{-1}$), and $W$ = weight of waste deposited ($t yr^{-1}$).

Mitigation of water impacts may be achieved through the following measures (Petts & Eduljee 1994, p. 185):
- Minimization of leachate generation,
- Containment of leachate within the landfill,
- Control over leachate quality,
- Collection and disposal of leachate as it is generated,
- Monitoring, and
- Contingency plans, in case the groundwater has been contaminated.

## 8.7 TRANSPORT OF CONTAMINANT SOLUTES IN AQUIFERS

### 8.7.1 *Theory of groundwater flow*

There are excellent books on the theme, e.g. Freeze & Cherry (1979), Bouwer (1978).

Way back in 1940, Hubbard developed the following physical law governing the steady state flow of groundwater.

$$\phi = gz + \frac{p - p_o}{\rho} \qquad (8.7)$$

where $\phi$ is the hydraulic potential at any point in a gravity field, $g$ is acceleration due to gravity, $z$ is the elevation of the point $P$ above the datum, $p$ is the pressure at the point $P$, $p_o$ is atmospheric pressure, and $\rho$ is the density of water.

Darcy's law for the flow of matter through a porous medium states:

$$v_x = K(\partial \phi / \partial x) \qquad (8.8)$$

where $v_x$ is the velocity in the $x$ direction ($m d^{-1}$), $K$ is the coefficient of permeability ($m d^{-1}$), and ($\partial \phi / \partial x$) is the hydraulic head or gradient (a fraction).

Darcy velocity ($m^3 s^{-1}$) is defined as specific discharge per unit area $A$ of the aquifer. Since only pore spaces have a role in the flow of water, it follows that the actual flow through the pores must be larger. The pore water velocity is given by $V = Q/(A - \varepsilon)$, where $\varepsilon$ is porosity.

The eminent hydrologist, Freeze, developed the following generalized equation for the groundwater system, i.e. two- or three-dimensional, steady-state or transient, all saturated, all unsaturated or saturated-unsaturated:

$$V \cdot [\rho K(\Delta \psi + \Delta Z] = \rho[S(\alpha + n\beta) + C](d\psi/dt) \qquad (8.9)$$

where $\psi$ is the pressure head, $\phi = z + \psi$; $\psi < 0$ refers to the unsaturated conditions; $\psi > 0$ refers to the saturated conditions; $\psi = 0$ refers to the water table; $K$ is the hydraulic conductivity of soil or geologic formation, $n$ is soil porosity (pore volume/total volume), $\alpha$ is the vertical compressibility of the soil, $\beta$ is the compressibility of water; $C$ is the specific moisture capacity of the soil.

The solution of the Freeze equation given above could lead to: plots of pressure head by which the water table is located; plots of the hydraulic head, $\phi$, from which the flow velocity at any given point can be determined; travel times along various flow paths; rate of recharge and discharge; moisture content fields for any desired cross-section at any point of time.

### 8.7.2   *Groundwater flows in homogeneous aquifers*

The term, piston flow, is a graphic description of vertically downward percolation of water in an unsaturated zone, where relief is moderate.

$$v = P/\varepsilon_w \qquad (8.10)$$

where $P$ is the precipitation surplus (m yr$^{-1}$), and $\varepsilon_w$ is water-filled porosity.

Let us consider a phreatic aquifer (of thickness, $D$) which is composed of homogeneous sediment, i.e. with the same porosity and permeability throughout (see the inset in Fig. 8.7; source: Appelo & Postma 1996, p. 333). Consequently, the flow velocities will be equal at all depths along a vertical line. Suppose water infiltrates at a point, $x_0$, upstream of $x$. It will percolate to depth $d$ after time $t$. The water that flows above $d$ is the water that is infiltrated between point $x_0$ and $x$. The water that flows below $d$ is the water that infiltrated upstream of $x_0$. Proportionality exists between the point along the upper reach where the water infiltrates and the depth in the aquifer, as per the following equation:

$$x/x_0 = D/(D - d) \qquad (8.11)$$

If $P$ is the precipitation surplus (m yr$^{-1}$) that enters the aquifer along its upper reach, and $\varepsilon$ is the porosity of the formation, the above equation could be written as follows (Appelo & Postma 1996, p. 332):

$$\ln (x/x_0) = P\, t/D\, \varepsilon \qquad (8.12)$$

Figure 8.7 (source: Appelo & Postma 1996, p. 333) shows the tritium profile of an aquifer (see Section 2.5.3 for the explanation of the use of tritium in the age dating of water in an aquifer). The age of water increases proportionately with increasing depth. The figure also leads to an important conclusion. It shows that if

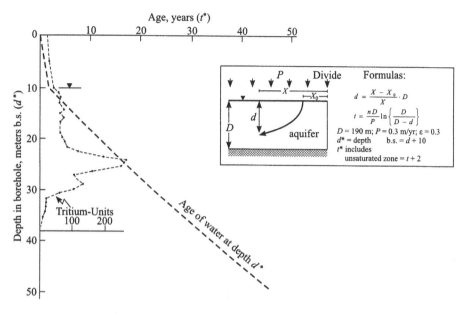

Figure 8.7. Tritium profile of an aquifer (source: Appelo & Postma 1996, p. 333).

the aquifer is homogeneous, and the water supply from above is uniform, the water at a given depth is characterized by the same age. In other words, the *isochrones* (lines connecting points of the same age) are horizontal.

Example 8.1 (adapted after Appelo & Postma 1996, 334)

Toxic effluents from a tannery (containing chlorides, sulfides and chromium) have contaminated an aquifer with the following characteristics: thickness ($D$): 40 m; precipitation surplus entering the aquifer at the upper reaches ($P$): 0.3 m yr$^{-1}$; porosity fraction ($\varepsilon$): 0.3. Polluted water infiltrates the aquifer at 400-600 m from the divide. A drinking water well located at 500 m from the divide, may be affected. Calculate:

– Thickness of the plume of the effluent,
– Its mean depth, and
– Expected arrival time of the pollution at the well.

Infiltration reach (R) = 600-400 = 200 m

$$\text{Thickness of the effluent plume} = \frac{\text{Infiltration reach (200)}}{\text{Distance of the well from the divide (500)}} \times \text{Thickness of aquifer (40 m)} = 16 \text{ m}$$

$$\text{Mean depth of the effluent plume} = \frac{\text{Mean distance from the divide (500 m)}}{\text{Distance of the well from the divide (500 m)}} \times \text{Thickness of aquifer (40 m)} = 40 \text{ m}$$

Expected time of arrival of the pollution at the well of water infiltrated at 400 m

$$= \ln (x/x_o) = Pt/D\varepsilon = \sim 9 \text{ years}$$

$$= \ln \frac{\text{Water infiltrated at (400 m)}}{\text{Distance of the well from the divide (500 m)}} = \frac{0.3 \times t}{40 \text{ m} \times 0.3}$$

$$= \ln (400/500) = (0.3 \times t)/(40 \times 0.3); \ 0.223 = (t/40); \ t = \sim 9 \text{ years}$$

### 8.7.3 *Consequences of inhomogeneities on groundwater flows*

Most aquifers are by no means, homogeneous – there may be large differences in permeability of an aquifer spatially. Also, the precipitation surplus may vary depending upon (say) the vegetation cover (forests intercept the rain more efficiently, and consequently have smaller precipitation surplus relative to lands with grass cover).

Figure 8.8 (source: Appelo & Postma 1996, 335) shows the consequences of two kinds of inhomogeneities:
– Differences in the precipitation surpluses ($P_1$, $P_2$) give rise to differences in the depth of infiltration $\Delta(d)$.

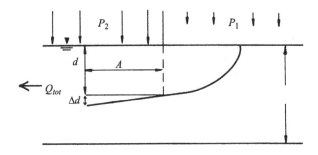

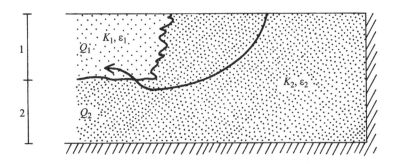

Figure 8.8. Effects of aquifer inhomegenieties on flow lines (source: Appelo & Postma 1996, p. 335).

$$\Delta\,(d) = (P_2 - P_1)AD/Q_{tot} \tag{8.13}$$

– The inhomegeneity in an aquifer may be deemed to be represented by two layers characterized by different thicknesses ($d_1,\ d_2$). permeabilities ($K_1,\ K_2$), porosities ($\varepsilon_1,\ \varepsilon_2$) and discharges ($Q_1,\ Q_2$). Assuming equal hydraulic gradient,

$$(Q_1/Q_2) = (K_1\,d_1)/(K_2\,d_2) \tag{8.14}$$

*Example 8.2 (adapted after Appelo & Postma 1996, 337)*
Excessive use of nitrogenous fertilizers in Haryana, India, have contaminated an aquifer ($D = 20$ m, $\varepsilon = 0.\,4$, $P = 0.5$ m yr$^{-1}$) which consequently contains 200 mg l$^{-1}$ of nitrate. Assuming that the use of fertilizer will be stopped with immediate effect, estimate the time ($t$) required to reduce the nitrate concentration to 10 mg l$^{-1}$ (US EPA prescribed Maximum Contaminant level in drinking water).
   Residence time of the nitrate in the aquifer ($t_R$) = $D\,\varepsilon/P = (20 \times 0.4)/0.5 = 16$ yr

$$10\ \text{mg l}^{-1}/200\ \text{mg l}^{-1} = 0.05 = \varepsilon^{-t}/t_R$$
$$t = -\,t_R\ \text{In}\ 0.05 = 16 \times 3 = 48\ \text{yrs}$$

### 8.7.4 *Retardation of organic micropollutants in groundwater*

Pollution of water resources by organic micropollutants from agricultural, industrial and household, etc. sources, has assumed serious proportions. It is therefore necessary to understand the dynamics of their movement in the groundwater.
   When a chemical enters the groundwater, part of it may be absorbed on the solid aquifer material. Henry's law governs the relationship between the adsorbed concentrations and the concentrations in the water.

$$s/C = K'_d \tag{8.15}$$

where $s$ the adsorbed concentration (in mg kg$^{-1}$ of dry soil, or ppm), $C$ is concentration in water (mg l$^{-1}$), and $K'_d$ is the distribution coefficient. If there is no adsorption, $K'_d$ will be zero, but this rarely happens as almost all chemicals do get adsorbed to some extent.
   The need to understand the adsorption process arises from two considerations:
– The more a chemical is adsorbed on the solid aquifer material, the less of it will be available for transportation by water,
– If a polluted aquifer is to be flushed and cleaned up, the part of the chemical that is bound to the solid aquifer material had to be desorbed and solubilized prior to flushing.
   Most of the organic micropollutants are virtually insoluble in water (a polar liquid), but are soluble in apolar liquids, like octanol. These pollutants are readily sorbed by the organic matter in the sediments. The tendency of a given organic chemical to get sorbed (i.e. the distribution coefficient of the chemical) on organic matter can be estimated from the distribution coefficient of the chemical vis-á-vis in an apolar liquid like octanol. If $C_w$ and $C_o$ are the concentrations of a given chemical

in the water and octanol phases, the distribution coefficient, $K_{ow} = C_o/C_w$. The coefficient, $K_{ow}$, is highly correlated with the distribution coefficient, $K_{oc}$, between organic carbon and water. Karickhoff (1981) reported $K_{ow}$ and $K_{oc}$ for a large number of organic micropollutants. He suggested the following simple relationship between $K_{oc}$ and $K_{ow}$, for these chemicals:

$$\log K_{oc} = \log K_{ow} - 0.35 \tag{8.16}$$

The $\log K_{oc}$ refers to partitioning between water and 100% organic carbon phase. The actual distribution coefficient for soil or sediment, $K'_d$, is obtained by multiplying $K_{oc}$ with the organic carbon fraction ($f_{oc}$) in the soil or sediment.

$$K'_d = K_{oc} f_{oc} \tag{8.17}$$

Table 8.11 gives the $\log K_{ow}$ and $\log K_{oc}$ values for a number of organic chemicals (source: Karickhoff, 1981).

The retardation of pesticides in soil water is studied in terms of the following parameters:
-   solubility (mg l$^{-1}$ or ppm),
-   coefficient of absorption on organic carbon ($K_{oc}$),
-   retardation factor, $R_f$, which varies from 0.00 (least mobile) to 1.00 (most mobile),
-   $K'_d$, coefficient of distribution.

The attenuation capacities of soils relative to various pesticides are estimated from laboratory investigations involving molecular diffusion or dispersion in soils, or leaching by percolation through synthetic soil samples or by chromatographic

Table 8.11. Partition coefficients for octanol – water ($K_{ow}$) and organic carbon – water ($K_{oc}$) (Karickhoff 1981).

| Compound | $\log (K_{ow})$ | $\log (K_{oc})$ |
|---|---|---|
| *Hydrocarbon and chlorinated hydrocarbons* | | |
| Tetracene | 5.90 | 5.81 |
| Anthracene | 4.54 | 4.20 |
| Napthalene | 3.36 | 2.94 |
| Benzene | 2.11 | 1.78 |
| 1,2 –dichloroethane | 1.45 | 1.51 |
| Tetrachloroethylene | 2.53 | 2.56 |
| ◊ (HCH (Lindane) | 3.72 | 3.30 |
| Methoxychlor | 5.08 | 4.90 |
| *Chloro-s-trazines* | | |
| Atrazine | 2.33 | 2.33 |
| Simazine | 2.16 | 2.13 |
| Propazine | 2.94 | 2.56 |
| *Carbamates* | | |
| Carbaryl | 2.81 | 2.36 |
| Carboturan | 2.07 | 1.46 |
| Chlorpropham | 3.06 | 2.77 |

methods. As the laboratory tests do not duplicate the conditions occurring in nature, the retardation factor, $R_f$, computed from laboratory data, should not be considered as sacrosanct, but only as indicative of the order of magnitude.

It can happen that some pesticides characterized by high solubility, may be immobilized because of their high affinity for absorption on organic carbon. For instance, Paraquat which is highly soluble, has virtually zero $R_f$, because of its high log $K_{oc}$ value (4.30). On the other hand, Chloroamben which has high solubility (700 mg l$^{-1}$) and low log $K_{oc}$ (1.11) is highly mobile.

Maximum concentrations of pesticides are found close to the surface, rapidly decreasing with depth due to absorption and degradation in the soil. The depth of penetration does not usually exceed 0.5 m, though there is evidence for the presence of some pesticides (e.g. carboform) at depths of two *m*.

The following guidelines may be kept in mind in estimating the contamination potential of pesticides (Calvet 1978):

1. $K_d$ values determined on laboratory soil columns may not be applicable to field cases,
2. $K_d$ would lead to proper estimates only in cases for which adsorption isotherms data are available.
3. There is a strong correlation between $K_d$ and organic content, but the correlation of $K_d$ with clay content of the soil is weak.

*Example 8.3 (adapted after Appelo & Postma 1996, 344)*
Calculate the retardation of methoxychlor with respect to groundwater flow in a sediment which contains 0.2% (0.002 fraction) organic matter.

From Table 8.11, the distribution coefficient $K_{oc}$ for methoxychlor is read as $10^{4.90}$

$$\log K'_d \text{(for methoxychlor)} = log\ K_{oc} + \log \text{ organic matter fraction}$$
$$= 4.90 + \log 0.002 = 4.90 - 2.70 = 2.20$$
$K'_d$ is antilog of 2.20 = 158 ml g$^{-1}$

$K_d = K'_d \cdot \rho_b/\varepsilon$ where $\rho_b$ is bulk density of the sediment, i.e. the density with air-filled pores, and $\varepsilon$ is porosity; $\rho_b/\varepsilon$ is generally taken as 6 kg l$^{-1}$ or 6 g ml$^{-1}$ = 158 × 6 = 948.

Thus, methoxychlor would have a velocity about 948 times less than water velocity.

### 8.7.5 *Sorption isotherms in relation to retardation*

The Retardation factor, $R$, is given by the following equation (Appelo & Postma 1996, 341):

$$R = 1 + K'_d \cdot \rho_b/\varepsilon \tag{8.18}$$

The simple linear relation used in the retardation equation earlier, may not yield correct results in all cases. The plot of absorbed versus solute concentration may turn out to be not a straight line but a curved line. The relationships between the solute and sorbed concentration, applicable to a single temperature, are called '*sorption isotherms*'. The isotherms that are widely used are those of Freundlich and Langmuir.

> *Freundlich* isotherm: $s = K_F C^n$
> *Langmuir* isotherm:   $s = S_m K_L C/(1+ K_L C)$

(*s* is absorbed concentration on solid, and *C* is concentration in water)

In the case of linear absorption, *q* (sorbed concentration on soil) is proportional to *C* (solute concentration), $q = K C$. In the case of Freundlich absorption, $q = K C^n$, where $0 < n < 1$. Unlike the linear isotherm and the Freundlich isotherm which show no maximum on adsorption sites, the Langmuir formula has a maximum, $S_m$.

### 8.7.6   *Macrodispersivity*

Particles of a chemical introduced into stagnant water, are spread by *diffusion*. On the other hand, particles of a chemical introduced into flowing water of an aquifer, undergo *dispersion*. Longitudinal or mechanical dispersion (symbolized by $D_L$) is far more dominant than transversal dispersion ($D_T$).

Both diffusion and dispersion are governed by Fick's law:

$$F = -D \cdot (\partial C/\partial x) \tag{8.19}$$

where $F$ = flux (in mol s$^{-1}$ m$^{-2}$), $D$ = Diffusion coefficient (in m$^2$ s$^{-1}$), and $C$ is concentration (say, in mol m$^{-3}$).

Field studies using tracers show that the depth integrated dispersivities in aquifers are much larger than those measured by column experiments in the laboratory. More importantly, dispersivity has been found to increase with distance. The longitudinal dispersivity ($\alpha_L$) is about 10% of the traveled distance, *x*. Substituting $\alpha_L = 0.1 x$, we get

$$\sigma^2 = 2 \alpha_L x = 2 \cdot 0.1 x \cdot x = 0.2 x^2 \text{ or } \sigma = \sqrt{0.2} \cdot x \tag{8.20}$$

where $\sigma^2$ is the variance of the pollutant front in an aquifer.

*Example 8.4   (adapted after Nkedi-Kizza et al 1983, and Appelo & Postma 1996, 349)*

Diuron is a herbicide whose dose of application is 1.8 kg ha$^{-1}$, and which is characterized by a long time scale of persistence (33-125 weeks) (source: Hassal 1987, p. 253). Nkedi-Kizza et al. (1983) showed that the sorption of diuron follows a Freundlich isotherm, $s = 6.0$, where *s* (sorption) is in µg g$^{-1}$ soil, and *C* (concentration) in µg ml$^{-1}$. Calculate the sorption of diuron on sand, given the following: $C = 10$ µg ml$^{-1}$; $\rho_b/\varepsilon = 6$ g ml$^{-1}$; organic content of the sand is 0.85% (i.e. 0.0085 fraction).

$$s = 6.0 \times 10^{0.67} = 6.0 \times 6.47 = 28 \text{ µg g}^{-1}$$

$q$ = absorbed concentration in mg $l^{-1}$ of pore water = $s$. $\rho_b/\varepsilon$ = 28 × 6 = 168 µg $ml^{-1}$

$$K_d = q/C = 168/10 = \sim 17$$

Retardation factor of diuron $(R) = 1 + K_d = 1 + 17 = 18$

$K_{oc}$ (distribution coefficient for diuron in the soil containing organic carbon) = $K_F/f_{oc} = 6.0/0.0085 = 706 \, l \, kg^{-1}$.

*Example 8.5 (adapted after Appelo & Postma 1996, 365)*
An aquifer has thickness $(D)$: 40 m; precipitation surplus entering the aquifer at the upper reaches $(P)$: 0.3 m $yr^{-1}$; porosity fraction $(\varepsilon)$: 0.3. A point source pollutant infiltrated the aquifer 2 km from the divide, after 2 km flow. Calculate:
1. The area which contains 60% of the pollutant, and
2. Estimate the time during which 60% of the pollutant mass with the highest concentration would pass, at 4 km from the divide (i.e. 2 km from the divide + 2 km. flow).

$$\sigma^2 = 0.2. \, x^2 \text{ or } \sigma = \sqrt{0.2}. \, x$$

where $\sigma^2$ = variance in the dispersivity parameters; and $x$ = distance of infiltration from the divide = 2000 m.

$$\sigma = \sqrt{0.2}. \, 2000 = 0.447. \, 2000 = 894 \text{ m}$$

Hence the dispersal of the pollutant will be between (4000 + 894) = 4894 and (4000 − 894) = 3106 m, from the divide.

Time period during which the highest concentration passes is the period between the arrival time of $(x + \sigma)$ and $(x - \sigma)$ at 4000 m.

$$x_1 = 4000/(1 + \sqrt{0.2}.) = 4000/1.447 = 2764 \text{ m}$$

The number of years in which the distance of 2764 m will be reached = thickness of the aquifer (40 m) In 2764/2000 = 40 In 1.382 = ~ 13 yr.

$$x_2 = 4000/(1 - \sqrt{0.2}.) = 4000/0.553 = 7233 \text{ m}$$

The number of years in which the distance of 7233 m will be reached = thickness of the aquifer (40 m) In 7233/2000 = 40 In 3.616 = ~ 51 yr.

Hence, the time period for 60% of the pollutant mass with highest concentrations, to pass the 4 km point is 51 − 13 = ~ 38 yrs.

*Example 8.6 (adapted after Johnson et al. 1989, and Appelo & Postma 1996, 365)*
A landfill is sealed with a clay barrier. Estimate the potential of the benzene in the leachate from the landfill to contaminate the drinking water resources, on the basis of the steady state of flux of benzene in waste site, given the following: size of the waste site 50 m × 40 m; thickness of the clay liner: 1 m; porosity of clay $(\varepsilon)$: 0.5; concentration of benzene in the waste site: 1 g $l^{-1}$; concentration of benzene

in the groundwater flow below the liner: 0.01 g $l^{-1}$; free water diffusion coefficient of benzene ($D_f$): 7. $10^{-6}$ cm$^2$ s$^{-1}$.

Effective diffusion coefficient ($D_e$) of benzene in water = $D_f$. $\varepsilon$ = Fick's Diffusion constant × porosity = 7. $10^{-6}$ cm$^2$ s$^{-1}$ × 0.5 = 3.5. $10^{-6}$ cm$^2$ s$^{-1}$

$$\text{Flux } (F) = -D_e.\varepsilon. (\partial C/\partial x) = -3.5. \ 10^{-6} \times 0.5 \times (1 - 0.01)/100$$
$$= \sim 1.8 \times 10^{-8} \text{ mg cm}^2 \text{ s}^{-1}$$

For one year (3.156 × $10^7$ s) and areal extent of the landfill (2000 m$^2$ = 2 ×$10^7$ cm$^2$), the flux would be (1.8 × $10^{-8}$)×(3.156 × $10^7$) × (2 ×$10^7$) = 11.4 × $10^6$ mg = 11.4 kg.

According to US EPA Drinking water standards, the Maximum Contaminant Level (MCL) allowable for benzene is 0.005 mg $l^{-1}$. Thus 11.4 kg of benzene has the potential to contaminate about 2.3 M m$^3$ of water resources.

### 8.7.7 *Movement of contaminant solutes in groundwater*

The chemistry of groundwater is affected by a large number of factors, such as, physiographical, geological, physico-chemical, physical, biological, anthropogenic, etc. The processes affecting the migration of solutes in an aquifer are summarized in Table 8.12 (after Schvarcev 1983, quoted by Sytchev 1988, p. 85)

The movement of the groundwater is generally slower and less turbulent than that of the surface water. Though chemical contamination of groundwater may sometimes be caused by natural sources (e.g. calcium-rich waters in a limestone country), most of the contamination occurs due to substances produced and introduced by man. These include synthetic chemicals (such as, solvents, pesticides and hydrocarbons), landfill leachates (i.e. liquids that have leached landfill dumps, and carry dissolved substances from them), organic wastes (e.g. bacteria and viruses from night soil dumps, hospital garbage), etc.

The issues concerned with the movement of contaminants in the groundwater have been dealt with by Petts & Eduljee (1994, p. 159), Reichard et al. (1990) and Devinny et al. (1990).

Table 8.12. Migration of solutes in an aquifer system.

| Action | Main process |
|---|---|
| Transfer of matter | Molecular diffusion; Convective diffusion mass transfer |
| Removal of solutes from groundwater | Hydrolysis; Sorption and ion exchange; Precipitation of compounds with low solubility; Filtration of coarsely dispersed matter; Bioaccumulation |
| Mobilization of material from rock matrix | Solution and leaching; Desorption |
| Combination of removal and mobilization processes | Formation of complexes; Oxidation – reduction and biogeochemical reactions; Radioactive decay |
| Interaction with water molecules | Hydration and dehydration of material; Underground evaporation and freezing; Dilution and concentration; Membrane effects |

When contaminated water moves through the soil column in the unsaturated zone, the contaminants present in the water tend to be removed through processes such as anaerobic deposition, filtration, ion exchange, adsorption, etc. On the other hand, the content of dissolved solids may increase as a consequence of the reaction between soil and water. The velocity of movement of the contaminant through and into the groundwater depends upon the physical and chemical characteristics of the soil, the nature of the contaminants and the flow system of the zone. It may be highly variable – ranging from nearly instantaneous to hundreds of years. If the aquifer concerned is composed of coarse-grained material like gravels, water may travel quickly, with residence times of a few days or weeks, before it is abstracted or reaches a river. On the other hand, if the aquifer is composed of fine-grained material such as chalk, water will move slowly. Consequently, water may reside in such aquifers for years, or even centuries, before it is abstracted or discharged.

The manner of spreading of a pollutant in an aquifer depends upon the nature of the pollutant and the aquifer. If the pollutant concerned is soluble (e.g. chloride or sulfate salts), it will get dissolved in the groundwater, and will move as a plume along the same path and with the same velocity as groundwater. On the other hand, if the pollutant is insoluble or poorly soluble (e.g. chlorinated solvents), its behavior will be markedly different from that of the soluble pollutant. The insoluble pollutant will tend to remain as a separate phase, and 'may sink below the water table, and flow separately along low permeability layers encountered at depth in the aquifer, or be trapped by capillary forces and act as a long-term contaminant source' (Petts & Eduljee 1994, p. 159).

Lyalko & Sakhatskii 1985, and Verigin 1979, quoted by Sytchev (1988, p. 86) generated the equation of mass transfer which can be made use of to model the leaching processes by groundwater (Fig. 8.9).

In order to model the sources of groundwater pollution and to recommend remedial action, we need to know the spatial, chemical, and physical characteristics of the source, and its temporal behavior. The spatial characteristics include information on the location, depth, areal extent, and whether it is a point source (e.g. septic tank), or a line source (e.g. sewage channel). The model could be used to predict the behavior of the contaminant plume (Fig. 8.10a, b, c and d). Models are increasingly being used to trace the source(s) of the existing pollution, by the technique of backward pathline tracking. Van der Heijde (1992) gave some case histories in this regard.

Appelo & Postma (1996) gave a detailed account of how to use the geochemical model, PHREEQE/PHREEQM, developed by Parkhurst et al. (1980), to find solutions to the problems that are commonly encountered, such as, Nitrate reduction in groundwater from agricultural land by organic matter, Recovery of freshwater injected in a brackish aquifer for storage purposes, Acidification of groundwater, Salinization of freshwater by seawater incursion, etc. Copies of the above program can be purchased from: US Geological Survey,

Books and Open-file Reports Section, Box 25425, Federal Center, Denver, Colorado, 80225-0425, USA.

(Readers interested in the derivation of the various equations, details of example calculations and modeling, used in this section, are referred to Chapter 9 of Appelo & Postma 1996).

In a general case, convective-diffusion transfer, accompanied by dispersion and rock-water interactions is described by the equation of mass transfer.

$$\Delta(DC) - \mathrm{div}(VC) = \frac{\partial(NC)}{\partial t} - \frac{\partial N}{\partial t} + W$$

Where   $\Delta = \dfrac{\partial^2}{\partial x^2} + \dfrac{\partial^2}{\partial y^2} + \dfrac{\partial^2}{\partial z^2}$ is the Laplace operator;

$\mathrm{div} = \dfrac{\partial}{\partial x} + \dfrac{\partial}{\partial y} + \dfrac{\partial}{\partial z}$ is the divergence operator;

$C$   is the mass concentration of substances
    in unit volume of fluid                                              $[\mathrm{kg} \cdot \mathrm{m}^{-3}]$;
$V$   is percolation rate                                                $[\mathrm{m} \cdot \mathrm{day}^{-1}]$;
$N$   is the concentration of matter in solid phase                      $[\mathrm{kg} \cdot \mathrm{m}^{-3}]$;
$n$   is effective porosity in fractions of unity;
$W$   is removal (input) of the studied component through the
    bottom and top of the aquifer                                        $[\mathrm{kg} \cdot \mathrm{m}^{-3} \cdot \mathrm{day}^{-1}]$;
$D$   is the generalized coefficient of dispersion                       $[\mathrm{m}^2 \cdot \mathrm{day}^{-1}]$;
$x,y,z$ are Cartesian coordinates;                                       $[\mathrm{m}^2 \cdot \mathrm{day}^{-1}]$;
$t$   is time in days

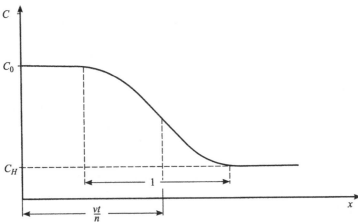

Groundwater dissolved solids content $C$ $[\mathrm{kg} \cdot \mathrm{m}^{-3}]$ variation along flow path x[m]; $C_0$ – initial concentration for time $t = 0$; $C_H$ – final concentration, $n$ – effective porosity, $V$ – percolation rate $[\mathrm{m} \cdot \mathrm{day}^{-1}]$

Figure 8.9. Modeling the mass transfer of groundwater and contained solute (source: Sytchev 1988, p. 85).

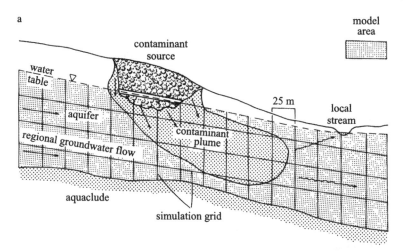

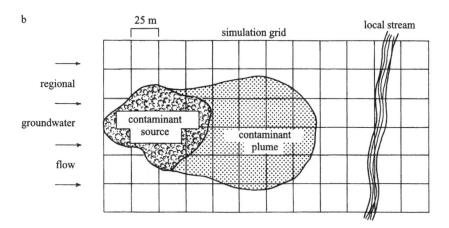

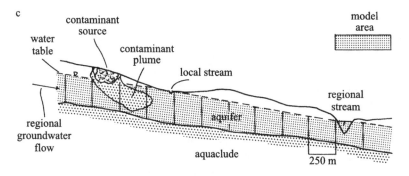

Figure 8.10. Groundwater model showing the disposition of the contaminant source and resulting plume in different perspectives (source: Van der Heijde 1992). (a) Vertical cross-section showing how a contaminant source initiates a contaminant plume, (b) Disposition in plan of the contaminant plume in relation to contaminant source, (c) Spatial relationship of the contaminant source and contaminant plume in relation to water table, regional groundwater flow, and aquiclude.

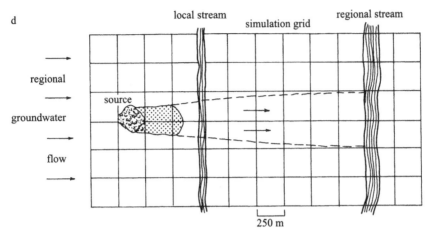

d

local stream
simulation grid
regional stream

regional

groundwater

flow

source

250 m

Figure 8.10. Groundwater model showing the disposition of the contaminant source and resulting plume in different perspectives (source: Van der Heijde 1992). (d) Contaminant plume in relation to local and regional streams.

## 8.8   FLUSHING AND AQUIFER CLEANUP

The clean-up of the an aquifer is technically complex, extremely costly and takes a long time to accomplish. But there might be circumstances where it may have to be undertaken nevertheless because of the present-day or future hazards (such as, pollution of drinking water wells, and seepage of the polluted groundwater into fresh surface water). An aquifer may have to be flushed many times (typically 5-10 times) in order to remove both solute and adsorbed pollutant.

Heavy metals and hydrophobic organic chemicals are flushed out from the aquifers by geochemical methods. Hydraulic or bioremediation methods are needed for cleansing of immiscible fluids, such as petroleum products (see, for instance, Abdul 1990; Karickhoff 1984; Weber & Miller 1988).

The critical step involved is the desorption of the chemical bound to the solid aquifer material. The greater the desorption, the more effective will be the flushing. The Freundlich absorption isotherm (referred to in Section 8.7.5) can be made use of to estimate the pore volumes needed for flushing.

In the Freundlich equation, $q = K_F C^n$, where exponent $n$ is less than 1.

A simple relationship has been found to exist between the slope of the desorption isotherm at concentration, $C_i$, and the number of pore volumes that are needed to flush the column in order to reach that concentration. The mass balance equation would therefore be: $\delta q = \delta C . V^*$, where $\delta q$ is the change in the sorbed amount that is transported as $\delta C$ in solution and $V^*$ is the flushing factor.

Experience has shown that 'mild' geochemical treatments give good long-term results.

For instance, an aquifer that has been accidentally contaminated by Cd Cl$_2$ spill, has been flushed with water with which common salt (Na Cl) has been mixed. When harsh techniques such as acid leaching are employed, flora and fauna have been found to have higher concentrations of pollutants, relative to untreated soils.

*Example 8.7 (adapted after Mehran et al. 1987, and Appelo & Postma 1996, 453)*
An aquifer has been contaminated by TCE (trichloroethylene), with the consequence that the concentration of TCE in groundwater rose to 10 mg $l^{-1}$. How many pore volumes need to be flushed through the aquifer to bring down the TCE concentration to 1 mg $l^{-1}$.

According to the Freundlich sorption isotherm for TCE given by Mehran et al. 1987:

$$q = 6 + 6 \ln C$$
$$\delta q = \delta C. \, V^*$$

where $V^*$ is the flushing factor. Thus:

$$V^*_{(c=1)} = \delta q / \delta C = 6/1 = 6$$

Hence $1 + 6 = 7$ pore volumes need to be pumped to lower the TCE concentration to 1 mg $l^{-1}$.

## REFERENCES

\* Suggested reading

Abdul, A.S., Th.L. Gibson & D.N. Ral 1990. Use of humic acids solution to remove organic contaminants from hydrogeologic systems. *Environ. Sci. Technol.* 24: 328-333.

Appelo, C.A.J. et al. 1982. *Controls on groundwater quality in the NW Veluwe catchment* (in Dutch). Soil Protection Series 11, Staatssuitgeverji, Den Haag 140 pp.

Appelo, C.A.J. 1985. CAC, computer aided chemistry, or the evaluation o groundwater quality with a geochemical computer model (in Dutch). *H₂O* 26: 557-562.

\*Appelo, C.A.J. & D. Postma 1996. *Geochemistry, Groundwater and Pollution.* Rotterdam: A.A. Balkema.

Archer, A.A., G.W. Luttig & I.I. Snezhko (eds) 1987. *Man's dependence on the earth.* Nairobi – Paris: UNEP – UNESCO.

Aswathanarayana, U. 1995. *Geoenvironment: An Introduction.* Rotterdam: A.A. Balkema.

Bagchi, A. 1990. *Design, Construction and Monitoring of Sanitary Landfill.* New York: John Wiley.

Böttcher, J., O. Strebel & H.M. Duynisveld 1985. Vertikale Stoffkonzentrationsprofile im Grundwasser eines Lockergesteins -Aquifers und deren Interpretation (Beispiel Fuhrberger Feld). *Z. dt. geol. Ges.* 136: 543-552.

Bouwer, H. 1978. *Groundwater Hydrology.* New York: McGraw-Hill.

Calvet, R. 1978. *Special Report to IHP Working Group 8.3.* INRA Versailles, France.

Devinny, J.S., L.G. Everett, J.C.S. Lu & R.L. Stollar 1990. *Subsurface Migration of Hazardous Wastes.* New York: Van Nostrand Reinhold.

Farquhar, G.J. & F.A. Rovers 1973. Gas production during refuse decomposition. *Water, Air and Soil Pollution* 2: 483-495.

Freeze, R.A. & J.A. Cherry 1979. *Groundwater*. Englewood Cliffs, N.J., USA: Prentice-Hall.

Hassal, K.A. 1987. *The Chemistry of Pesticides*. ELBS Ed. Basingstoke, UK: Macmillan.

Johnson, R.L., J.A. Cherry & J.F. Pankow 1989. Diffusive contaminant transport in natural clay: a field example and implication for clay-lined waste disposal sites. *Environ. Sci. Technol.* 23: 340-349.

Karickhoff, S.W. 1981. Semi-empirical estimation of sorption of hydrophobic pollutants on natural sediments and soils. *Chemosphere* 10: 833-846.

Karickhoff, S.W. 1984. Organic pollutant sorption in aquatic systems. *J. Hydraulic. Eng.* 110: 707-735.

Klimentov, P.P. 1983. *General Hydrogeology*. Moscow: Mir Publishers.

*Laconte, P. & Y.Y. Haimes (eds) 1982. *Water Resources and Land-use Planning: A Systems Approach*. The Hague: Martinus Nijhoff Publishers.

Manahan, S.E. 1991. *Environmental Chemistry* (5th ed.), Chelsea, Michigan USA: Lewis Publishers.

Mehran, M., R.L. Olsen & B.M. Rector 1987. Distribution coefficient of trichloroethylene in soil-water systems. *Ground Water* 25: 275-282.

Nkedi-Kizza, P., P.S.C. Rao & J.W. Johnson 1983. Absorption of diuron and 2,4,5 T on soil particle-size separates. *J. Env. Qual.* 12: 195-197.

O'Neill, P. 1985. *Environmental Chemistry*. London: George Allen & Unwin.

Parkhurst, D.L., D.C. Thorstenson & L.N. Plummer 1980. *PHREEQE – A computer program for geochemical calculations*. US Geol. Surv. Water Resour. Inv. 80-96: 210 pp.

Petts, J. & G. Eduljee 1994. *Environmental Impact Assessment for waste treatment and disposal facilities*. Chichester: John Wiley.

Reichard, E., C. Cranor, R. Raucher & G. Zapponi 1990. *Groundwater Contamination Risk Assessment: A Guide to understanding and managing uncertainties*. Wallingford, UK: Int. Assn. Hydrol. Sci.

Salcedo, R.N., F.L. Cross & R.L. Chrismon 1989. *Environmental impacts of Hazardous waste treatment, Storage and Disposal facilities*. Lancaster, Penn., USA: Technomic Publ.

*Sytchev, K.I. 1988. *Water Management and Geoenvironment*. Paris – Nairobi: UNESCO – UNEP.

*Van der Heijde, P.K.M. 1992. Computer modeling in groundwater protection and remediation. In P. Melli & P. Zannetti (eds), *Environmental Modeling*: 1-21. Amsterdam: Elsevier.

Vartanyan, G.S. (ed.) 1989. *Mining and Geoenvironment*. Paris – Nairobi: UNESCO – UNEP.

Weber, W.J. & C.T. Miller 1988. Modeling the sorption of hydrophobic contaminants by aquifer materials. *Water Res.* 22: 457-474.

CHAPTER 9

# Economics and management of water supplies

## 9.1 INFORMATION MANAGEMENT FOR WATER RESOURCES ASSESSMENT

The International Conference on Water and the Environment (ICWE), Dublin, Ireland, 1992, stressed the need for reliable information about water resources. The *Expert Group meeting on Strategic Approaches to Freshwater Management* (Harare, Jan. 1998) made the following recommendation about the Information Management:

'There is need to finance, establish and maintain effective data collection and dissemination, information management system and research in order to provide a sound basis for policy formulation, planning and investment decisions and operational management of freshwater resources. The collection of all freshwater resource and related socio-economic and environmental data and information needed for policy decisions, planning and management action and monitoring, should have a high and continued priority'.

### 9.1.1 *Components of the Integrated Information Management System about water*

The Integrated Information Management System involves the following components (Paul 1998):
– Various dimensions of water: Surface and groundwater, quality and quantity, ability to support the ecosystem,
– Interaction between freshwater and other environmental systems, on land surface, atmosphere, coastal and offshore waters and terrestrial ecosystems, etc.
– Interrelationship between water and social and economic activity (irrigated agriculture, hydropower generation, navigation, fisheries, etc.)
– Institutions and legal instruments for the management of water.
The impetus for the integrated information management arises from the following considerations:
– Environmentally sustainable development of natural resources,
– Accountability to the public,
– Pressure on financial resources, requiring cost-benefit analysis of various options,

– Advances in information technology (computers, sensors, communication systems).

The Water Information System has to take into account the benefits, costs and risks to the 'stakeholders', i.e. contributors and users of water information. A stakeholder may be a family, a community, a company or a government department or corporation. A successful conclusion needs a consultatitive approach based on open communication and negotiation with all the concerned stakeholders.

Based on the procedure recommended by Paul (1998), the following scenario in respect of (say) farmers who would be using irrigation water from a canal, can be envisaged:

– Identification of all the stakeholder farmers,
– Definition of the purpose (irrigation) for which each stakeholder requires information,
– How much water, of what quality is needed and when,
– Identification of policy issues, such as cost-recovery requirements,
– Preparation of an inventory of existing information, and existing practices, and identification of gaps and overlaps in coverage,
– What the information should do, and how it should do it,
– Survey of successful approaches in place in other fields or in other countries,
– Development of options, and their cost-benefit and risks to the stakeholder,
– Finally, decision whether or not to proceed and if so, with what option.

### 9.1.2 *Pressure – State – Response model*

The role of Integrated Information Management could be visualized by applying the Pressure – State – Response model to (say) environmental monitoring.
– The stress or pressures under which a resource (say, water) is placed due to human activity,
– What is the present state of the resource (say, quantity and quality, in terms of various descriptors),
– How does the Society respond to the observed state of the resource (say, legislative, technical, regulatory, etc. means),
– Perception of the society about the resource (depending upon the society's values and priorities).

The model is schematically shown in Figure 9.1 (source: New Zealand Ministry for Environment, as quoted by Paul 1998, p.107).

### 9.1.3 *Decentralized and centralized network of databases*

In most countries, it is not uncommon for water issues to be handled by a variety of agencies, each handling a particular kind of information and database. Advances in information technology have made it possible to link them in such a manner that that every unit is aware what information is available with other units and can draw information from other units. This is analogous to CD-ROM infor-

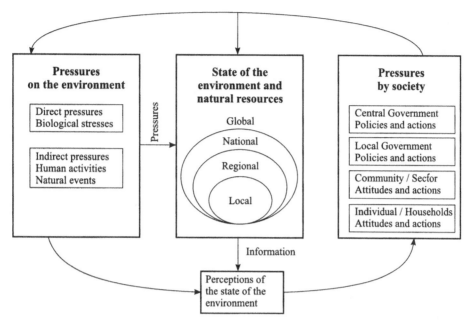

Figure 9.1. Pressure – State – Response model (source: Paul 1998, p. 107).

mation system that is maintained by libraries in a country. By punching the key words about the area in which a person is interested, the computer will show what books are available in that area, and which particular library (libraries) hold them. In the case of water resources, free (i.e. without paying money) and unrestricted downloading of desired information is not always possible, as security and proprietary considerations may be involved. Figure 9.2 shows the Network of decentralized databases (source: New Zealand Ministry for the Environment, quoted by Paul 1998. CRI in the figure stands for Crown Research Institutes, which collect environmental information in New Zealand).

The same set of databases as may be organized in different ways (Fig. 9.3: Network of databases with a central 'hub'; source: Paul 1998). In the centralized system, a 'hub' will receive information from individual agencies and will provide information to contributors and users as requested by them. The hub will store metadatabases of data summaries, rather than complete records.

## 9.2 WATER RESOURCE INVENTORIES

Before we develop models for the management of water resources, we need to know how much water is available in a given watershed, and its temporal and spatial distribution in various reservoirs. De Backer (1982) gave a good account of what he calls as pre-management water resources study.

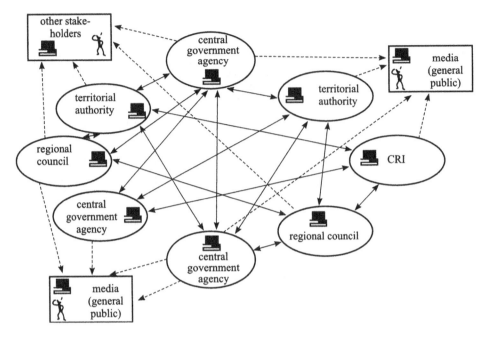

Figure 9.2. Network of decentralized databases (source: Paul 1998, p. 110).

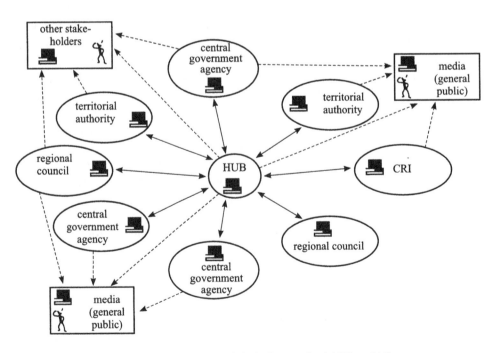

Figure 9.3. Network of databases with a central 'hub' (source: Paul 1998, p. 110).

### 9.2.1 *Water resource balance in a watershed*

The water resources balance equation for a watershed or a region may be written as follows:

$$S = (P - E) - (Q + W) \tag{9.1}$$

where $\Delta S$ is the water storage variation during a given time period, $\Delta t$, and $P$ is precipitation, $E$ is evapotranspiration, $Q$ is the total volume of discharge at the outlet of a watershed, and $W$ is the leakage between the aquifers (all expressed in water thickness per unit area).

When the study covers a longer period (i.e. the greater the $\Delta t$), the variations in water storage smoothen out (i.e. the less the $\Delta S$). Also, the leakage term may be ignored. Thus, the water resource equation would be concerned with three parameters, $P$, $E$ and $Q$. The actual situation is far more complex: large amounts of flood water may flow during a short time interval or water shortages may occur during the dry season. These problems may be addressed by groundwater recharge of the flood water, or supplementing the surface water with withdrawal from groundwater. Evidently, the water quality undergoes change along the path of the water cycle. In humid, tropical regions, the groundwater tends to be purer than surface water. In the arid regions, the groundwater tends to be salty, whereas the surface runoff water may be better.

The difference between the precipitation and evapotranspiration either runs off into streams and rivers as 'base flow', and/or infiltrates into the soil to be stored in the soil, subsoil or as groundwater. Thus, the total storage variation, $\Delta S$, during the time time interval, $\Delta t$, is the sum of the variations in the storage of the various reservoirs.

$$\Delta S = \Delta S_1 + \Delta S_2 + \Delta S_3 + \Delta S_4 + \Delta S_5 \tag{9.2}$$

where $\Delta S_1 =$ water storage variation for the *soil surface*, $\Delta S_2 =$ water storage variation for the *soil profile*, $\Delta S_3 =$ water storage variation for the *subsoil*, $\Delta S_4 =$ water storage variation for the *aquifer*, $\Delta S_5 =$ water storage variation for the *river bed*.

Figure 9.4 (source: De Backer 1982, p.58) gives the block diagram of the five reservoirs described above. Gravity drives the movement of water between the various reservoirs. It is necessary to measure the various parameters described above in order to be able to evaluate, control and forecast the resources of surface and groundwater in a given area, and also to forecast potential floods and droughts. But the tricky job is to determine the periodicity of measurement of a given parameter. If the time unit chosen is too small, too many details gathered that may be irrelevant. On the other hand, if the time unit is too long, some short-period phenomena (e.g. floods) may be obscured by averaging. So a convenient periodicity of measurement of a given parameter is chosen.

The time periods, $\Delta t$, between the maxima and minima for each reservoir, are of the order of a few months. A given reservoir tends to reach its maxima in about a month.

In general, the ground filling phenomena $\Delta t_f$ takes about 3 months, and the ground emptying phenomena $\Delta t_E$ takes about six months.

The $\Delta t$ and fluxes (inputs and outputs) of the five reservoirs are summarized as follows (Table 9.1)

The water storage capacity of a reservoir which is the volume of water that can be held in a unit volume of the reservoir, depends not only on the geometric dimensions of the reservoir, but also on the hydrodynamical properties and conditions. Take for instance, the groundwater reservoir, ($S_4$). A rise in the water table does not change its water storage capacity, $C\,S_4$, but its total water storage. An in-

Table 9.1. Reservoirs and their time-frames and fluxes (source: De Backer 1982).

| Reservoir and $\Delta t$ | Fluxes |
|---|---|
| Soil Surface ($S_1$) $\Delta t_1 = 4$ months | The input is precipitation ($P$), and the outputs are evaporation ($E$), infiltration ($I$) and rapid runoff ($R$) |
| Soil profile ($S_2$) $\Delta t_2 = 5$ months | The input is precipitation ($I$) and the outputs are evaporation ($E$), drainage ($D$) and slow runoff ($L$) |
| Subsoil ($S_3$) $\Delta t_3 = 6$ months | The input is drainage ($D$) and the output is groundwater supply ($A$) |
| Aquifer ($S_4$) $\Delta t_4 =$ over 8 months | The input is groundwater supply ($A$), and the outputs are baseflow ($B$) and possible leakage ($W$) |
| River bed ($S_5$) | The inputs are rapid runoff ($R$), slow runoff ($L$), and base flow ($B$), and the output is the total volume discharge ($Q$) |

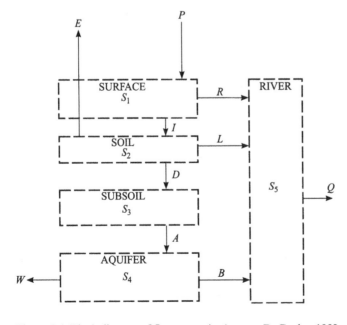

Figure 9.4. Block diagram of five reservoirs (source: De Backer 1982, p. 58).

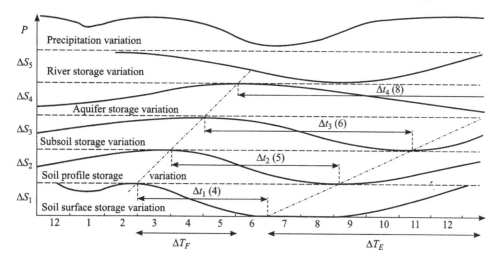

Figure 9.5. Temporal variation of water storage in different reservoirs (source: De Backer, 1982, p 63).

crease in $C\,S_4$ will either increase the groundwater leaks in the form of springs, or base flow in the river. The storage capacities can be derived either from flow equations, or empirically through observations, preferably a combination of both.

The temporal variation in precipitation, and the variation in storage (maxima and minima) in the five reservoirs are shown schematically in Figure 9.5, in a yearly cycle of 12 months, with $\Delta t_f$ of about three months, and $\Delta t_E$ of about six months (source: De Backer 1982, p. 63). Depending upon when the aquifer/river storage reach their maxima, decision may be taken when to pump water from the aquifer or from the river.

The time variation in the storage capacity ($CS$) of a reservoir, is given by the following equation:

$$\frac{CS_{ij}}{t} = \frac{\Delta(L_{ij}S_{ij})}{L_{imx}\Delta t} + \frac{\Delta f_i}{\Delta t} \tag{9.3}$$

where, $C$ is the storage capacity of the reservoir, $L$ is the geometric dimensions of the reservoir, $S$ is the degree of saturation and $V$ is the flux. The indices $i, j$ identify the reservoir and the considered time interval, and $mx$ refers to the highest value reached during the time interval of measurement.

## 9.3 ECONOMIC VALUE OF WATER

### 9.3.1 *General principles*

The economic value of water – whether water is a 'free good' or 'economic good' – has been the subject of vigorous debate during the last three centuries (see the excellent review of Janusz 1998). The so-called *Paradox of Value* refers to the fact that water which is so essential to life should be so cheap whereas diamonds which are not all needed, should be so expensive !

The International Conference on Water and Environment, Dublin 1992, enunciated the following guiding Principles regarding the economic value of water, which are eminently reasonable: 'Water has an economic value in all its competing uses and should be recognized as an economic good.... Within this principle, it is vital to recognize first the basic right of all human beings to have access to clean water and sanitation at an *affordable price*... Managing water as an economic good is an important way of achieving an efficient and equitable use, and of encouraging conservation and protection of water resources..'

There are many ways of estimating the value of water:

– *Utility*: the economic benefit yielded by the use of a given amount of water with specific characteristics, and the time and location of its availability – for instance, the value of the additional crop production that has been achieved by the use of (say) 1000 m$^3$ of irrigation water over a period of four months,
– *Exchange*: the quantity/value of some other commodity for which a given amount of water can be exchanged – a vendor may sell 100 l of water in exchange for certain quantity of fuelwood or cash,
– *Scarcity and marginal utility:* a commodity is said to be scarce if people want it, and the quantity available is limited. 'The more scarce the water is, the higher will be its economic value, and the more it will be economized' (Janusz 1998, p. 409). In desert lands, water is scarce. For this reason, people will use it for the most important purpose, such as drinking and preparation of food. In humid regions where water is not that scarce, there is no economic compulsion to choose the best use. Only scarce things have marginal utility and therefore economic value. If people would be willing to pay for water, then it is scarce. If water is freely available in any quantity, then it has zero marginal utility and no economic value.

Experience has shown that the willingness of people to pay for water is very high for some selected uses, such as for drinking and preparation of food. As prices go up, they tend to economize on less essential uses, such as washing and gardening. In one case of a municipality in USA, when water prices went up by 10%, the consumption went down by 3% only.

### 9.3.2   *Water transfers*

Water as an input may be transformed into a variety of outputs. As water becomes more and more scarce, the value of the possible outputs will have to be taken into account in allocating the water to the competing uses. In USA, it has been estimated that one acre-foot of water typically yields only about USD 400 on the farm versus USD 400,000 in manufacturing. In such a situation, it is inevitable there would be transfer of water from lower economic returns (e.g. agricultural) to higher economic returns (municipal and industrial). In countries, where the State owns the water rights (as for instance, in India), the State can allocate the water by an executive order. The position is complicated in countries like USA, where the water rights are held by the landowners (e.g. farmers). In such a situation, water transfers should be based on three principles: voluntarism, infrastructure and third party interests (Bouwer 1994).

Voluntarism means that the State cannot confiscate the water resources belonging to a farmer. The farmer should be induced to transfer his water. If the farmer finds that selling the water rights is financially more advantageous to him than using the water for irrigation, he will voluntarily sell the groundwater under his farm (this is a common practice in California – the district of Devils Den sold its water rights to Santa Clarita, a suburb of Los Angeles about ten years ago, and the process continues). The availability of infrastructure (e.g. conveyance systems) would ensure that water transfer is cost-effective.

Third party interests can be quite complicated. When a farmer sells his water rights, instead of farming, jobs based on agriculture would be lost. There may be serious environmental implications. A city may draw water from a stream and return the water to the stream in the form of sewage effluent. A farmer who would be using the stream water in the downstream side may find that he cannot use the water for irrigation either because it contains too much nitrogen or it contains pathogens. The presence of the pathogens would also adversely affect the recreational use of the stream, or the use of water for fisheries, and so on. A good approach would be for all the concerned users of stream water to sit together, and decide upon the water transfer which is fair to all concerned.

An innovative system of water transfer is water banks that have been set up in California. Just as we put money in the bank, and the bank lends it to a borrower on interest, farmers who have water to spare will place their water at the disposal of the bank. The bank will then sell the water to the needed persons/ communities/industries, and pay the farmer. During the California drought of 1987-1993, the water bank was handling water at the rate of about a billion $m^3$ per year.

### 9.3.3  Total Economic Value of water

The concept of Total Economic Value (TEV), developed by Rogers, Bhatia & Huber (1997), is based on the premise that economic analysis should include all aspects of water uses whether or not they are reflected in the market. Direct use refers to the contribution that water makes to production of goods and services (e.g. crop production due to irrigation). Indirect uses refer to value of benefits such as return flows in irrigation. Societal values take into account issues such as poverty alleviation and food security. Non-use values refer to pleasure derived from amenities, such as 'water view' (scenic beauty). Janusz (1998, 415) gave a schematic depiction of the categories of economic values of water.

Water is a mobile resource which an be used and reused many times (e.g. municipal water supply $\rightarrow$ waste water $\rightarrow$ bio-ponds $\rightarrow$ irrigation water $\rightarrow$ groundwater recharge → withdrawal from wells, etc.). The kind of changes that water undergoes in terms of its quality, quantity, location and timing because of a given use, would determine its value for the succeeding uses.

In the market economies of the Industrialized countries, the economic value of water is determined by the market. But the market prices do not reflect the true worth of water, as they do not take into account social and environmental factors and as they may get distorted due to income disparities, subsidies, etc. On the other hand, a water-supply system which ignores the market in deciding about water prices, will not be sustainable. Most public institutions use three sets of prices: market prices (based on demand and supply), administered prices (decided upon by the government for specific purposes, on socio-economic and political grounds, such as domestic water supply or irrigation) and accounting prices ('shadow' prices that reflect the true economic value of water).

There are different levels of considering the value of water. An individual user (e.g. householder/farmer/industry) would look at water from a narrow perspective of their particular net revenues or satisfaction. A regional perspective would addition-ally take into account some aspects which are not take into consideration by individ-ual user, such as employment generation, new economic activities, poverty allevia-tion, improved sanitation and health of the community, tourism, etc. A national perspective would take a broader view of integrating water resources planning in the context of national income, national productivity, etc. Conflicts can arise between these perspectives, because of NIMBY (Not In My Back Yard) approach. A farmer knows that canals need to be dug to distribute water from the reservoir, but he wants the canal to be located in somebody else's farm. Similarly, the beneficial effects of water use may accrue to users in one part of the region (e.g. irrigated agriculture), whereas another part of the region is exposed to the detrimental effects (e.g. drainage water). Besides, in order to protect the natural water system, the State may have to enforce some limits (such, as maintaining minimum streamflow rates). The various conflicts have to be reconciled in a fair, equitable and ecologically-sound manner.

Based on the technique of decomposition of linear programs developed by Dantzig (1963), Stephenson (1989) developed a layered model for use in the de-velopment of water resources in a country at different levels (national, departmen-tal, river basin and project levels). 'Each successively lower model is optimized on its own, but uses shadow values on output imposed by the successively higher models. The lower models in turn feed back optimal plans and technical aspects, such as output, to the higher model which is further able to refine planning' (Ste-phenson 1989, p. 63)

### 9.3.4   *Estimates of water value for different uses*

*Environment and Ecosystem*

The complexity and importance of impact of water use on environment and eco-systems is increasing steadily. As Global Water Partnership puts it, 'Until realistic values can be placed on ecological services, economic planning techniques will ignore or marginalize the role of the environment as provider and user of water resources'. For instance, the economic value of a tree is not confined to the market value of its timber – it should include the cost of growing such a tree, the environmental benefit that the tree has been providing (e.g. absorbing carbon dioxide, protecting against erosion, habitat for birds, biodiversity) and the adverse effects that would arise when the tree is felled.

Turner & Adger (1996) developed innovative methods of economic valuation of non-market issues involved in the assessment of economic value of water (Table 9.2):

*Crop irrigation*
Roughly half the water used in irrigation is lost through evaporation and transpiration. Whereas evaporation from the surface is a waste, transpiration by crop plants is necessary and productive (techniques of moisture conservation have been described in Section 6.8). The value of water depends upon the climate, soil, crop grown (high-valued vegetables and fruits versus low-valued crops and forages), stage of growth of the crop and the techniques of application of water. The value of water may be estimated at the point of delivery to the farmer, or for the watershed as a whole.

*Municipal water use*
Water for municipal use is given the highest priority, as water is essential for life. Two aspects of municipal water use need special mention: the water has to be of the highest quality, and the quantities involved are small.

The criteria used for the estimation of the economic value of water for productive use are not applied in the case of municipal use of water.

*Industrial water use*
Water is used in the industry for cooling, transportation and washing, and as a solvent. In some industrial processes, water enters the composition of the finished

Table 9.2. Economic valuation of non-market issues (source: Turner & Adger 1996).

| Effects category | Valuation method option |
|---|---|
| Productivity | Market valuation via prices or surrogates; Preventive expenditure; Replacement cost/shadow projects/ cost-effective analysis; Defensive expenditure |
| Health | Human capital or cost of illness; Contingent valuation; Preventive expenditure; Defensive expenditure |
| Amenity (e.g. lake) | Contingent valuation/ranking; Travel cost; Hedonic property method |
| Existence values (eco-systems; cultural assets) | Contingent valuation |

product. The principal water uses are in thermal power production, chemical and petroleum plants, ferrous and non-ferrous metallurgy, wood pulp and paper industry, etc. Cooling accounts for about 80% of the water use. In some countries, withdrawals of water for industry are more than 50% of all withdrawals. In thermal power generation, only about 0.5-3% of the water intake is actually consumed, and recycling is the norm. But some industrial processes (e.g. food industries) consume 30-40% of the water intake.

The value of water for industrial use is estimated on the basis of the 'cost of alternative processes that will produce the same product while using less water' (Janusz 1998, p. 413). The costs of internal recycling of water are usually low. But it is generally more expensive to remove the toxic dissolved materials which the industrial wastewater may be carrying. The costs vary greatly depending upon the extent of degradation in water quality occurring in the process. In general, the costs of water supply and wastewater treatment for industry are usually about 2-3% of the production costs. The total water consumption in the industrial sector is stabilizing or even decreasing, as more and more industries are switching over to water-free or dry technologies.

### Waste assimilation

The economic value of water for waste assimilation is estimated on the basis of the alternative cost of treating the effluent. A given flow of water can only assimilate a certain quantity of effluent with certain characteristics (such as, TDS, SS, BOD, SAR, etc.) without exceeding the quality standards. The treatment costs needed to bring down the effluent characteristics in order to stay within the standards, gives a good estimate of the value of water.

### Navigation

The economic value of water for navigation is the difference between the costs of water transport vis-á-vis the lowest cost of alternative mode of transportation. It is known that transport of goods through river barges (in terms of USD per ton-km) is cheaper than other modes, such as rail or truck (towed barges plying over the Mississippi river in USA routinely carry large quantities of ores, various kinds of construction materials, scrap, etc.). If, however, a navigational facility has to be constructed anew, the value of water for navigation may be zero or even negative.

### Hydropower generation

The value of water for hydropower generation is the difference between the costs of hydropower (including transmission) and other lowest-cost means of power generation, say, coal-fired thermal power plants (including transmission).

### Recreation

The value of water for recreation depends upon a number of factors such as location, accessibility, scenic setting, water quality, etc. The valuation is done on the basis of the behavior of tourists by means of a survey – how much extra travel costs, admission price, etc. they are willing to bear to visit a particular kind of recreational facility.

### 9.3.5 *Water value in system context*

Since many combinations of use and reuse of water are possible simultaneously or in sequence, it is critically important to know how the various uses of water combine and interact in space and time. Consideration has therefore to be given to return flow and reuse. A given physical unit of water can be used over and over again if water is used fast and returns large quantities of clean or well-treated water. Even though the value addition for each use of water is small, a pattern of multiple use can generate large values. The location and timing of water use becomes important in the system context of water use. The system gains more value when the waste-generating users are located as far downstream as possible.

Wollman (1962) made a pioneering study of water value in the system context on the value of water in the San Juan and Rio Grande basins of New Mexico, USA. The demand of water far exceeded the supply for water use categories, such as, irrigation, municipal and industrial uses, recreation and others. The study evaluated the effects on the economy of New Mexico state as a consequence of the use of water in several different ways. A knowledge of the value addition through the use of an acre-foot (1 acre-ft = 1235 m$^3$) of water for alternative uses permitted informed decision about the directions in which the economy could move in order to make the best use of all its resources, as the scarcity of water increases.

For populist reasons, a government may take an ad hoc decision about the upstream use of water, and this may create serious problems and civil unrest in the country on the downstream side. Such conflicts could be avoided if informed decisions could be made on the basis of different of techno-socio-economic options that are available.

## 9.4 PRICE COORDINATION OF WATER SUPPLIES

Often, a given region may have a complex system of supplies and demands. Price coordination methods have to be developed to make the system techno-socio-economically efficient. Guariso et al. (1982) adapted the classical methods of price coordination to water resource systems. The objective of the exercise is to maximize the total regional net benefit. A disaggregated approach is used whereby 'the marginal benefit at each demand point is equal to the marginal cost of delivering water to that point'.

Traditionally, water use demands are projected in terms of fixed water requirements of the user (e.g. so many liters of drinking water per capita per day, so many ha cm of irrigation quality water for irrigating a crop, so many liters of water for refining one barrel of crude, etc.). Supplies of water are sought to be organized on the basis of the projected demand. But the picture is changing all over the world for two reasons. As population grows and industries and irrigated agriculture expands, several countries find that known supplies of good quality fresh water are grossly inadequate to provide all the requirements of the users. Water use has to be optimized, using pricing as a control. Also, research and development is leading to more efficient use of water, use of inferior quality of water and reuse and recycling of water. In other words, water demand cannot be assumed to be inelastic, but would depend in part on the availability and pricing of water of a particular quality.

As Zimmerman put it, '*Resources are not, they become*'. Inferior quality water, drainage water and waste water which were unusable earlier, can be used now. For instance, acid mine water can be treated for use in agriculture and for drinking. Brackish water can be used for irrigation with gypsum amendment, or made use of to grow salt-tolerant plants. It therefore follows that the supplies and demands of water should be treated in dynamical terms. Thus, instead of investing more money in developing new supplies of water, it is generally cost-effective to use the same money to develop methods of efficient use of water and recycling and reuse of water.

There are two approaches for price coordination – aggregated and disaggregated. In the aggregated approach, a large mathematical model is formulated to represent all the components of supply and demand for the whole region. The equations are solved to maximize the net benefit from water use for the entire system. When the supplies and demands are complex and are managed centrally, the aggregated approach would, of course, the proper one to use, but the model would be too cumbersome and too expensive in terms of computer time to be practicable.

In the disaggregated approach, supply and demand are treated independently. They are coordinated sequentially by a supervisor. In effect, the supervisor would be engaged in a dialogue with the various supply and demand entities. The process will go on until it converges to produce an optimal balance or equilibrium between supply and demand.

The coordination algorithm is aimed at maximizing the total regional net benefit, but instead of the usual practice of computing the total benefits and costs, the algorithm of Guariso et al. (1982) seeks to achieve maximization of benefit on the basis of marginal benefits and costs ('… if a certain flow is to be transferred from a supply to a demand, the cost of delivering the final unit of water, which is the marginal cost of this flow, must be equal to the benefit generated by this final unit, or the marginal benefit'). This is analogous to determining the equilibrium price in a market.

The application of the price coordination method of the market to water resources management suffers from a short-coming, namely, that water supplies and demands are not independent. For instance, the upstream use of water affects the use of water downstream, and extensive withdrawal of groundwater leads to the reduction of surface water, and so on. The model of Guariso et al. (1982) does take this matter into account, but still involves the making of some assumptions, such as, that water of requisite quality is available, that all flows are made available at the same time, and that all supplies have the same reliability, and so on.

### 9.4.1 *Principles of optimization*

A demand unit may be a farm, a factory or a residence or a community or a region. The benefit (*B*) accruing to a demand unit is a function of the amount, *Q*, of water used. Thus, $B = B(Q)$. Where the demand unit is, say, a farm, it is critically important to know the timing and reliability (say, 95%, 90%) of supply of the irrigation water during the critical weeks of the growing season. Some irrigation models take into account the total amount of water needed for the whole season, rather than particular flow levels during particular periods. The assumption of a constant average flow in the present algorithm may be consistent with the output of these models.

If the water is paid for at a price p, the profit of the unit is $[B(Q) - pQ]$. Assuming that the unit is profit maximizing, the amount of water concerned can be obtained by solving the following optimization problem:

$$\text{Max}_{Q}\left[B(Q) - pQ\right] \tag{9.4}$$

If the Equation 9.4 is solved for all values of the parameter *p*, we can derive the following demand function, which gives the amount demanded by the unit as a function of the price of water:

$$Q = Q^{D}(p) \tag{9.5}$$

The necessary condition for optimality can be written as:

$$\frac{d}{dQ}B(Q) = p \tag{9.6}$$

The demand function (Eq. 9.5) may be interpreted as the marginal benefit of the unit.

A supply unit may be a reservoir, a pumping station, a runoff storage pond, or a desalination plant, etc. At a cost of $C(Q)$, the supply unit supplies *Q* amount of water at price *p*. The optimization problem for the unit is solved by

$$\text{Max}_{Q}\left[pQ - C(Q)\right] \tag{9.7}$$

The corresponding solution for the supply function is obtained by:

$$Q = Q^{S}(p) \tag{9.8}$$

We assume that the demand and supply structures have been designed for optimum performance. Thus, the benefit and cost functions, $B(Q)$ and $C(Q)$, are expected to have optimization built into them.

When a supply and demand unit are connected, the value of water exchanged can be simply obtained from Figure 9.6 (source: Guariso et al. 1982, p. 377). The Equilibrium Point *E* in the plot corresponds to the following equation:

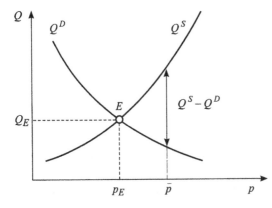

Figure 9.6. Supply and demand units and the equilibrium point (source: Guariso et al. 1982, p. 377).

$$Q^S(p_E) = Q^D(p_E) \tag{9.9}$$

$Q_E$ corresponds to the equilibrium flow which must be exchanged between the supply and demand units in order to maximize the total net benefit of the system, such that

$$\underset{Q}{\text{Max}}\, B(Q) = C(Q) \tag{9.10}$$

Similarly, *pE* corresponds to the equilibrium price which is the price 'which leads the supply and demand units not only to exchange the same amount of water but also to select the particular value $Q_E$ which maximizes the total net benefit of the system' (Guariso et al. 1982, p. 377). One way of solving the problem is to assume a particular value for p and compute the corresponding imbalance, $Q^S - Q^D$ (Fig. 9.6) and the reiterating the price until $Q^S - Q^D$ is zero.

### 9.4.2 *A typical irrigation system*

The price coordination method is applied to a simple situation (Fig. 9.7; source: Guariso et al. 1982, p. 378). An irrigation system has two sources:
– A pumping station ($S_1$) where energy has to be spent for pumping groundwater, and from where water is transferred through an artificial channel to irrigate the area ($D_1$), and
– A reservoir ($S_2$) which has already been built, and from where water flows out by gravity through a natural channel (i.e. at no recurring expense) to an artificial channel to irrigate the area ($D_2$). Though the economics of capital and operating costs of the sources and transfer units are different, irrigation water has to be delivered at the same price to farmers in the areas $D_1$ and $D_2$.

The same irrigation system can be depicted in the form of interaction graph (Fig. 9.8; source: Guariso et al. 1982, p. 381). A two-level decision-making process is explained as follows (Fig. 9.9; source:Guariso et al. 1982, p. 386): The central block represents the supervisor, and the external block represents the seven

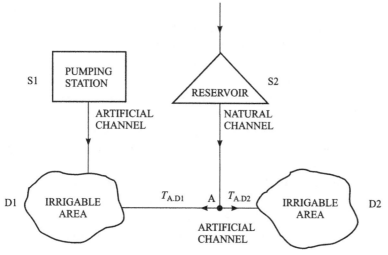

Figure 9.7. Example of the methodology of price coordination (source: Guariso et al. 1982, p. 378).

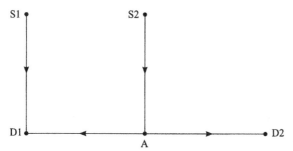

Figure 9.8. Interaction graph for price coordination (source: Guariso et al. 1982, p. 381).

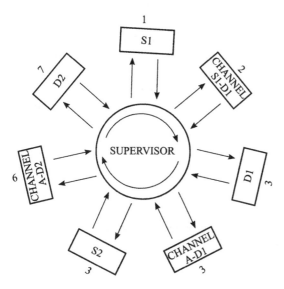

Figure 9.9. A two-level decision-making process (source: Guariso et al 1982, p. 386).

steps of 'dialogue' between the supervisor and the source units (S1 and S2), demand units (D1 and D2), and the channel units, (A-D1, A-D2 and S1 – D1). Thus
the algorithm envisages a recursive sequence of questions and answers between
the supervisor and each component. The maximization of the total net benefit of
the system is obtained by maximization of each sub-problem or component. The
analogy of the procedure with recursive trade negotiations should be evident.

Guariso et al. (1982) applied the above algorithm to the Northwest Water Plan
in Mexico. The system involved a groundwater supply, three surface water supplies from rivers, and four irrigation areas which would be interconnected through
an interbasin water transfer. Convergence was achieved in 12-15 iterations. The
model indicates a 6% increase in the total net benefit due to crop production. It
also shows that the market price of the water supplied is double the existing
(probably subsidized) price if water is to be treated as a market commodity. The
model can be further refined by bringing in parameters such as the quality of water, and the reliability of supplies.

## 9.5 OPTIMIZATION METHODS IN WATER MANAGEMENT

This section is largely drawn from the excellent and lucid summary by Holy
(1982).

Systems engineering concepts are applied to the solution of the water management problems in relation to the human environment, namely, protection of water
resources, multiplying the number of water resources, and improving the efficient
utilization of water resources (Holy 1982, p. 81). The objective of the water production process is to collect and store water having the required utility properties
(quantity, quality, place and time) which may be required for final consumption
(e.g. water for domestic use, irrigation water) or a means of production (e.g. water
for transportation, water for power production). Water is thus an economic good.
But unlike other economic goods like coal and oil, water cannot be treated in
purely economic terms, since human life itself depends upon it. Water resources
are sought to be protected and augmented by providing for well-head protection,
protection around water springs, and wastewater reuse. The efficient use of water
involves, among other things, the reduction in conveyance losses, and minimization of evaporation.

### 9.5.1 *Water reservoirs*

The technical management of water reservoirs has been dealt with under Section
6.2.

There may be many reasons for building water reservoirs, such as:
– Providing water for domestic purposes (potable water), industry (utility water),
  and agriculture (irrigation)

- Flood control,
- Power production,
- Inland water transport,
- Recreation, etc.

The construction of a reservoir perturbs the natural environment and affects the geographical, climatic and social conditions. There may be beneficial impacts on the economy and way of life of the people through the development of new economic activities (agriculture, power production, inland transport, recreation, etc.). There may be adverse effects on the health and productivity of men and animals (e.g. incidence of waterborne diseases, schistosomiasis, etc.), and there is the potential danger arising from the failure of the dam due to torrential rains or an earthquake.

Reservoirs affect the following components of the biophysical environment (Holy 1982, p. 73):

- Solar radiation and the thermal balance of the accumulated water,
- Temperature of the air, and fog formation,
- Air currents (similar to those near the sea), and
- Air humidity. As the water level of the reservoir fluctuates, muddy areas with shallow water develop at the edges of the reservoirs. Water weeds and swamp plants grow in these areas, and form breeding grounds for mosquitoes.

Construction of reservoirs have the following adverse impacts:

- Large populations may have to be moved and resettled away from the areas that are going to be flooded,
- Silting reduces the life-span of the reservoir,
- Buildings, cultural monuments, soils, forests, mineral deposits, etc. may be irretrievably lost,
- Ecosystems in the impounded water may be severely disturbed (changes in thermal and chemical stratification, reduction in DO (Dissolved Oxygen) in water, movement of fish, etc.) and
- Problems of eutrophication, and the development of thick, and extensive mats of water hyacinth (*Eichhornia crassipes*) which seriously impede fluvial transport.

The favorable impacts of the reservoirs in the downstream side are the decrease in the flood flow, and the improvement in the low minimal rates in water courses. Reservoirs trap silt and insoluble pollutants. So the water quality improves in the discharge of the reservoir.

### 9.5.2   *Water management systems*

Water management systems are of crucial importance to the quality of life and national economy. They are often complex and costly. Most water projects are multipurpose. It is therefore necessary to identify the principal purpose of the project (water supply, power generation, recreation, etc.) and use the secondary purposes as normative limitations.

Economic activities which are dependent on water can be divided into two categories: water users and water exploiters. Water users remove water from its source (reservoir, pumping station, runoff pond, etc.) and use it for some purpose (agricultural, industrial, domestic, etc.) and then discharge it at a different place, reduced in quantity, and most often of degraded quality. On the other hand, water exploiters do not remove water from its source, but use it where it is and as it is, for production of hydropower, for river transport and fisheries, for recreation, etc. This division is, however, arbitrary. For instance, when large quantities of water are stored for purposes of power generation, there is considerable loss of water due to evaporation from the surface of water. Agriculture and forestry operations, draining of swamplands, construction of new townships, etc. have a great impact on the quality and quantity of water resources.

The objective of optimization is to attain the optimal relation between the purposes for which the system is being built and the means available for building it. The following summary is largely drawn from Holy (1982).

The standard function $\phi$ for the optimization of a water management system contains three groups of economic variables:

$N$ – Unit costs of the individual equipments of the system for different purposes, $P$ – Unit benefit of the system for water supply for individual purposes, $Z$ – Unit losses that arise due to failure to supply water for individual purposes.

The model assumes that

– The system contains $\underline{k}$ reservoirs (1,2, 3....$n$),
– $N$, $P$, and $Z$ vary with time $\underline{t}$. Value $\underline{T}$ corresponds to the assumed physical or economic lifespan of an installation or the whole system.

It therefore follows that

$$N = f(k, t) \tag{9.11}$$

$$P = f(k; i; t) \tag{9.12}$$

$$Z = F(i; t) \tag{9.13}$$

A general standard economic function which simulates a multi-purpose water management system, may be written as follows:

$$\phi_{\text{opt}} = \sum_{k=1}^{m} \sum_{i=1}^{n} \sum_{t=1}^{T} \phi(N_{kt}; P_{kit}; Z_{it}) = \max; \min; 0 \tag{9.14}$$

Optimization cannot be based on economic criteria alone. Many times, non-economic criteria, such as social priorities and environmental concerns, may have to be taken into account or may be given priority over purely economic criteria. In a multi-purpose water management system, this process is accomplished by the method of matrix of contradiction of purposes. We identify the principal purpose of water use, for which the system is to be optimized, and consider other purposes as limitations or contradictions.

The following criteria for contradiction may be considered:
1. Demand on reservoir volume,
2. Demand on water quality,
3. Demand on water level fluctuation.

There is bound to be contradiction in demands in most cases. There are a few exceptions – for instance, there may be no conflict involved if water is to be used for navigation and recreation. There can be movement of boats, and water sports in the water in a river or lake. The most frequently occurring contradiction is contradiction 1, followed by 3 and in some cases, 2.

Evidently, the priority of purpose would determine would determine the principal consideration that has to be taken into account.

| Priority of purpose | Principal consideration |
|---|---|
| Drinking water | Water quality |
| Hydropower production | Water level in the reservoir |
| Irrigation | Reservoir volume |

Table 9.3 gives the matrix of contradiction of purposes (source: Holy 1982, p. 87):

For instance, if priority is to be given for the use of water in irrigation, maintenance of adequate storage of water would be the prime concern. Limited importance will be given to other uses, such as, water supply, power, etc.

## 9.6 ALLOCATION OF WATER TO COMPETING USERS

### 9.6.1 *General considerations*

In most parts of the world, there is intense competition among the users for scarce water resources. The competitors may be individual users (e.g. factories) in an area or communities or provinces within a country (violent confrontation between

Table 9.3. Matrix of contradiction of purposes.

| Purpose | Irrigation | Water supply | Power | Navigation | Flood control | Recreation |
|---|---|---|---|---|---|---|
| Irrigation | – | 1, 2, 3 | 1, 3 | 1, 3 | 1 | 3 |
| Water supply for population | 1, 2, 3 | – | 1, 3 | 1, 3 | 1 | 2, 3 |
| Hydropower production | 1, 3 | 1, 3 | – | 1, 3 | 1 | 3 |
| Navigation | 1, 3 | 1, 3 | 1, 3 | – | 1 | – |
| Flood control | 1 | 1 | 1 | 1 | – | – |
| Recreation | 1 | 2, 3 | 3 | – | – | – |

the provinces of Tamilnadu and Karnataka in India about the sharing of the waters of the Cauvery River), or even countries themselves (Syria, Turkey, Iraq about sharing of the waters of Euphrates – Tigris). While the value of the possible output should have primacy in allocating the water to the competing users, it is not possible to ignore some cultural, socio-economic, political and technical factors. Innnovation and practicability should be the key words in addressing such problems.

How an innovative approach could provide a way-out of the water competition problems can be illustrated with two examples from the Indian sub-continent (Suresh 1998).

The province of Tamilnadu in south India makes use of 95% of the surface water potential through 55 major reservoirs and more than 300 medium and small anicuts, barrages and diversion structures. On the other hand, the province of Kerala to the west has plenty of water (43 flowing rivers), but does not have enough land to construct the reservoirs. Thus, it makes economic sense to transfer the waters of Kerala towards the east, and construct the reservoirs in Tamilnadu where land is available. The Parambikulam Aliyar Project (PAP) seeks to divert 863 M $m^3$ of Kerala waters to Tamilnadu, in a way which will be mutually beneficial to both the states.

River Ganges ranks next only to Amazon in terms of the volume of water, with surface water availability of 446 million acre-ft (MAF) or about 551 $km^3$. The headwaters of the major tributaries of the Ganges lie in Nepal which contribute to about 40% of the annual flows and 71% of the dry season flows of the Ganges available at Farakka (in the West Bengal province of India). But Nepal does not have the resources to develop these water resources, and sell water and energy to the downstream countries, namely, India and Bangladesh. Techno-economic solutions do exist to solve these problems. What is needed is the political will and economic means to implement them.

### 9.6.2 *Case histories*

Two case histories are given to illustrate the use of linear programming methodology (Lee 1976) to decide about the allocation of water to competing users.

Barandemaje (1988) tried to develop a policy aimed at optimizing the allocation of water to competing industrial users in Burundi (Africa). A finite amount of water (26,000 $m^3$ $d^{-1}$) has to be allocated to five competing industrial users (a cigarette company, a brewery, a textile mill, a paper mill and a dairy). A linear mathematical model was developed to conceptualize the five firms. It consists of an objective function and constraint equations. Each equation takes into account the water-use/net-profit relationship for each firm. The constraints are based upon the lower and upper water consumption capacities and the non-negativity conditions on the value of the decision variables. Optimization methods were applied to two kinds of situations: involving only competitive criteria (i.e. which kind of allocation would yield the maximum profits), and involving both social criteria (the needs of the dairy being given the highest priority and the cigarette company, the lowest) and economic competitive criteria. Even though the dairy may provide less profit per unit of water used relative to the cigarette company, the community needs milk more than it needs the cigarettes, and hence the higher priority for the dairy in water allocation. On the basis of this kind of analysis, the decision-makers could make their choices.

Singh et al. (1987) made use of goal programming (a case of linear decision problems which may have more than one objective) to allocate water to the winter crops (wheat, *ahu* paddy, pulses, oil seeds and potato) in Garufella catchment in Assam in northeast India. Garufella is a typical monsoon land – though the rain-fall is very heavy (3710 mm), 90% of it occurs in the monsoon period (June to October).

On the basis of data (water availability data, crop yield data, food requirement), the various parameters are maximized, taking into account various constraints and priorities.

Under this study, three objective functions are formulated for maximizing the net return, protein content, and the calorific value of crop, under various socio-economic constraints.

*Maximization of net return:*

$$\text{Max } Z_1 = \sum_{i=1}^{n} A_i N_i \tag{9.15}$$

where $A_i$ = area under $i$-th crop activity in hectares, $N_i$ = net return from hectare from $i$-th crop activity in rupees, $n$ = number of crops being considered.

*Maximization of calorific value:*

$$\text{Max } Z_2 = \sum_{i=1}^{n} A_i Y_i C_i \tag{9.16}$$

where $Y_i$ = yield of $i$-th crop activity (in kg ha$^{-1}$), $C_i$ = calorie of $i$-th crop (calories kg$^{-1}$)

*Maximization of net protein value:*

$$\text{Max } Z_3 = \sum_{i=1}^{n} A_i Y_i P_i \tag{9.17}$$

where $P_i$ = nutrient value of $i$-th crop activity (in g kg$^{-1}$), $C_i$ = calorie of $i$-th crop (calories kg$^{-1}$).

The three optimization models given above are subject to the following con-straints:

*Water availability constraint:*
This refers to the amount of surface water available for crop production.

$$\sum_{i=1}^{n_j} A_i R_{ij} \leq S_j \qquad (9.18)$$

where $S_j$ = surface water available in $j$-th month, $R_{ij}$ = water requirement per unit area in excess of effective rainfall for the $i$-th crop in $j$-th month, $n_j$ = total number of crops which are grown in $j$-th month.

*Land availability constraint:*
The extent of land used for various crops cannot exceed the total available land. Also, the land allocated to a given crop has to remain unchanged from sowing to harvesting.

$$\sum_{i=1}^{n} A_i \leq TA \qquad (9.19)$$

where $TA$ = Total available land (in ha).

*Minimum area constraint:*
In order to meet the minimum food requirements of the population, a given crop (say, wheat) needs to be grown in a minimum area.

$$A_i \geq T_i \qquad (9.20)$$

where $T_j$ = minimum area allocated to $j$-th crop.

*Protein requirement constraint:*

$$\sum_{i=1}^{n} A_i Y_i P_i \geq PR \qquad (9.21)$$

where $PR$ = total protein requirement (in g).

*Calorie requirement constraint*

$$\sum_{i=1}^{n} A_i Y_i C_i \geq CR \qquad (9.22)$$

where $CR$ = total calorie requirement (in calorie units).

Each goal constraint (nutritional requirement constraint, Net return constraint, Production constraint based on the food habits of the people) may be assigned a positive or negative deviation variable or both.

The priorities assigned are as follows:

$P_1$: The highest priority is assigned to the maximization of net return.

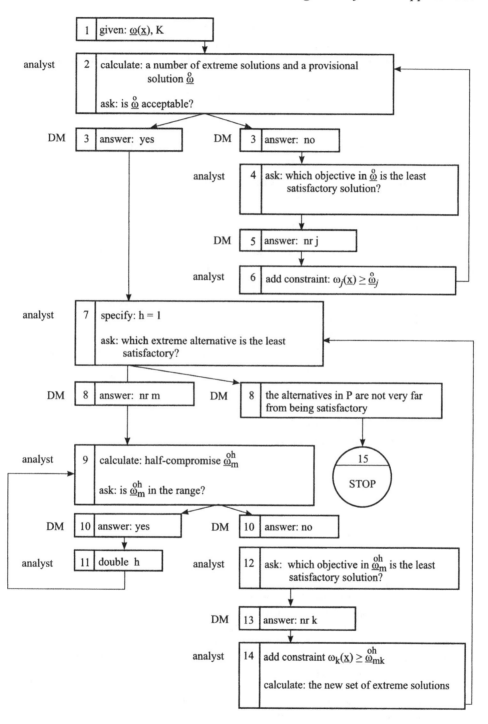

Figure 9.10. Flow-chart of decision-making process (source:Nijkamp & Rietveld 1982).

$P_2$: The second priority is assigned to protein and calorific value. Between the two, a higher weight is given to the calorific value.

$P_3$: The production of different crops should be adequate to meet the actual food requirements of the population. Wheat is assigned a higher weight than paddy which is also cultivated during the monsoon season.

The existing situation (EXT) is compared with ten models, involving the following parameters: Total area (ha), Total water utilization (ha m), Net return (millions of rupees), Total protein ($\times 10^8$ g), and Total calories ($\times 10^{10}$ calories). From among these, the model which yielded the maximum net return, while providing adequate nutrition in tune with the food habits of the people, was chosen.

### 9.6.3   *Decision-making process*

Nijkamp & Rietveld (1982) gave a good account of the conceptual basis of the decision-making process. The process has the following elements:

1. DM (Decision Maker) has at his disposal $I$ instruments, $\underline{x} = (x_1, x_2 .... x_I)$.
2. The vector of the instruments $\underline{x}$ is an element of the convex set $K_I$, being a subset of $\mathfrak{R}^I = (\underline{x} \in K \subset \mathfrak{R}^I)$,
3. The DM considers $J$ objectives: $\underline{w} = (w_1, w_2 ... w_j)$, which he wants to maximize,
4. For each combination of instruments, $\underline{x}$, the effect on the set of objective $\underline{w}$ can be determined with certainty. Hence, a set of $J$ concave objective functions $\underline{w}$ ($\underline{w} = w_1, w_2 ... w_j$) is assumed to exist, each mapping $\underline{x} \in \mathfrak{R}^I$ to $w_j \in \mathfrak{R}^I$.

Figure 9.10 gives the flow chart of the decision-making process. In this arrangement, the analyst comes up with a variety of solution. The Decision Maker (DM) evaluates them. By iteration, he will eliminate the least satisfactory solutions and in due course, arrives at the most satisfactory solution.

## REFERENCES

* Suggested reading

Barandemaji, D. 1988. Optimum allocation of water to competing users. *Proc. Int. Training Course on Environment Management in Developing counties*. Leipzig.

Bouwer, H. 1994. Role of geopurification in future water management. In *Soil and Water Science: Key to Understanding Our Global Environment*. Soil Soc. Amer. Sp. Publ. 41: 73-81.

Dantzig, G.B. 1963. *Linear Programming and Extensions*. Princeton, N.J.: Princeton Univ. Press.

De Backer, L.W. 1982. Pre-management of water resources study. In P. Laconte & Y.Y. Haimes (eds), *Water Resources and Land-use Planning: A Systems Approach*. The Hague, Netherlands: Martinus Nijhoff Publishers. p. 53-67.

*Guariso, G., D. Maidment, S. Rinaldi & R. Soncini-Sessa 1982. In P. Laconte & Y.Y. Haimes (eds), *Water Resources and Land-use Planning: A Systems Approach*. The Hague, Netherlands: Martinus Nijhoff Publishers. p. 373-392.

*Holy, M. 1982. Environmental aspects of water management. In P. Laconte & Y.Y. Haimes (eds), *Water Resources and Land-use Planning: A Systems Approach*. The Hague, Netherlands: Martinus Nijhoff Publishers. p. 69-91.

Janusz, K. 1998. Economic value of water. In H. Zebidi (ed.), *Water: A Looming Crisis?* IHP-V, Tech. Doc. no. 18. Paris: UNESCO. p. 407-416.

Lee, Sang M. 1976. *Linear optimization for management.* New York: Petrocelli/Charter.

Nijkamp, P. & R. Rietveld 1982. Selecting a range of alternatives by individual or group decision-makers. In P. Laconte & Y.Y. Haimes (eds), *Water Resources and Land-use Planning: A Systems Approach.* The Hague, Netherlands: Martinus Nijhoff Publishers. p. 41-45.

*Paul, M.M. 1998. Integrated Information Management for Water Resources Assessment. In H. Zebidi (ed.), *Water: A looming crisis?,* p. 103-112. Paris: Unesco.

Rogers, P., R. Bhatia & A. Huber 1997. Water as a social and economic good: How to put the principle into practice. Paper presented at the meeting of the Technical Advisory Committee of the Global Water Partnership in Namibia.

*Shiklomanov, I.A. 1998. *World Water Resources: A New Appraisal and Assessment for the 21st. Century.* Paris: UNESCO.

Singh, R., B. Soni & A.K. Changkakoti 1987. Optimal utilization of irrigation water in Garufella catchment in Assam, India. In T.H. Anstey & U. Shamir (eds), *Irrigation and Water Allocation.* IAHS Publ. no. 169.

Stephenson, D. 1989. Planning model for water resources development in developing countries. In D.P. Loucks & Uri Shamir (eds), *Closing the Gap between Theory and Practice,* IAHS Pub. no. 180.

Suresh, S. 1998. Intersectoral competition for land and water policy between users and uses in Tamilnadu, India. In H. Zebidi (ed.), *Water: A Looming Crisis?* IHP-V, Tech. Doc. no. 18. Paris: UNESCO. p. 441-445.

Turner, R.K. & W.N. Adger 1996. *Coastal Zone Resources Assessment Guidelines.* LOICZ Reports and Studies. No.4. Texel, The Netherlands: LOICZ.

Wollman, N. 1962. *The value of Water in Alternative Uses.* Albuquerque: The University of New Mexico Press.

CHAPTER 10

# Etiology of diseases arising from toxic elements in drinking water

A caveat should be entered straight-away about the scope of the chapter – *this chapter covers only those diseases (such as, arseniasis and fluorosis) where drinking water is the principal route of intoxication.* The chapter does not discuss numerous water-based, parasitic, bacterial, viral, etc. diseases, such as, gastrointestinal disorders, cholera, typhoid, malaria, schistosomiasis, etc.

## 10.1 ESSENTIALITY/TOXICITY OF TRACE ELEMENTS

*All things are poison, and nothing is without poison, and the dosis alone decides that a thing is not poison* – Paracelsus of Hohenheim (1493-1541).

Though trace elements are present in very small quantities, and *in toto* comprise only 0.1% of animal matter, they play a crucial role in the health and disease of humans and animals. Essential elements are trace elements which have importance in biological processes. Inadequate intake of essential elements leads to the impairment of the relevant physiological functions. Supplementation of the deficient element will prevent or ameliorate the impairment.

The partitioning of elements between rocks and plants through soils and water is critically dependent upon the ability of the element to form soluble complexes in the soil environment. With the exception of selenium, the essential elements are sufficiently abundant in the crust (more so, in mafic rocks) to meet the physiological functions of human beings.

In the ultimate analysis, all essential elements that humans need are derived from the geoenvironment (rocks, soils, water, and air). Diet is the principal source of essential elements, in about 80% of the cases. Fluoride is an exception in that the bulk of the intake is through water. Seaspray is an important source of iodine for people who live in the coastal zones, which accounts for the lower rate of goiter incidence in such areas. The quantity of trace element that reaches humans through food depends upon

– Geochemical availability (depending on the leachability of the element and its mode of distribution, and availability in the rock or soil), and
– Bioavailability (the fraction of the element in the food of plant or animal origin available to humans).

Trace elements exist in nature in different chemical forms, and their speciation influences their toxicity (methyl mercury is more toxic than metallic mercury, and chromium as Cr (III) is an essential element needed by humans, whereas Cr (VI) is toxic to them). Besides, trace elements may undergo biotransformations in the human body and may form metallobiocomplexes. This may involve a change in the oxidation state and hence in the toxicity of the concerned trace element. Two or more elements may act synergistically, accentuating their combined effect, or antagonistically, suppressing the effect in part. For instance, zinc can moderate the toxicity of cadmium.

The physiological function and the Recommended Dietary Intake of Essential Elements is given in Table 10.1.

The US National Research Council recommended certain levels of intake of essential elements (RDA) to maintain normal physiological functions from nutritional point of view. On the other hand, the US EPA recommended a reference dose (RfD) which is an estimate of the daily exposure levels to protect from the deleterious effects of toxicity. Conflicts have arisen from the two sets of recommendations when the RDA is higher than RfD.

In order to enable people to make appropriate intake decisions that will help them 'to live longer and better', Nielsen (2000) proposed that the Dietary Reference Intakes (DRIs) be spelt out in terms of Nutritionally beneficial, Pharmocologically beneficial and Conditionally essential elements. He makes out a case for the inclusion of boron and chromium to the essential elements. He makes the point that in the case of arsenic, the US EPA prescribed an RfD of 0.3 μg/body weight/day i.e. 21 $\mu g\ d^{-1}$ for a 70 kg person, whereas nutritionists believe that a safe upper limit of arsenic intake could well be 140-250 $\mu g\ d^{-1}$.

The joint FAO/IAEA/WHO Expert Consultations on Trace Elements in Human Nutrition (1996) defined the concerned terms as follows:

Table 10.1. Recommended dietary intake of essential elements.

| Element | Physiological function | Recommended daily Intake (mg $d^{-1}$) |
|---------|------------------------|------------------------------------------|
| Calcium | Enzyme activator, electrolyte | 800 |
| Phosphorus | Nucleic acid element | 800 |
| Magnesium | Enzyme activator; | 350 (males); |
| | Electrolyte | 300 (females) |
| Iron | Metallo-enzyme; | 10 (males), |
| | enzyme activator | 18 (females) |
| Zinc | – same | 15 |
| Manganese | – same | 2.5-5.0 |
| Fluorine | Enzyme inhibition (glycolysis) | 1.5-4.0 |
| Copper | Metallo-enzyme | 0.15-0.5 |
| Molybdenum | – same | 2.0-3.0 |
| Chromium | | 0.05-0.2 |
| Selenium | Enzyme protein | 0.05-0.2 |
| Iodine | Thyroid element | 0.15 |

*Requirement*: This is the lowest continuing level of nutrient intake that, at a specified efficiency of utilization, will maintain the defined level of nutriture in the individual.

*Basal requirement*: This refers to the intake needed to prevent pathologically relevant and clinically detectable sign of impaired function attributable to inadequacy of nutrient.

*Normative requirement*: This refers to the level of intake that serves to maintain a level of tissue storage or other reserve that is judged (by the expert consultation) to be desirable'.

*Recommended or safe level* of intake is taken as the average requirement + 2 SD in requirement.

The Acceptable Range of Oral Intake (AROI) is defined as protecting 95% of an unselected human population from even minimal adverse effects of deficiency or toxicity. Figure 10.1 (source: Nordberg et al. 2000) is a theoretical model describing the distribution of intakes to meet the nutritional requirements (left) and distribution of intakes giving rise to toxicity (right), with acceptable range of oral intakes (AROI) in between. Special population subgroups, such as A, may exhibit toxicity at intakes lower than the acceptable range (e.g. Wilson's disease and copper intake). In contrast, some population subgroups, such as B, may have requirements higher than the upper limit of the acceptable range (e.g. zinc intakes in subjects with acrodermatitis enteropathica).

The medical consequences arising from the deficiency or excess of some elements are summarized in Table 10.2 (Aswathanarayana 1995, p. 227-251, gave a brief account of the concerned diseases and their etiology).

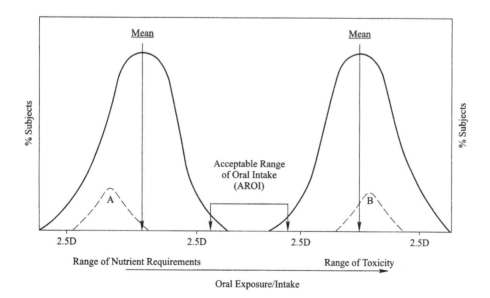

Figure 10.1. Acceptable range of oral intake (AROI) (source: Nordberg et al. 2000).

Table 10.2. Medical consequences of excess or deficiency of elements.

| Element disorder | Medical consequence |
| --- | --- |
| Excess of Hg | Minamata disease (Japan) |
| Excess of Cd | Itai-itai disease (Japan) |
| Deficiency of Se | Keshan disease (China) |
| Lack or excess of elements (Se?), possibly mycotoxins | Kashin-Beck disease, bone and joint degeneration, deformation of children (China) |
| Deficiency of I | Goiter (Tanzania) |
| Deficiency of F | Dental caries |
| Excess of F | Dental mottling, skeletal fluorosis (Tanzania) |
| Deficiency of P | Osteomalacia |
| Deficiency of Zn | Impeded growth, decreased immune function, hypogonadism (Egypt, Iran) |
| Deficiency of Mg | Nervous disease, depression |
| Deficiency of Cr | Cr- diabetes, cardiovascular diseases |
| Deficiency of Cu + Zn + Se | Arthritis due to overproduction of peroxidase |
| Excess of Cd | Degeneration of kidney and bones |
| Copper accumulation | Wilson's disease |

### 10.1.1 *Routes and consequences of exposure*

The potential exposure routes through which toxic substances reach humans are given in Table 10.3 (source: US EPA 1989).

Drinking water may be contaminated with radionuclides near nuclear reactors or nuclear facilities or as a consequence of nuclear accidents (such as, Chernobyl). The Maximum Permissible Concentrations (MPC) in terms of (Ci ml$^{-1}$) of selected radionuclides in drinking water are given in Table 10.4 (source: Laconte & Haimes 1982).

Leachates and effluents from industrial activities could contaminate the drinking water resources. Three case histories are given about health effects arising from the drinking of such contaminated groundwater (Table 10.5; source: Marsh & Caplan 1987)

## 10.2 ARSENIASIS

The writing of this section has been much facilitated by the opportune appearance of two excellent reviews (Chappell et al. 1999; Hindmarsh 2000). The UN agencies have put out a Synthesis Report on Arsenic in Drinking Water, which is due to be released in late 2000 as hard copy and on the web (http://www.who.int/ water_sanitation_health/water_quality/arsenic.htm).

### 10.2.1 *Introduction*

Arseniasis (also called arsenicosis or arsenicism) has emerged as an environmental issue of global concern. Arseniasis (manifested in the form of skin lesions,

Table 10.3. Potential exposure routes of contaminants from various sites (Source: US EPA 1989).

| Exposure route | Domestic | Commercial/industrial | Recreational |
|---|---|---|---|
| *Food* | | | |
| Ingestion | L | – | L |
| *Groundwater* | | | |
| Ingestion | L | A | – |
| *Surface water* | | | |
| Ingestion | L | A | L, C |
| Dermal contact | L | A | L, C |
| *Sediment* | | | |
| Dermal contact | C | A | L, C |
| *Air* | | | |
| Inhalation of vapor phase chemicals | | | |
|   Indoors | L | A | – |
|   Outdoors | L | A | L |
| | | | |
| Inhalation of particulates | | | |
|   Indoors | L | A | – |
|   Outdoors | L | A | L |
| *Soil/dust* | | | |
| Incidental ingestion | L, C | A | L, C |
| Dermal contact | L, C | A | L, C |

L: Lifetime exposure, C: Exposure in children may be significantly greater than in adults, A: Exposure to adults (highest exposure is likely to occur during occupational activities), –: Exposure of this population via this route is unlikely to occur.

Table 10.4. Nuclear and health physics data for selected radionuclides (source: Paper 1.1 in Laconte and Haimes 1982).

| Radionuclide | Half-life (years) | Major radiation | Critical organ | Biological half-life | MPC * ($\mu Ci\ ml^{-1}$) |
|---|---|---|---|---|---|
| $^{3}H$ | 12.26 | $\alpha$ | Total body | 12 days | $3 \times 10^{-3}$ |
| $^{90}Sr$ | 28.1 | $\beta$ | Bone | 50 years | $3 \times 10^{-6}$ |
| $^{129}I$ | $1.7 \times 10^{7}$ | $\beta, \gamma$ | Thyroid | 138 days | $6 \times 10^{-8}$ |
| $^{137}Cs$ | 30.2 | $\beta, \gamma$ | Total body | 70 days | $2 \times 10^{-3}$ |
| $^{226}Ra$ | 1,600 | $\alpha, \gamma$ | Bone | 45 years | $3 \times 10^{-8}$ |
| $^{239}Ra$ | 24,400 | $\alpha$ | Bone | 200 years | $5 \times 10^{-6}$ |

* MPC = Maximum Permissible Concentration for water consumed by the general public without readily apparent ill effects (these figures are revised as the toxicity of the radionuclides is better understood).

vascular damage, cancers of bladder, lung, liver and kidney, etc.) arises from the ingestion of excessive quantities of arsenic, through drinking water, and inhalation. Chen et al. (1999) summarized the present situation about the incidence of arseniasis in Asia (Table 10.6).

Table 10.5. Health effects due to drinking of contaminated groundwater.

|  | Love Canal, New York, USA | Woburn, Massachusetts, USA | Hardeman County, Tennesse, USA |
|---|---|---|---|
| Nature of the contaminant | Largely hydrocarbon residues from pesticide production, chemical waste leachates, and toxic vapors | Lead, arsenic and organic contaminants from industrial wastes | Diverse chemicals, including chlorinated organic compounds |
| Health measurements | Ill health, and birth defects: Birth weight, birth height | Cancer incidence | Urine analysis for selected organic compounds, blood analysis for liver profile |
| Exposure measurements | Distance from site | Reported in interview | Household water and air samples |
| Health results | Significantly lower birth weight and birth height | Greater than expected incidence of childhood leukemia and renal cancer | Significant differences in liver profiles |

Table 10.6 Incidence of arseniasis in Asia (source: Chen et al. 1999).

| Area | Source | Population at risk | Non-cancer manifestations* | Cancer manifestations ** |
|---|---|---|---|---|
| *Bangladesh* | Well water | 50,000,000 | M/K, D, G, B | S |
| *China* |  |  |  |  |
| Guizhou | Burning of high-As coal | 200,000 | M/K, G, P | S, Li |
| Inner Mongolia | Well water | 600,000 | M/K, G, P | A, S, Lu, Li, U, K |
| Shaanxi | Well water | 1,000,000 |  |  |
| Xinjiang | Well water | 100,000 | M/K, G, P | A, S, Lu |
| Yunnan | Metal smelting | 100, 000 |  |  |
| *India* |  |  |  | S |
| West Bengal | Well water | 1,000,000 | M/K, D, G, B,P |  |
| *Japan* |  |  |  |  |
| Toroku/Matsuo | Metal smelting | 217 patients | M/K, D, G, B,P | A,S, Lu, U, K |
| Nakajo | Contaminated water | 44 patients | M/K, G, B, P | A, Lu, Li, U, K |
| *Taiwan* |  |  |  |  |
| Southwest coast | Well water | 100,000 | M/K | A,S,N, Lu,Li, U, K, P |
| Northeast coast | Well water | 100,000 | M/K | A,S,N, Lu, Li, U, K |
| *Thailand* |  |  |  |  |
| Ronpibool | Tin mining | 1000 patients | M/K | A |
| *Philippines* |  |  |  |  |
| Mindanao | Geothermal drilling | ? |  |  |

*Non – cancer: M/K – melanosis/keratosis; D – dermatitis; G – gastroenteritis; B – bronchitis; P – polyneuropathy, ** Cancer: A – all sites; S – skin; Nasal cavity – N; Lung – Lu; Liver – Li; Urinary bladder – U; Kidney – K; Prostate – P. Besides these, the following cancers have been reported from the regions noted against them: esophagus – Inner Mongolia, cervix uteri – Nakajo, and stomach - northeast coast of Taiwan.

The natural sources of arsenic are: volcanic exhalations and products, forest fires, and weathering of As-bearing minerals, while anthropogenic sources are: fertilizers and pesticides, slag piles and mining wastes, combustion of fossil fuels and smelting of non-ferrous metals (Fig. 10.2; source: Piver 1983). In terms of geologic setting, the natural sources of arsenic causing arseniasis are: estuarine sediments in West Bengal and Bangladesh, Quaternary volcanics in Chile, black shale in Taiwan, etc. Anthropogenic sources causing arseniasis are: coal burning as in China, and mining and smelting as in NWT, Canada, etc.

Arsenic has a long history of use in the Indian system of medicine (*Ayurveda*). There are stories of how some Indian kings got young girls fed with small, regular doses of arsenic over many years, to be used as live homicidal weapons against their enemies (legends of *visha kanya* or poison maiden). In England, Fowler's solution containing arsenic was widely used in the eighteenth and nineteenth centuries, for the treatment of psoriasis and asthma. Its use was discontinued only recently, when the toxicity of arsenic became widely known. The use of arsenical pesticides and fungicides has also declined.

In the medieval times in Europe, arsenic was a commonly used homicidal poison. It was said that the British murdered Napoleon Bonaparte by slow-poisoning him with arsenic. According to Hindmarsh (2000), this allegation is unfounded, and that Napoleon actually died of stomach cancer.

When Clare Booth Luce of *Time-Life* fame, was US Ambassador to Italy, she suffered from chronic arsenic poisoning. She was convinced that she was being poisoned by her enemies. Later it turned out that she got intoxicated by the arsenic in the pigment used in the wall paper in her villa in Rome! Earlier, her mission in Italy began with a diplomatic *faux pas*. The State Department advised her that the then President of Italy was a reputed scholar in *etymology* (the study of word origins). She mistook it as *entomology* (the study of insects). So at the time of the presentation of her credentials, she ceremoniously handed over to the President a box of insects. The President froze – he could not understand why he was being presented with insects, of all things!

Arsenic compounds are currently used as a clarifier in glass industry, as a wood preservative (copper arsenite) and in the manufacture of semiconductors (as gallium arsenate), and as desiccant and defoliant in agriculture. Arsenic is byproduct of smelting of non-ferrous metals, particularly of gold and copper.

### 10.2.2 *Stability of the dissolved species of arsenic*

Arsenic (atomic number: 33; electronic configuration: 2-8-18-5) can exist in three oxidation states: metalloid (0), trivalent (+3, –3) and pentavalent (+5). The toxicity of arsenic is highly dependent upon speciation. It increases in the following order: elemental < organic < arsenates < arsenites < arsines. This implies that the toxicity of arsenic could be reduced by manipulating a change in speciation from (say) arsenites to arsenates, or from arsenates to organo-metallic compounds. A lethal human dose in the form of arsenic trioxide is about 100-200 mg.

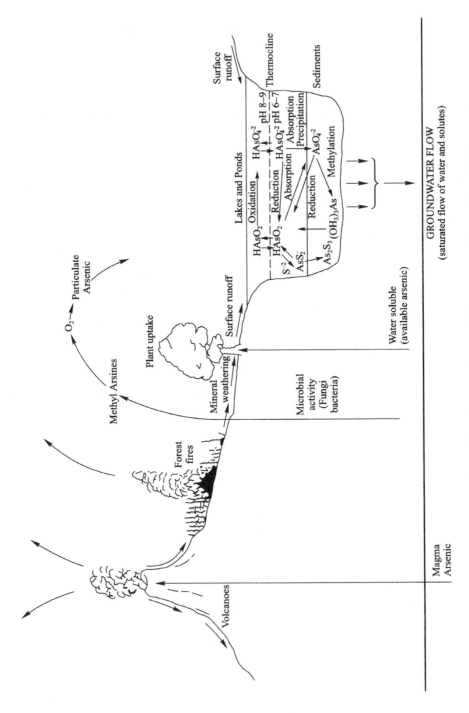

Figure 10.2. Natural sources and speciation of arsenic in various environmental compartments (source: Piver 1983).

The following are the important arsenic compounds that are relevant to human toxicity (Hindmarsh, 2000): Arsenic trioxide – $As_2 O_3$; Arsenous acid – $H_3As O_3$; Arsenite – $H_2As O_3^{1-}$, $H As O_3^{2-}$, $As O_3^{3-}$; Arsenic pentoxide – $As_2 O_5$; Arsenic acid – $H_3As O_4$; Arsenate – $H_2As O_4^{1-}$, $HAs O_4^{2-}$, $As O_4^{3-}$; Arsanilic acid – $C_6 H_4 N H_2As O (OH)_2$; Arsenobetaine – $(CH)_3 As^+ CH_2 COOH$, Dimethylarsinic acid $(CH_3)_2 As O (OH)$; Methylarsonic acid – $CH_3 AsO(OH)_2$.

Since the toxicity of arsenic is strongly species-dependent, a knowledge of the pE – pH stability range of various arsenic species helps us to understand the etiology and mitigation of arseniasis. The following account, including Figure 10.3, is largely drawn from Cherry et al. (1979) and Appelo & Postma (1996, p. 249).

pE is redox potential expressed as – [log $e^-$], where [$e^-$] is 'activity 'of electrons.

$$pE = Eh/0.059 \text{ at } 25°C$$

The stability of some dissolved species of arsenic depend on pH only.

As (V) and As (III) are the principal arsenic species found in natural waters. Both the species are protolytes, which may release protons step-wise.

$$H_2 As O_4^- \leftrightarrow H_2 As O_4^{2-} + H^+ \tag{10.1}$$

The corresponding mass action equation is:

$$\log [H_2 As O_4^{2-}] - pH - \log [H_2 As O_4^-] = -6.9 \tag{10.2}$$

Thus, below the pH of 6.9, $H_2 As O_4^-$ dominates, whereas above this pH, the dominant form would be $H_2 As O_4^{2-}$. In the pE – pH diagram, the stability fields of the two species are separated by a vertical line at pH 6.9. It may be noted that the boundary is independent of pE.

Similarly,

$$H_3 As O_3 \leftrightarrow H_2 As O_3^- + H^+ \tag{10.3}$$

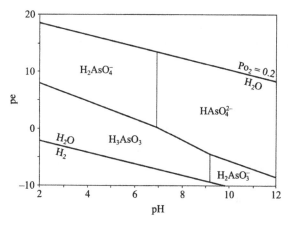

Figure 10.3. A partial pE – pH stability diagram for dissolved arsenic species. Boundaries indicate equal activities of both the species (source: Cherry et al. 1979; Appelo & Postma 1996, p. 249).

The corresponding mass action equation is:

$$\log [H_2 \, As \, O_3^-] - H - \log [H_3 \, As \, O_3] = -9.2 \tag{10.4}$$

A vertical line at pH 9.2 indicates equal activities of both the species.

The boundaries between As (V) species of $H_2 \, As \, O_4^-$ and As (III) species of $H_3 \, As \, O_3$ are dependent upon both pE and pH, as follows:

$$H_3 \, As \, O_3 + H_2O \leftrightarrow H_2 \, As \, O_4^- + 3 \, H^+ + 2_e^- \tag{10.5}$$

$$\log [H_2 \, As \, O_4^-] - 3 \, pH - 2 \, pE \, \log [H_3 \, As \, O_3] = -21.7 \tag{10.6}$$

Thus, equal activities of both species exist at

$$2 \, pE = -3 \, pH + 21.7 \tag{10.7}$$

For

$$H \, As \, O_4^{2-} / H_3 \, As \, O_3 \text{ boundary, } 2 \, pE = -4 \, pH + 28.5 \tag{10.8}$$

### 10.2.3   *Geochemical distribution of arsenic species*

The arsenate $(AsO_4)^{3-}$ species are strongly sorbed onto clays, Fe-Mn oxides/ hydroxides, and organic matter. The amount of sorption is determined by the concentration of arsenic species, time and Fe-Mn content of the soil. Arsenite salts are 5-10 times more soluble than arsenate salts. Reducing conditions, such as those occurring in the flooded soils (e.g. paddy soils, such as those in Bangladesh and West Bengal), enhance the proportion of As (III), in which form it is more available and more toxic. Soil bacteria can accelerate the oxidation of arsenite to the less soluble arsenates. They can also bring about methylation, in the form of As (III) methyl derivatives.

Natural remediation or self-cleansing processes tend to reduce the bioavailability of arsenic. It is necessary to understand these processes, to enable us to use them to mitigate arseniasis. For instance, the soils and mine smelter wastes in the historical arsenic mining areas in southwest England are very heavily contaminated with arsenic. Yet, there are no manifestations of adverse health effects among the people living in these areas. This is so because ferric iron in the soil/mine waste combines with arsenic species to form insoluble ferric arsenate, which has a low bioavailability to plants, and animals/man (Thornton 1999).

Surface waters tend to have high Eh and slightly acid pH, as a consequence of their exposure to oxygen and carbon dioxide in the atmosphere. For this reason, arsenic exists in the pentavalent form in the surface waters. Iron present in the surface waters either in a dissolved form or in the form of suspended particles reacts with pentavalent species of arsenic to form insoluble salts with arsenate and arsenite, which then get precipitated onto the sediments. For instance, the dissolved arsenic content of the waters of Lake Michigan is only 0.5-2.4 µg l$^{-1}$, whereas the As content of the lake sediments is 10-15 times higher (7-29 µg g$^{-1}$). The conditions associated with groundwaters (low Eh, absence of suspended particles containing iron, near neutral pH) do not allow the formation and precipitation of insoluble arsenic species. Thus, arsenic species tend to remain in a soluble form in groundwaters. Self-cleansing of arsenic due to iron may take place to some extent if the aquifer concerned contained iron and manganese. These considerations explain why arseniasis became endemic in Bangladesh and the West Bengal province of India during the last few decades, when people switched from surface water to borewell water for their supply of drinking water.

The movement of arsenic in groundwater (in the case of bank filtration with pumping wells in parallel with the bank), can be estimated from the one-dimensional advection – dispersion equation for the saturated zone (Park et al. 1994):

$$\frac{\partial C}{\partial t} = -\frac{v \partial C}{R \partial x} + \alpha_L \frac{v \partial^2 C}{R \partial x^2} - k \frac{v C}{R} - \lambda C \tag{10.9}$$

where $C$ = Concentration (m$^{-3}$), $R$ = Retardation coefficient; $t$ = time (day); $\alpha_L$ = dispersivity (m), $x$ = distance (m), $k$ = filtration coefficient (m$^{-1}$); $v$ = pore velocity (m d$^{-1}$); $\lambda$ = inactivation rate (d$^{-1}$).

Schijven & Rietveld (1997) annotated the terms in the above equation as follows: The first term on the right hand side describes transport of a particle by advection with the flow of groundwater. The second term describes the longitudinal dispersion of the particle, where $\alpha_L$ is scale-dependent dispersivity. The third term refers to filtration term for irreversible adsorption with filtration coefficient, $k$. $R$ refers to retardation due to reversible adsorption, and $\lambda$ stands for the inactivation rate and is assumed to be dependent on groundwater temperature.

### 10.2.4 *Pathways of arsenic to humans*

The following are the pathways of arsenic to man, through:
- Drinking water – as in Bangladesh, West Bengal (India), Inner Mongolia, Shaanxi, Xinjiang (China), southwest and northeast coasts of Taiwan, western USA, etc.
- Inhalation of arsenic-containing aerosols – coal burning, as in Guizhou (China), copper smelting and arsenic mining, as in Yunnan (China), NWT (Canada), etc.
- Dermal uptake: Although the human skin does act as a barrier, it cannot completely block the entry of environmentally harmful substances from entering the body. For instance, arsenic-induced skin cancers resulting from the use of arsenical pesticides have been reported among the wine growers of Beaujolais. Similarly, rice farmers, fishermen and salt workers whose feet and legs are exposed to arsenical waters for long periods, have been found to develop arseni-

cal skin lesions. The dermal pathway explains the prevalence of 'Blackfoot disease' in Taiwan where rice farming is practiced, but not in arid Chile, though in both the cases the water has high arsenic content.

– Diet: The daily dietary intake of inorganic arsenic ranges from 8-14 $\mu$g d$^{-1}$ in USA, and 5-13 $\mu$g d$^{-1}$ in Canada. It is probably 50 $\mu$g d$^{-1}$ in the case of Taiwan. Though the content of arsenic in fish (0.1-64 mg As kg$^{-1}$) and shellfish (0.2-126 mg As kg$^{-1}$) is high, it is present in a relatively non-toxic form (e.g. arsenobetaine: $(CH)_3$ As$^+$ $CH_2COOH$), which is readily excreted. Herbal medicines of Indian and Chinese origin which are becoming increasingly popular in the western countries, have high content (25-107,000 mg As kg$^{-1}$) of arsenic, all of which is in inorganic form. No significant incidence of arseniasis has been reported any where in the world based on diet alone.

10.2.4.1 *Route of Rock – soil/sediment – water – humans:*
This is, by far, the most important pathway. As against 50 $\mu$g l$^{-1}$ prescribed by US EPA and 10 $\mu$g l$^{-1}$ recommended by WHO, the tubewell water in west Bengal has 60-3700 $\mu$g As l$^{-1}$. In Pabna district in northern Bangladesh, the pumped groundwater has an exceptionally high Arsenic content of 14 mg or 14,000 $\mu$g As l$^{-1}$ (Rott & Friedle 1999). Estuarine sediments are the most likely sources of arsenic in West Bengal (India) and Bangladesh. Estuaries act as sinks of heavy metals, such as As, coming from the rivers. Though the normal range of As in the sediments is not high (5-12 $\mu$g g$^{-1}$), instances are known of the existence of very high concentrations (up to 900 $\mu$g g$^{-1}$) in some estuaries. Authigenic pyrites (Fe S$_2$) is probably the most important host mineral for inorganic arsenic in estuarine sediments. The mobilization of arsenic from pyrites and its entry into the groundwater depends upon the mode of association and speciation of As in pyrites. For instance, arsenic existing as intergranular films or 'paint' on pyrite, gets mobilized more readily than (say) arsenic occupying a lattice site.

Within the conterminous USA, high arsenic levels in groundwater are attributed to the following sources:
1. Upflow of geothermal water,
2. Dissolution of, or desorption from, iron oxide,
3. Dissolution of sulfide minerals, and
4. Evaporation of concentration. Arsenic concentrations exceeding 10 $\mu$g l$^{-1}$ are more common in the western USA than in the east (Welch et al. 1999). Significantly, thermal springs and high evaporation are, among other things, responsible for the high fluoride content of the waters in northern Tanzania. It therefore stands to reason that on the analogy of western USA, thermal springs and evaporation might be contributing arsenic to the waters in northern Tanzania, though very little data is available on the subject.

Compounds of arsenic enter the human body through gastrointestinal tract and the lung. They are transported in the blood, and reach the liver and kidney. Most mammals have a built-in mechanism to detoxify arsenic. They do so through the process of methylating inorganic arsenic to methylarsonic acid (MMA: $CH_3$ $AsO(OH)_2$) or dimethylarsonic acid (DMA: $(CH_3)_2$ As O (OH)). The methylated arsenites are less reactive, less toxic and more readily excreted in urine. About 60-70% of the intake of arsenic is thus excreted within 48 hours. The component of arsenic that is not excreted gets sequestrated in the skin, hair and nails. This explains the manifestation of arseniasis in the form of arsenic corns and hyperkeratosis on the hands and feet, loss of hair, and transverse striations or Mees lines on the nails.

Since arsenic is rapidly cleared in urine and blood in 1-3 days, arsenic concentration in these media cannot be used for retrospective diagnosis of As toxicity. On the other hand, urine analysis can be relied upon for identifying *ongoing* exposure. Hair and nails have the same kind of affinity for As, but hair is more convenient to handle, and is therefore preferred. Hair arsenic is a good indicator of arsenic toxicity, but only when external contamination can be excluded, which is extremely difficult to achieve. Arsenic content of hair (in terms of $\mu g\ g^{-1}$) is < 1 in normal hair, < 3 in polluted areas, and $\geq$ 10 in the case of chronic poisoning. A practical way to diagnose chronic arsenic poisoning is on the basis of characteristic clinical features, such as skin lesions, debility, weight loss and neuropathy, with hair arsenical levels providing supportive evidence (Hindmarsh, 2000).

If the body is deficient in methionine due to (say) nutritional inadequacies, this process of detoxification of arsenic may not take place. The less the excretion of arsenic in urine, the more the arsenic that is retained in the body, and the greater the severity of arseniasis. This explains as to why the gangrene of the extremities (Blackfoot disease) is more prevalent in the Developing countries, particularly among the poor.

Avoidance of drinking high-As water is the first step in the mitigation of arseniasis endemic in Developing countries, such as Bangladesh and India. Three possibilities exist:

1. An eminently practical way for countries like Bangladesh and India, is to go in for roof-harvesting of rain water in a big way. Assuming a roof area of 40 $m^2$, annual rainfall of 1000 mm, and collection efficiency of 50%, the amount of rainwater that could be harvested from rooftop of even a modest dwelling in a village, would be about 20,000 l. Thus, a family can have about 50 l $d^{-1}$ of potable, arsenic-free water for drinking and cooking purposes. For instance, Bermuda depends for its supplies of water (80 l $d^{-1}$ per capita) almost exclusively on harvesting of precipitation (1430 mm) (see Section 4.1.1).

2. In West Bengal (India) and Bangladesh, a number of low-cost methods have yielded good results:
   - community supply based on alumina column,
   - domestic supply based on three-pitcher (locally known as 3-kalshi) filtration assembly (uppermost containing iron chips and coarse sand, the middle one with wood charcoal and fine sand, with the lowermost receiving filtered water): this filtration can bring down the total arsenic to below 10 ppb level,

even through the input water may be having 1100 ppb. The daily capacity of the system is about 40-140 l d$^{-1}$ (Alauddin 2000 – Abstract, p. 62, Fourth Int. Arsenic Conf., San Diego, June, 2000).

3  Removal of arsenic: Sancha (1999) described a method of treating high-As water with an oxidant (such as, Cl$_2$) and a coagulant (such as, Fe Cl$_3$) to bring down the As content of raw water from 0.400-0.450 mg As l$^{-1}$ to 0.040-0.050 mg As l$^{-1}$. The arsenic removal facility at Salar de Carmen has a capacity of 520 l s$^{-1}$ (about 45,000 m$^3$ d$^{-1}$), costs about USD 20 million, and produces water with residual As of 0.040 mg As l$^{-1}$ at a cost USD 0.04 cents m$^{-3}$. Joshi & Chaudhuri (1996) developed a technique for the removal of arsenic from groundwater through the use of iron-oxide coated sand. Clifford et al. (1999) used type 2 polystyrene divinylbenzene, a strong base resin, for the removal of arsenic from water.

A novel approach has been proposed by Rott & Friedle (1999) for the subterranean removal of arsenic from groundwater. In effect, the aquifer is used as a natural, biochemical reactor, whereby the water treatment that is normally performed above ground is performed in the aquifer itself. Oxygen-enriched water is injected under pressure into the aquifer with a water jet air pump. A number of physical, chemical and biological processes get initiated by the introduction of oxygen. Autotrophic micro-organisms utilizing the energy from the oxidation process, become active. Dissolved iron and manganese are adsorbed on the bacteria sheaths by the bio-film. The introduced oxygen oxidizes As (III) to Arsenic (v), and facilitates the precipitation of arsenic and its absorption on ironhydroxide and manganese-oxide. Field experiments using the above principle were conducted to make the aquifer permanently capable of delivering water which satisfy the requirements of potability, namely, 10 µg As l$^{-1}$, 200 µg Fe l$^{-1}$, 50 µg Mn l$^{-1}$. While the removal of arsenic and iron normally starts after a few treatment cycles, the removal of manganese has been found to take longer (a few months). The method has the following advantages:

– It is a low-cost technique, using a naturally occurring process; it involves the use of only oxygen as a reagent. No other chemicals are involved,
– No sludges or waste products are produced,
– No above-ground buildings are necessary. The technique has to be tested in the context of different aquifer characteristics, and the chemistry of the groundwater.

Microbial remediation of community water supplies through the use of bacterial cultures of *Thiobacillus acidophilus* is an affordable alternative.

### 10.2.4.2  *Route of Mineral – aerosol – humans:*

This route is based upon an entirely anthropogenic source. Low-rank coals invariably contain pyrite. Arsenic may substitute in pyrite (FeS$_2$) or may be found in the form of a separate mineral, arsenopyrite (Fe As S). In Guizhou province of China, coals have very high arsenic content (9600 mg kg$^{-1}$). When such high-As coal is used for cooking, keeping warm, and drying of grain, arsenic content of the ambient kitchen air rises to 0.003-0.11 mg m$^{-3}$. Exposure to this environment leads to the absorption of arsenic by the respiratory tract, skin, and digestive tract (Zheng, B.S. et al. 1994, quoted by Sun et al. 1999).

About 30,000 workers in the copper smelting and arsenic mining industries are exposed to high-As aerosols. In the case of tin mine workers in the Yunnan province, where the workers are exposed to high-As aerosols, the cancer incidence among these workers is 716.9 per 100,000 which is 82 times that of the controls. The average As content in the lungs of the cancer patients was 43.33 mg kg$^{-1}$.

Carcinoma of the lung is associated with inhalation of arsenic dusts. Instances are known from Southeast Asia where lung cancer is attributed to As in drinking water.

In the Xinjiang province of China, both arsenic and fluoride contents are high in drinking water as well as the coal used for burning. This led to the concurrent endemicity of arseniasis and fluorosis among the populations.

Coal is the principal source of energy in China. China is the largest producer of coal in the world (1235 Mt in 1998). With increased industrialization, and with people aspiring for a higher standard of living, consumption of coal-fired thermal energy as also the use of coal in home heating, has been growing rapidly. There is a price for this. It has been said that nine out of ten most polluted cities in the world are in China, and one out of three deaths in China is due to contaminated air and water (*Time*, USA, Nov. 8, 1999).

### 10.2.4.3 *Possible mitigation measures*

The following technological options are available to reduce the emissions of the particulate matter and harmful emissions containing arsenic:

1. Coal-cleaning technologies to bring down the non-combustible ash content, and sulfur content (which is usually present in the form of pyrite),
2. Mechanical and electrical devices which can remove the particulates to the extent of 99%,
3. Flue gas desulfurization technologies ('scrubbers') which can remove sulfurous emissions by 90%, and
4. Fluidised bed combustion technologies, which can bring about abatement of sulfur dioxide emissions by 90%, besides reducing the nitrogen oxide emissions significantly. An innovative technology of integrated coal gasification combined-cycle technologies with fluidised bed combustion, holds great promise (this does not require pre-cleaning of coal, has high thermal efficiency and brings about significant reduction in the emissions of sulfur dioxide and nitrogen oxide. The cost-effectiveness of these technologies could be evident from the fact that fluidised bed combustion which completely eliminates the particulate matter, reduces the SO$_2$ emissions by 90%, and NO$_x$ emissions by 56%, has thermal efficiency of 33.8% – and all this at an added cost (as a percentage of generation costs) of less than 2% (*Development and the Environment*, World Bank Report 1992, p. 119).

### 10.2.5   *Etiology of arseniasis*

Etiology means the study of causation of diseases.

It has been established that ingestion of water containing inorganic arsenic ($As_i$) is associated with increase in skin lesions, cardiovascular diseases and cancers of skin, kidney, liver and bladder. The new carcinogen risk assessment paradigm of US EPA is based on Mode of Action (MOA) hypotheses which 'relate carcinogenicity to obligatory precursor effects, link cancer and non-cancer responses through common pathways, and predict dose-response relationships via biologically-based dose-response (BBDR) models... $As_i$ – induced carcinogenesis involves impaired DNA repair, altered DNA methylation, increased growth factor synthesis, and increased oxidative stress' (Andersen et al. 1999). Arseniasis has been studied most intensively in Taiwan. Tumour incidence increases when the $As_i$ content of water is more than 600 $\mu$g $l^{-1}$. Incidence of bladder cancer is highly non-linear, with sharp increases occurring when $As_i$ level of water is more than 500 $\mu$g $l^{-1}$. A non-cancer end-point is 'Blackfoot disease '(gangrene of the extremities). The incidence of Blackfoot in Taiwan is 9/1000, which increases with age, and dose of $As_i$. Dose-dependent relationship has been found in regard to diabetes mellitus, ischemic heart disease, and peripheral vascular disease. Cancer incidence has been found to be high among those with Blackfoot disease and skin disease, though it has not been established whether these non-cancer effects are direct precursors of tumors. The pathways of arsenate metabolism are extremely complex. Which of the intermediate metabolites (such as, MMAA, DMAA, glutathione conjugates), singly or jointly, are involved in molecular interactions, toxicity and carcinogenicity of $As_i$, and at what point the normal cells are converted to malignant phenotype, are poorly understood (Andersen et al. 1999).

### 10.2.6   *Regulatory process*

*How much arsenic is too much?* There is no simple answer to this question (vide compilation by Grissom et al. 1999, and Thornton 1999).

*Drinking water*: Arsenic level in surface water is usually less than 10 $\mu$g As $l^{-1}$. Uncontaminated groundwater has about 2 $\mu$g As $l^{-1}$, but in most cases, the figure is much higher. The US EPA is in the process of revising downward the Maximum Contaminant Level (MCL) in drinking water of 50 $\mu$g As $l^{-1}$. WHO (1993) has already introduced a provisional guideline of 10 $\mu$g As $l^{-1}$.

*Soil*: The arsenic content of 'unpolluted soil' is 1-40 mg As $kg^{-1}$. In areas with sulfide deposits or geothermal activity, the figure may go up to 8000 mg As $kg^{-1}$. Mean levels may be taken as 5 mg As $kg^{-1}$. The regulations in respect of soil are still tentative. In 1987, the Interdepartmental Committee on the Redevelopment of Contaminated Land (ICRCL) of UK prescribed 'tentative trigger concentrations' for contaminant chemicals at various sites. It is implied that it is safe to use the site below this 'trigger' level of concentration. Above this level, remedial action is recommended. Above the 'Action' level, remedial action is mandatory or land-use must be changed. ICRCL prescribed 10 $\mu$g As $g^{-1}$ for soil sites under urban development, and 40 $\mu$g As $g^{-1}$ for parks and amenity areas. In the case of concentrations above these levels, the matter has been left to the judgment of the local authorities (on its own initiative, one local authority in UK prescribed an 'action' value of 140 $\mu$g As $g^{-1}$). The trigger values have been criticized as being unrealistic, since their rigid application would prohibit redevelopment of large areas of Devon and Cornwall. Australia followed a different approach, based on site-specific assessment of 'fitness for the purpose'. Accordingly, the following values (in terms of $\mu$g As $g^{-1}$) have been prescribed for the soils for purposes shown against them, to which UK also may adhere: Residential with garden – 175; Residential without garden – 300; Allotment: 250; Parks, open spaces – 300; Commercial/Industrial: 1000.

*Air*: Ambient air levels are usually about 0.1 $\mu$g As $m^{-3}$. Kitchen air with 110 $\mu$g As $m^{-3}$ have been reported in China. Levels of arsenic in indoor dust could be as high as 2000 mg As $kg^{-1}$. There do not appear to exist any NOEL (No Observed Effect Level) data in this regard, presumably because this is a hazard restricted only to some Developing countries which use soft, high-As coal in the kitchen for cooking and drying (such as, Guizhou province of China, and to some extent, India). But there is an urgent need to provide the guide-lines (in terms of $\mu$g $m^{-3}$) to enable the concerned governments come up with mitigation measures in view of the large populations at risk.

### 10.2.7  *Where do we go from here?*

The important issues in the case of arseniasis are posed in the form of four questions.

1. *Is arsenic an essential element?*
   Animal experiments conclusively show that arsenic has a physiological role affecting methionine metabolism. The essentiality of arsenic for human beings is, however, not established. The central question is whether arsenic is like selenium (i.e. an essential element, with a Recommended Daily Intake, though toxic in high doses) or whether arsenic is like cadmium which is toxic at any level (gun kills – there is no such thing as a benevolent gun).

2. *Is there a threshold for arsenic exposure for non-cancer and cancer endpoints?*
   There are sound reasons to believe that the dose – response relationship in the case of arsenic is non-linear. For instance, Guo et al. (1994) found the incidence of bladder cancer in Taiwan increases sharply above 500 $\mu$g $l^{-1}$. On the

basis of the statistical analysis of data from Taiwan and elsewhere, Stöhrer (1991) concluded that there is a common threshold (400 μg l$^{-1}$ (?)) for skin cancers, internal cancers and non-cancer endpoints. While the incidence of arseniasis above 200 μg l$^{-1}$ is well established, there is much controversy about the risks of intake of less than 100 μg l$^{-1}$ (Hindmarsh 2000).

3. *How do nutrition/socio-economic status, genetics, etc. modify the risk?*

There is a clear correlation between the susceptibility for arseniasis (blackfoot disease in Taiwan and arsenical melanosis in West Bengal) and poverty and inadequate nutrition. That the victims having diffuse melanosis (preliminary stage of skin lesions) recovered when they discontinued the use of contaminated water and got nourishing food, demonstrates the essential correctness of the above surmise (SOES 1996).

Chile is a spectacular example of the role that genetics could play in modifying the risk. Arroyo of Chile (quoted by Brown 1999) raised two pertinent questions:

– Why is there no arseniasis among the people of Atacameño who have been drinking water with high levels of arsenic (600 μg l$^{-1}$) for many decades, and

– Why do some people in Antafagosta develop arsenic-related diseases and others do not, though they are exposed to the same levels of arsenic?

4. *What is the best treatment for chronic arsenic intoxication?*

Therapy with d-penicillamine did not prove effective. On the other hand, oral treatment with retinoids and the use of selenium as an antioxidant nutrient, hold promise to mitigate the cutaneous arsenicism and other forms of arsenic toxicity. Use of antidepressants for pain relief in regard to peripheral neuropathy, and the application of topical keratolytics for palmar-plantar hyperkeratoses, may offer short-term symptomatic relief (Kossnet 1999). Field experiments in Chile, Inner Mongolia (China) and Romania showed that treatment with DMPS (2,3-dimercatopropane-1-sulfonate) increases the urinary excretion of arsenic in humans, without any adverse effects (Aposhian 2000 – p. 54, Abstract in the Fourth Arsenic Conference, San Diego, June, 2000).

## 10.3 FLUOROSIS

Fluoride disorders and fluorosis are caused by excessive ingestion of fluoride, mainly through drinking water route.

### 10.3.1 *Geochemistry of fluorine*

Fluorine (at. wt.: 9; electron configuration: 2-7) has only one oxidation date (F$^{-1}$).

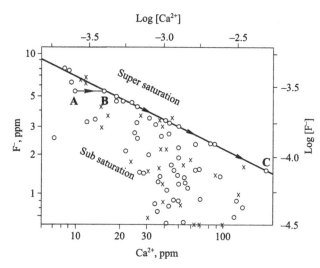

Figure 10.4. The stability of fluorite and the saturation of groundwaters from Sirohi, Rajasthan, India (source: Handa 1975; Appelo & Postma 1996, p. 46). The changes in water chemistry that take place on the addition of gypsum (points A, B and C) are described in the text.

Fluorite ($Ca F_2$) is the principal mineral of fluorine. The dissolution of fluorite takes place in the following manner:

$$Ca F_2 \leftrightarrow Ca^{2+} + 2 F \tag{10.10}$$

The solubility constant for the reaction is:

$$\log K_{\text{fluorite}} = \log [Ca^{2+}] + 2 \log [F^-] = -10.57 \tag{10.11}$$

Thus, on a log – log plot of $Ca^{2+}$ (ppm) and $F^-$ (ppm), the equilibrium condition between fluorite and solution is given by a straight line (Fig. 10.4; source: Handa 1975, Appelo & Postma 1996, p. 46). All combinations of $Ca^{2+}$ and $F^-$ that plot below the equilibrium line are subsaturated, and those plotting above the line are supersaturated. This plot can be made use of to place an upper limit to $F^-$ concentrations in the groundwaters. It also shows that groundwaters with higher natural concentrations of $Ca^{2+}$ are characterized by lower concentrations of $F^-$, and vice versa. Handa (1975) verified this observation in the field in the case of fluorous waters associated with Sirohi granites, Rajasthan province, western India. Similarly, in northern Tanzania, groundwaters in low-ca, alkali basalt aquifers are invariably more fluorous than groundwaters in high-ca, tholeiitic basalt aquifers. On the basis of this consideration, it is possible to identify broadly the areas where the groundwater is likely to be relatively safer (i.e. low fluoride concentration).

The $Ca^{2+}$ vs $F^-$ plot could also be used to understand and apply the principle behind the use of gypsum ($CaSO_4 \cdot 2 H_2O$), or dead-burnt magnesite ($MgCO_3$), or combinations of alum, and lime, to defluoridate the fluorous waters.

Gypsum is more soluble than fluorite.

$$CaSO_4 \leftrightarrow Ca^{2+} + SO_4^{2-} \tag{10.12}$$

$$K_{gypsum} = [Ca^{2+}][SO_4^{2-}] = 10^{-4.60} \text{ at } 25°C \qquad (10.13)$$

When gypsum is added to a water corresponding to composition, $A$, the water composition will change to composition, $B$, on the equilibrium line. Further addition of gypsum will move the composition down the equilibrium line, until point, $C$, when the saturation of gypsum is also reached (Appelo & Postma 1996, p. 47).

### 10.3.2 *Geochemical distribution of fluorine*

Fluorine in surface and ground waters is derived from the following natural sources:
– Leaching of rocks in the area: fluorine is concentrated in granites (750 ppm), and phosphatic fertilizers (3-3.5%),
– Dissolution of fluorides from volcanic gases by precipitation,
– Fresh or mineral springs,
– Marine aerosols and continental dusts, etc.
Anthropogenic sources of fluoride are:
– Industrial emissions such as freons, organo-fluorine compounds produced by the burning of fossil fuels, and from dust in the cryolite factories,
– Industrial effluents, and
– Runoff from farms using phosphatic fertilizers extensively. In areas like northern Tanzania, the high fluoride content of the waters arises entirely from natural sources, and hence it is a case of 'natural pollution'. On the other hand, fluorine-bearing cryolite dust (as in Denmark in the past) is wholly anthropogenic. The high fluoride contents in parts of Greenland is attributed not only to the leaching of alkalic rocks by the stream waters, but also to the effluents from an industrial unit processing the fluorine-bearing ores. This is a good example of a situation where the 'natural' and anthropogenic 'pollutions' cannot be separated.
Surface waters generally contain about 0.2 mg l$^{-1}$ of fluoride, whereas the fluoride content of the groundwaters depend upon the nature of the aquifer. It may vary from 0.4 mg l$^{-1}$ in aquifers composed of shales, to 8.7 mg l$^{-1}$ in areas of alkalic rocks. Effluents from phosphate mining may contain upto 100 mg l$^{-1}$ of fluoride. Plants have high fluoride content in the range of 0.1 to 10.0 ppm (dry weight). Some plants accumulate excessive quantities of fluoride – for instance, the fluoride content of tea could be 100-760 ppm. Among food items, fish concentrates may contain 20-760 ppm of fluoride. A saline encrustation, trona (Na$_2$ CO$_3$. Na HCO$_3$ 2 H$_2$O), called 'magadi' in Swahili language, is used in the cooking of beans, and as salt. 'Magadi' has very high fluoride content, ranging from 200 to 6000 ppm (Jan. 2000 issue of *National Geographic* of USA carries a spectacular photograph of huge blocks of trona forming in Lake Natron, a rift lake in northern Tanzania). Poor people use trona in lieu of sea salt because trona improves the cookability of the beans, and is collectible at no cost.

While fluorosis arising from the drinking of fluorous waters continues to persist in northern Tanzania, fluorosis arising from the inhalation of cryolite dust in Denmark has been eliminated.

Groundwaters in several parts of India (provinces of Andhra Pradesh, Madhya Pradesh, Rajasthan, etc.) are fluorous. The problem is got over either through the defluoridation of water using alum-lime treatment or by passing the water through activated charcoal made from coconut shell. In the outlying parts of Hyderabad city in southern India, the tubewell water is used for most purposes (bathing, washing, gardening, etc.), and potable water drawn from Manjira river and supplied through tankers on payment basis, is used for drinking. This is an eminently sensible approach.

### 10.3.3 *Pathways of fluorine to man*

Fluorine is an essential element with Recommended Daily Intake (RDI) of 1.5 to 4.0 mg d$^{-1}$. Health problems may arise from deficiency (dental caries) or excess (dental mottling and skeletal fluorosis). Unlike other elements for which food is the principal source to the extent of (say) 80%, water is the principal source of fluoride. The diadochic relationship between F$^-$ and OH$^-$ ions in many silicates and phosphates arises from the close similarity in their ionic charge and radius (F$^-$: 1.36 Å; and OH$^-$: 1.40 Å). This property explains the replacement of hydroxylapatite in teeth and bones by fluorapatite in the case of persons who are exposed to higher levels of intake of fluoride.

### 10.3.4 *Fluorosis in northern Tanzania – a case study*

Aswathanarayana (1990) gave a detailed account of the etiology and epidemiology of fluorosis in northern Tanzania. Figure 10.5 shows the areas of endemic goiter and fluorosis in Tanzania.

The natural waters in northern Tanzania are characterized by very high contents of fluoride, which are among the highest in the world: Maji ya Chai river (12-14 mg l$^{-1}$) (in Swahili language, maji means water, and chai means tea; the color of the river water looks like tea decoction!), Engare Nanyuki river (21-26 mg l$^{-1}$), drinking water pond of the Kitefu village (61-65 mg l$^{-1}$) (as should be expected, every child in the village has mottled teeth!), thermal springs of Jekukumia (63 mg l$^{-1}$), and the soda lakes of Momella which are the habitat for hippos (upto 690 mg l$^{-1}$). The high fluoride content of water of several streams is attributed to their being fed by thermal springs.

Aswathanarayana et al. (1986) proposed a geochemical model to account for the high fluoride contents, of natural waters of northern Tanzania (Fig. 10.6). Fluoride is derived from two sources:
1. Steady influx of fluoride into the surface and groundwaters by the leaching of the Quaternary volcanics of the East African Rift whose fluoride content ranges from 0.029 to 0.49%,

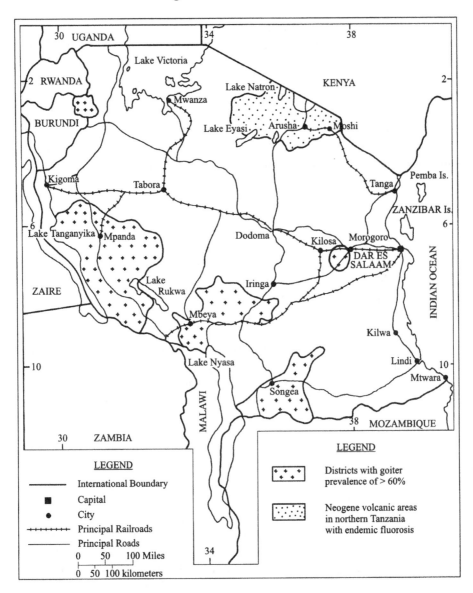

Figure 10.5. Areas of endemic fluorosis and goiter in Tanzania (source: Aswathanarayana 1995, p. 242).

2. Episodic, massive influx of fluoride which arose due to the leaching of the highly soluble villiaumite (Na F) present in the volcanic ash, exhalations and sublimates related to Miocene to Recent volcanism (for instance, the ash of the unique carbonatitic volcano of Oldoinyo Lengai in the region which erupted as recently as 1960, has fluoride content of 2.7%; minor eruptions of this volcano have continued, and as the area is remote and uninhabited, the eruptions are known only from the satellite images).

Figure 10.7 is a plot of the $\delta D$ vs $\delta^{18}O$ of the fluoride waters of the northern Tanzania. Its slope of 5.5 is indicative of excessive evaporation, and is sharply different from the slope of 8 which is characteristic of meteoritic waters (see Section 2.5.1). Whereas the cluster related to brackish springs and rivers is characterized by negative $\delta D$ and negative $\delta^{18}O$ values, the cluster pertaining to the saline lakes (with high F and Na, and low Ca) shows positive $\delta D$ and positive $\delta^{18}O$ values. One freshwater lake with low F and Na and high Ca does not plot with the saline lakes cluster, but has similar slope and isotopic characteristics as other lake waters.

It has been estimated that the rate of ingestion of fluoride in northern Tanzania is about 30 mg d$^{-1}$, contributed as set out below.

This is about 15 times more than the estimated intake of 2 mg d$^{-1}$ of fluoride per capita per day in temperate countries (1-2 l d$^{-1}$ of water with 1-1.5 mg l$^{-1}$ of flouride).

| Source | Contribution |
|---|---|
| 3 l of drinking water, with 8 mg l$^{-1}$ | 24 mg d$^{-1}$ |
| 10 g of locally grown tea, with 200 ppm of chloride | 2 mg d$^{-1}$ |
| 5 g of 'magadi' with 1000 ppm of fluoride | 5 mg d$^{-1}$ |
| Miscellaneous (through diet) | 2 mg d$^{-1}$ |
| Total | 33 mg d$^{-1}$ |

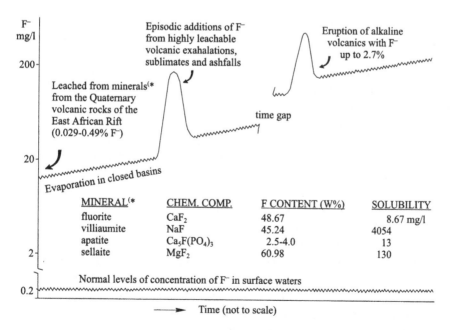

Figure 10.6. Schematic diagram of the geochemical model to account for the high fluoride content of natural waters in northern Tanzania (source: Aswathanarayana 1995, p. 244).

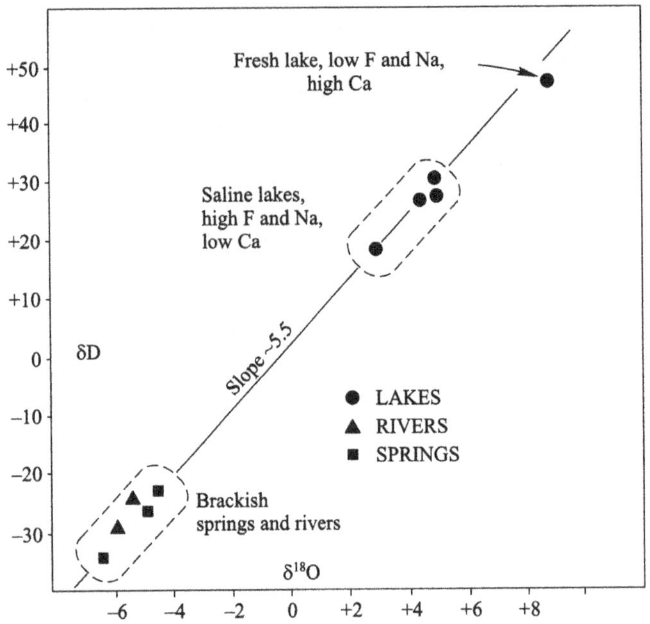

Figure 10.7. Plot of δD vs δ$^{18}$O of the fluoride waters of northern Tanzania (source: Aswathanara-yana 1995, p. 244).

The endemicity of fluorosis in northern Tanzania is a consequence of prolonged ingestion of more than 8 mg d$^{-1}$ of fluoride. Fluorosis may be manifested as dental mottling, and in more severe cases, as skeletal fluorosis (genu valgum, character-ized by bow legs, knock-knees, stiffness of trunk, impeded movement of limbs, severe joint pains, etc.). Variability in the severity of fluorosis in the population whose drinking water is drawn from the same source, is attributed to the follow-ing factors, acting singly or jointly:

– Fluorosis may be aggravated by the inadequate intake of nutrients, such as, pro-teins, ascorbic acid, calcium, etc. Thus, low-income groups who are unable to afford nutritious food, are at higher risk than affluent people who can afford nutritious food,
– Children are more affected than adults, probably because their teeth and bones (in which F$^-$ substitutes for OH$^-$) are still forming,
– Men, particularly those involved in doing manual labor in the sun, are more af-fected than women, probably because such men tend to drink more water, and therefore ingest more fluoride. There is evidence to suggest that drinking of water with 8 mg l$^{-1}$ of flouride protects post-menopausal women from osteoporosis (this is probably because fluorapatite resists resorption better than hydroxylapatite).

To prevent dental caries, some communities fluoridate their drinking water sup-plies in order to ensure about 1 mg l$^{-1}$ of fluoride. Recent studies indicate that the substitution of OH$^-$ by F$^-$ in some metabolites, could render them carcinogenic.

On this ground, some authorities have raised doubts about considering fluoride as an essential element.

### 10.3.5 *Concurrent endemicity of fluorosis and arseniasis*

Waters in Chile have high As content (400-600 µg $l^{-1}$) but normal fluoride content (0.30-0.50 mg $l^{-1}$), and hence while arseniasis is endemic in Chile, fluorosis is not. On the other hand, both arseniasis and fluorosis are endemic in Inner Mongolia (China) and possibly in the Kilimanjaro region of northern Tanzania. In the case of northern Tanzania, drinking water is the only route of ingestion of fluoride and arsenic. In the case of Inner Mongolia, the ingestion of arsenic and fluoride is not only through the drinking water but also through the inhalation of smoke from the burning of arsenious coal.

As and F form complexes in the aqueous environment – As $F_3^0$ (aq) and As $F_5^{2-}$ (aq) – but the pathways of these complexes to man, and their metabolic consequences are poorly known. The behavior of these complexes in the aquatic environment is of particular relevance in the understanding of the concurrent endemicity of fluorosis and arseniasis in Inner Mongolia, and possibly in northern Tanzania.

Alumina-based, metal oxide adsorption media are commercially available to remove both arsenic (arsenite and arsenate) and fluoride simultaneously.

## 10.4  RISK ASSESSMENT

'Risk Assessment (RA) is a process in which prediction and evaluation are combined to estimate the probability or frequency of *harm* (risk) for a given *hazard* (an event which has the potential to be harmful). Manifestation of a risk requires an event, and a pathway for transport, and a receptor which could be harmed at the exposure point' (Petts & Eduljee 1994, p. 116). Risk management process includes not only risk analysis and risk evaluation but also risk reduction through various mitigation measures.

Risk is estimated on the basis of Poisson statistics. Haas et al. (1993) found that a beta-Poisson model best describes the probability of virus infection. They used this model to estimate the risk of infection, clinical disease and mortality due to hypothetical levels of viruses in drinking water, and calculated the annual and lifetime risks.

Risk assessment, as applied to arseniasis, involves the following four steps (Nat. Acad. Sci. 1983):
1. Hazard identification: to define human health effects associated with the hazard (ingestion of excessive quantities of arsenic, present in water/dust/diet, etc.),
2. Dose – response assessment: to characterize the relationship between the dose administered, and the incidence of the health effect (ingestion of arsenic above the threshold, leading to arseniasis, in the form of skin lesions, vascular damage, cancers, etc. in the exposed population),

3. Exposure assessment: to determine the size and nature of the population exposed, and the route, amount and duration of the exposure,
4. Risk characterization: an integration of the above three steps in order to estimate the magnitude of the public health problem. In the case of Taiwan, there were 18 times as many cases of skin signs (hyperpigmentation or keratosis) as skin cancer, skin signs were present in 94% of the skin cancer cases, and skin cancer was 70 times more prevalent among those with skin signs than those without skin signs. The above information can be schematically shown in the form of a Venn diagram with areas proportional to the size of the total population, number of persons showing skin signs in the population, and the number of persons showing skin signs, and suffering from skin cancer (Tseng et al. 1968, Yeh 1973, quoted by Brown 1999).

Chen et al. (1992) studied the relationship between the content of inorganic arsenic in drinking water, and the incidence of liver cancer, lung cancer, bladder cancer and kidney cancer among the population at risk in Taiwan. For each cancer, the number of deaths during a 13-year period were analyzed, in terms of four age categories (25, 40, 60 and 80 yrs.) and four water categories (0.02, 0.2, 0.45 and 1.0 mg As $l^{-1}$). The data for males and females were modeled separately. Two lag periods (0 and 15 yrs) were considered.

The number of cancer deaths in the $i$-th age and $j$-th exposure category were modeled as a Poisson variable with the expected value

$$Py_{ij} * a_i [ 1 + \beta* (t_i - \text{Lag}) * e_j ] \tag{10.14}$$

where $Py_{ij}$ is the number of person-years of observation in the $ij$-th cell, $t_i$ is the mid-point of the $i$-th age interval, $e_j$ is the assumed average inorganic arsenic concentration in the $j$-th exposure category, and $\beta$ is the parameter that quantifies the carcinogenic potency of inorganic arsenic. The model provided an acceptable fit ($p > 0.05$) for the various cases.

The risk of cancer from the lifetime consumption of inorganic arsenic in drinking water is estimated from the following parameters:

$Ed_{10}$: Concentration of inorganic arsenic in water (mg $l^{-1}$) and in air ($\mu g\ m^{-3}$) from which a person exposed constantly over a lifetime would have a 10% increased risk of dying of a particular cancer.

$LED_{10}$: 95% statistical lower bound on the $Ed_{10}$.

Clewell et al. (1999) gave the following consolidated comparison of risk estimates for inorganic arsenic for oral and inhalation routes:

| Route | $LED_{10}$ | MCL* | $LED_{10}$ – Linear unit risk | Current USEPA unit risk |
|---|---|---|---|---|
| Oral | 10 $\mu g\ l^{-1}$ | 50 $\mu g\ l^{-1}$ | 0.01 $(\mu g\ l)^{-1}$ | $5 \times 10^{-5}\ (\mu g\ l)^{-1}$** |
| | 144 $\mu g\ l^{-1}$ | | $7 \times 10^{-4}\ (\mu g\ l)^{-1}$ | |
| Inhalation | 168 $\mu g\ m^{-3}$ | N.A. | $6 \times 10^{-4}\ (\mu g\ m^{-3})^{-1}$ | $4.3 \times 10^{-3}\ (\mu g\ m^{-3})^{-1}$ |

*Maximum Contaminant Level in drinking water, prescribed by US EPA, **Based on skin tumor incidence (all others based on internal cancers)

The conclusions of Clewell et al. (1999) are as follows:

1. Though it is generally agreed that the carcinogenicity of inorganic arsenic is non-linear, the exposure level at which non-linearity might occur is not known,
2. There is a significant risk involved in the consumption of water with MCL of US EPA, and
3. Inhalation does not appear to entail significant risks of cancer.

## REFERENCES

*Suggested reading

Andersen, M.E. et al. 1999. Mode of Action studies for assessing carcinogenic risks posed by inorganic arsenic. In Chappell, W.R., C.O. Abernathy & R.L. Calderon (eds), *Arsenic Exposure and Health Effects*, p. 397-406. Amsterdam: Elsevier.

Appelo, C.A.J. & D. Postma 1996. *Geochemistry, Groundwater and Pollution*. Rotterdam: A.A. Balkema.

Aswathanarayana, U., P. Lahermo, E. Malisa & J.T. Nanyaro 1986. High fluoride waters in an endemic area in northern Tanzania. In I.Thornton (ed.), *Environmental Geochemistry*: 243-249. London: Royal Society.

Aswathanarayana, U. 1995. *Geoenvironment: An Introduction*. Rotterdam: A.A. Balkema.

Brown, K.G. 1999. Observation of arsenic exposure and health effects. In Chappell, W.R., C.O. Abernathy & R.L. Calderon (eds), *Arsenic Exposure and Health Effects*, p.407-413. Amsterdam: Elsevier.

*Chappell, W.R., C.O. Abernathy & R.L. Calderon 1999. *Arsenic Exposure and Health Effects*. Amsterdam: Elsevier.

Chen, C.J. et al. 1999. Emerging epidemics of arseniasis in Asia. In Chappell, W.R., C.O. Abernathy & R.L. Calderon (eds), *Arsenic Exposure and Health Effects*, p.113-121. Amsterdam: Elsevier.

Chen, C.J. et al. 1992. Cancer potential in liver, lung, bladder and kidney due to ingested inorganic arsenic in drinking water. *Br. J. Cancer* 66: 888-892.

Cherry, J.A. et al. 1979. Arsenic species as an indicator of redox conditions in groundwater. *J. Hydrol.* 43: 373-392.

Clewell, H.J. et al. 1999. Application of risk assessment approaches in the US EPA proposed cancer guidelines to inorganic arsenic. In Chappell, W.R., C.O. Abernathy & R.L. Calderon (eds), *Arsenic Exposure and Health Effects*, p. 99-111. Amsterdam: Elsevier.

Clifford, D.A., G. Ghurye & A.R. Tripp 1999. Development of an anion exchange process for arsenic removal from water. In Chappell, W.R., C.O. Abernathy & R.L. Calderon (eds), *Arsenic Exposure and Health Effects*, p. 379-387. Amsterdam: Elsevier.

Grissom, R.E. et al. 1999. Estimating total arsenic exposure in the United States. In Chappell, W.R., C.O. Abernathy & R.L. Calderon (eds), *Arsenic Exposure and Health Effects*, p. 51-60. Amsterdam: Elsevier.

Guo, H.-R. et al. 1994. Arsenic in drinking water and urinary cancers: A preliminary report. In Chappell, W.R., C.O. Abernathy & C.R. Cothern (eds), *Arsenic Exposure and Health*, p. 119-128. Northwood, England: Science & Technology Press.

Haas, C.N. et al. 1993. Risk assessment of virus in drinking water. *Risk Analysis*. 13: 545-552.

*Handa, B.K. 1975. Geochemistry and genesis of fluoride-containing ground waters in India: *Ground Water* 13: 275-281.

*Hindmarsh, J.T. 2000. Arsenic, its clinical and environmental significance. *J. Trace Elem. in Experim. Medic.* 13: 165-172.

Joshi, A. & M. Chaudhuri 1996. Removal of arsenic from groundwater by iron oxide-coated sand. *J. Environ. Eng.* 122(8).

Kosnett, M.J. 1999. Clinical approaches to the treatment of chronic arsenic intoxication: from chelation to chemoprevention. In Chappell, W.R., C.O. Abernathy & R.L. Calderon (eds), *Arsenic Exposure and Health Effects*, p. 349-354. Amsterdam: Elsevier.

Laconte, P. & Y.Y. Haimes (eds) 1982. *Water Resources and Land-use Planning: A Systems Approach*. The Hague: Martinus Nijhoff Publishers.

Marsh, G. M. & R.J. Caplan 1987. Evaluating health effects of exposure at hazardous waste sites: a review of the state-of-art, with recommendations for future research. In Andelman, J.B. & D.W. Underhill (eds), *Health effects from hazardous waste sites*. Michigan, USA: Lewis Publishers.

Nielsen, F.H. 2000. Importance of making dietary recommendations for elements designated as nutritionally beneficial, pharmacologically beneficial and conditionally essential. *J. Trace Elem. in Experim. Medic.* 13: 113-129.

Nordberg, G. et al. 2000. Essentiality and toxicity of trace elements: principles and methods for assessment of risk from human exposure to essential trace elements. *J. Trace Elem. in Experim. Medic.* 13: 141-153.

Park, N.S., T.N. Blandford & P.S. Huyakorn 1994. *CANVAS 2.0 A composite and analytical-numerical model for viral and solute transport simulation*. Documentation and User's Guide. Hydrogeologic Inc., Herndon, Va., USA.

Petts, J. & G. Eduljee 1994. *Environmental Impact Assessment for waste treatment and disposal facilities*. Chichester: John Wiley.

Piver, W.T. 1983. Biological and environmental effects of arsenic. *Top. Environ. Health*. New York: Elsevier.

Rott, U. & M. Friedle 1999. Subterrenean removal of arsenic from groundwater. In Chappell, W.R., C.O. Abernathy & R.L. Calderon (eds), *Arsenic Exposure and Health Effects*, p. 389-396. Amsterdam: Elsevier.

Sancha, A.M. 1999. Full-scale application of coagulation processes for arsenic removal in Chile. In Chappell, W.R., C.O. Abernathy & R.L. Calderon (eds), *Arsenic Exposure and Health Effects*, p. 373-378. Amsterdam: Elsevier.

Schijven, J.F. & L.C. Rietveld 1997. How do field observations compare with models of microbial removal. In *Under the Microscope: Examining microbes in Groundwater*. Denver, Co., USA: AWWA.

SOES (School of Environmental Studies) 1996. *Bangladesh's arsenic calamity may be more serious than West Bengal*. A Report from the School of Environmental Studies, Jadavpur University, Calcutta, India.

Stöhrer, G. 1991. Arsenic, opportunity of risk assessment. *Arch. Toxicol.*, 65, 525-531.

Sun, G.F. 1999. The present situation of chronic arsenism and research in China. In Chappell, W.R., C.O. Abernathy & R.L. Calderon (eds), *Arsenic Exposure and Health Effects*, p. 123-125. Amsterdam: Elsevier.

*Thornton, I. 1999. Arsenic in the global environment: looking towards the millennium. In Chappell, W.R., C.O. Abernathy & R.L. Calderon (eds), *Arsenic Exposure and Health Effects*, p. 1-8. Amsterdam: Elsevier.

US EPA (1989) *Exposure Factors Handbook*. Washington, D.C.: US EPA.

Welch, A.H. et al. 1999. Arsenic in groundwater supplies of the United States. In Chappell, W.R., C.O. Abernathy & R.L. Calderon (eds), *Arsenic Exposure and Health Effects*, p. 9-17. Amsterdam: Elsevier.

WHO (1993) *Recommended guidelines for drinking water*. Geneva: WHO.

*WHO/FAO/IAEA. 1996. Trace elements in human nutrition and health.* Geneva: WHO.

CHAPTER 11

# Climate change impacts on water resources management

*'What we perceive as the present is the bright crest of an ever growing past, and what we call the future is a looming abstraction ever coming into concrete appearance'*

Vladimir Nabokov

This chapter deals with three issues related to climate change:
- Causes and projected consequences of climate change, and proxy reconstructions of climate shifts,
- Likely impact of climate change on water resources, aquatic ecosystems, and human health, and
- Adaptation strategies to mitigate its adverse consequences with respect to water resources.

## 11.1 CLIMATE CHANGE FORCING

While the pre-twentieth century temperature variability has been attributed to both solar and volcanic forcing, there is unequivocal correlation between the temperature trend and atmospheric $CO_2$ in the twentieth century. It is therefore not an accident that the year 1998 turned out to be the warmest year of the millennium in the Northern Hemisphere (Mann et al. 1999).

Fossil fuels (coal, oil, and natural gas) account for 88% of the world's commercial primary energy. Burning of fossil fuels leads to the production of climate-relevant emissions of carbon dioxide ($CO_2$), methane ($CH_4$), nitrogen oxides ($NO_x$), carbon monoxide (CO) and volatile organic compounds (VOC). Atmospheric $CO_2$ concentrations have risen from a base of 280 ppmv (parts per million, volume ) before the Industrial Revolution to 360 ppmv at present; they are projected to rise to 560 ppmv ($2 \times CO_2$) by 2065. These emissions cause climate change, particularly in the form of anthropogenic greenhouse effect, or global warming as it is popularly known.

The extent of contribution of emissions from various sources to the greenhouse effect is as follows:
- Energy industries – about 50%, out of which $CO_2$ accounts for 40%,

- Chemical products, particularly CFCs: about 20%,
- Destruction of tropical rain forests and related causes: about 15%,
- Agriculture and others (e.g. waste deposit sites, etc.): about 15%.

### 11.1.1 *Role of water vapor in climate change*

A succinct account of the role of the water vapor in climate has appeared in *Eos* (AGU) of April 11, 2000.

The study of the temporal and spatial distribution of water vapor in the atmosphere is essential for the understanding of climate, atmospheric chemistry and hydrology. Water vapor is the most important atmospheric greenhouse gas – but for the water vapor, the earth would have become a cold desert. It is also the prime source of the hydroxyl radical (OH) which regularly cleans up the atmosphere by oxidizing many of the pollutants in air. The large amounts of latent energy associated with the phase changes of water drives the atmospheric processes, such as the storm systems.

The atmospheric concentration of water vapor has an enormous range – from a few percent near the surface of the earth to parts per million in the lower stratosphere. Designing instruments which can measure water vapor accurately across four orders of magnitude has been a difficult task. Consequently, the temporal and spatial distribution of water vapor in the troposphere (from the surface upto 10-15 km) and the stratosphere (about 15-55 km above the surface) is not known with sufficient accuracy.

'Water vapor absorbs infrared radiation notably in the 6.3 micron (1200-2000 $cm^{-1}$) vibrational band, and in the rotational band (40-900 $cm^{-1}$) as well as a continuum in the 'atmospheric window' (800-1200 $cm^{-1}$)' (*Eos*, Apr. 11, 2000). Although the continuum absorption is weaker, the continuum needs to be known accurately as it is a significant player in a number of climate-relevant processes (such as, strength of the hydrological cycle, and the way the changes in the carbon dioxide concentrations affect the climate, etc.).

As should be expected, increase in the temperature at the surface and in the troposphere leads to greater evaporation, and higher concentration of water vapor in the atmosphere. Since water vapor is a greenhouse gas, increased content of water vapor in the atmosphere leads to further warming. This kind of water vapor feedback is easy to understand. What is not clear is how the temporal and spatial distribution of water vapor in the upper troposphere is affected by the increase in surface temperatures.

Water vapor feedback also influences the hydrological cycle. Increased surface and tropospheric temperatures could lead to acceleration in hydrological cycle, increased rates of precipitation and more vigorous storms. Some authorities hold that the hydrological cycle has already started accelerating, but this cannot be determined one way or the other in the absence of research-quality, long-term water vapor data for various segments of the atmosphere.

### 11.1.2 *Evidence of global warming*

The most compelling evidence for recent global warming is found in the rapidly melting icecaps in the alpine regions of the tropics and subtropics, such as those of the Peruvian Andes and eastern Tibet. The Himalayas have about 5000 glaciers, but only a small number of them have been subjected to regular monitoring. The 5 km long Dokriani Bamak glacier in Garhwal Himalayas had retreated by 20 m in 1997, the highest annual retreat thus far.

Most of the Industrialized countries happen to be characterized by temperate and cold climates. A 2°C rise in temperature in these regions may have a number of beneficial consequences, such as:

– The areas in which crops can be grown will enlarge substantially,
– The growing season will be extended (say, by about three weeks), making it possible to grow crops of longer duration, and
– The energy required for home heating will be reduced (say, by about 15% ).

On the other hand, the island nations (e.g. Maldives) among the developing countries are expected to suffer the greatest economic losses from global warming. These island countries have long coastlines, tend to be strongly dependent upon tourism, and have small, undeveloped economies. For such countries, global warming is no longer an academic issue. Most of them stand to lose a good chunk of their territories, and some islands may even disappear altogether (for instance, the Marshall Islands in the Pacific lie almost entirely within three metres of the sea level. The rise of sea level by (say) half a metre would reduce their areas, and render non-potable about half of their groundwater). When a western delegate, in the course of UN deliberations, remarked that global warming has not been conclusively proved, Ambassador Robert F. Van Lierop, Permanent Representative of the UN from Vanuatu (a small island in the Pacific), is reported to have retorted, '*The proof, we fear, will kill us*'. By the time global warming is proved beyond a shadow of doubt, his country probably may not be around!

### 11.1.3 *IPCC Models*

The Intergovernmental Panel on Climate Change (IPCC 1996) generated five climate models taking into account greenhouse gases and aerosols. They predict that by 2100 the mean surface temperature could rise 1 to 3.5°C, depending on the assumed factor of climate sensitivity (defined as the equilibrium change in annual surface temperature due to doubling of atmospheric concentrations of $CO_2$ or equivalent doubling of other greenhouse gases). Some non-IPCC models, however, project a much higher global temperature rise of 3 to 9°C, most probably by 5°C, by 2100. Generally, high latitudes are projected to warm more than the global average.

*Example 11.1*
It is projected that by 2065, the atmospheric concentration of $CO_2$ will be double that of the pre-industrial concentration of $CO_2$. Assuming that the pre-industrial

and present day concentrations of $CO_2$ in the atmosphere are 280 ppmv and 360 ppmv respectively, calculate the expected temperature rise (due to radiative forcing of climate by $CO_2$), by
−  Doubling, and
−  Trebling,
of $CO_2$ atmospheric concentrations.

$\Delta T$ (°C) = 1.9 ln (C/C$_o$) (Shine et al. 1990, Climate Change, IPCC, p.41-68) where T = temperature rise, C$_o$ = base conc. of $CO_2$, C = Projected conc. of $CO_2$
−  Doubling: $\Delta T$ (°C) = 1.9 ln (560/360) = 1.9 ln 1.555 = 1.9 × 0.441 = ∼ 0.8°C.
−  Trebling: $\Delta T$ (°C) = 1.9 ln (840/360) = 1.9 ln 2.33 = 1.9 × 0.846 = ∼ 1.6°C.

IPCC models (1996, p. 22) suggest the following consequences of climate change:

1. Climate warming will enhance evaporation. There will be an increase in global mean precipitation and the frequency of intense rainfall. These consequences are, however, unlikely to be universal. Some land regions may not experience an increase in precipitation, and even those that do may experience decreases in soil moisture because of enhanced evaporation. There may be increased precipitation at high latitudes during winter and a decrease in soil moisture in mid-latitudes during summer.
2. Severe droughts and floods are likely to occur in some places. It is unclear what effect the climate change would have on the frequency of extreme weather events, such as tropical storms, cyclones, and tornadoes.
3. The climate change will have potential adverse effects on physical and ecological systems, human health, economy, and quality of life.
4. Gerstle (1992) gave the following estimates of sea level rise in the next century:

| Estimate | Year 2050 | Year 2100 |
|----------|-----------|-----------|
| High | 0.5 m | 1.6 m |
| Low | 0.1 m | 0.4 m |
| Median | 0.3 m | 0.8 m |

## 11.2   CLIMATE SHIFTS DURING THE HOLOCENE

We are naturally most concerned with what is going to happen to us. Consequently, much interest is evinced about climate changes in the recent past as indicators of future climatic changes (Holocene corresponds to the last about 10,000 years).

Our ability to predict future climatic change is critically dependent upon our understanding of the specific causes for past and present changes. We must bear in mind the fact that climate change is influenced not only by greenhouse gases, but also by changes in solar irradiance and major volcanic eruptions. On the basis of the analyses of Northern Hemisphere temperature variations over the past 600 years, it was found that

- Up to the start of this century at least, changes in recorded temperatures are more closely linked to changes in solar irradiance than to either volcanic or greenhouse gas variations, and
- But, during this century, increases in greenhouse gas emerged as the dominant forcing mechanism. Three of the past 8 years were the warmest across the Northern Hemisphere since at least 1400 A.D. (Mann et al. 1998).

An examination of the trends in temperature and sea-level pressure anomalies since 1976 and especially during the 1990s, is suggestive of the possibility that the frequency and intensity of ENSO (El Niño Southern Oscillation) events may be a response to global warming. This theory is being checked through a study of the 'natural archives', such as, tree rings, annually banded corals and ice cores from tropical glaciers, and the record of flood events preserved in lake sediments.

Two key facts emerge from the latest drillings in Antarctica: firstly, the unprecedented rise in atmospheric methane and carbon dioxide concentrations during this century are attributable to human activities, and secondly, there is a close coherence between the changes in temperature and the greenhouse gases during the last 400,000 years. The coherence is particularly marked during the periods of rapid warming at the end of each period of glaciation. Intensive research is ongoing to reconstruct as quantitatively as possible, the sources, sinks, and fluxes of carbon within the Earth system as a whole during periods of rapid change, on the basis of the records preserved in ice and sediment cores.

### 11.2.1 *Palaeoclimate modeling*

Historical Geology is based on the principle, 'The Present is the Key to the Past'. If we observe a particular type of ripple-mark in a low-tide situation, and if we find a similar type ripple mark (say) in a Devonian sandstone, we conclude that that sandstone must have been formed in a low-tide environment. In the case of climate change studies, we reconstruct the palaeoclimate record in order to predict what is going to happen. Another difference between the two approaches concerns the resolution. In the case of Historical Geology, a resolution of the order of a million years is usually adequate. In the case of climate studies, people are not worried about the weather in the next millennium or century but during the next year or next decade.

The instrumental record of climate change is too short to allow us to project the climatic variability on a decadal or century timescales. Proxy records of palaeoclimate have been reconstructed in the following manner (Alverson 1999; Alverson et al. 2000):

- Records of temperature and precipitation with a resolution of one year or better, have been built on the basis of geochemical ratios (e.g. Sr/Ca) and stable isotopic abundances (e.g. $\delta D$, $\delta^{18} O$, $\delta^{13}C$) in ice cores, tree rings, spelothems, varved lake sediments and corals (see Section 2.5 about the principle of iso-topic tracing; the concentration of $^{18}O$ varies by about 0.7 of a part per thou-sand for each degree change in the average temperature at the surface. Thus, from a study of $^{18}O$ distribution in the ice cores from polar icecaps, it is possi-ble to reconstruct the palaeotemperature record for the period when the ice core was laid down).

- North Atlantic Oscillation as well as Southern Oscillation Index have been re-constructed with monthly resolution extending back to 1675 (Luterbacher et al. 1999, quoted by Alverson 1999),
- Greenhouse gas levels in the past are measured in air pockets trapped in ice cores, and gaseous inclusions in volcanic rocks.
- Measurements of $^{10}Be$ in ice cores to build proxy records of variability in inso-lation ($^{10}Be$ is a cosmic-ray produced isotope of beryllium, and thus could be used to reconstruct the record of the cosmic ray flux, and thus the insolation).

Stalagmite growth layers from the Cold Air Cave in Makapansgat Valley in South Africa have been used to build a 3000-year, high resolution record of palaeocli-mate. This record has been compared with the reconstruction of the annual growth rings in an African tree species (*Pterocarpus angolensis*, from Sukumi forest in Zimbabwe), pertaining to the period 1762-1831 A.D. In both the cases, the growth rings respond primarily to moisture availability and are thus indicative of ambient climate. The existence of a good correlation between these two com-pletely different approaches, strengthens our confidence in the essential correct-ness of the proxy reconstructions (Holmgren et al. 1999; Stahle 1998).

Sea Surface Temperatures (SSTs) calculated from Sr/Ca ratios of corals, are compared with $\delta^{18}O$ data of modern as well as mid-Holocene (5350 yrs old) Porite corals from Orpheus Island, central Great Barrier Reef, Australia. They show that mid-Holocene water temperatures were warmer by 1.2°C than the pre-sent (Gagan et al. 1998, quoted by Alverson 1999).

The proxy records do show that rapid climate changes did take place in the past – in other words, strong linearity exists in the dynamics of the climate sys-tem. A spectacular example is the termination of the Younger Dryas cold event (so called because it was marked by the spread of the arctic flower, *Dryas oc-topetala*) about 11,600 years ago, when the temperature rose by 15°C, and annual precipitation doubled in less than a decade (Alley et al. 1993). This abrupt climate shift has a global signature, though it has been initially recorded in the ice cores of central Greenland. The Younger Dryas event is not an isolated event. During the last glacial period, a number of rapid and climatic shifts occurred. What trig-gered these climatic shifts? A generally accepted explanation is that when conti-nental icesheets over Greenland and eastern Canada melted, the resulting fresh water pulses into the North Atlantic induced changes in the pattern of thermoha-line circulation.

It should not, however, be understood to mean that rapid climatic shifts are confined to glacial times and that large ice sheets should necessarily be present to cause them. For instance, the desiccation of sub-Saharan Africa during mid-Holocene appear to have been caused by changing insolation modulated by non-linear oceanic and continental biospheric processes and the rapid monsoon climate shifts. When these processes went too far, and the threshold was crossed, an irreversible desiccation has set in (DeMenoccal et al. 2000).

Valdiya (1999) gave an excellent account of critical Holocene developments in the Indian subcontinent. The warming up following the termination of the Younger Dryas (11,600 yr BP) event is clearly manifested in the Indian subcontinent. The climate became progressively wetter and warmer, as the SW monsoon got intensified. Several lines of evidence (Stone-Age settlements, pollen studies, Himalayan lake levels., etc.) indicate that a warm and wet climate existed in a large part of the Indian subcontinent and Tibet, 10,000 to 4000 yr BP. Suddenly, around 3500 yr. BP, the SW monsoon weakened (the tell-tale slackening of the upwelling currents is indicated by the decline in plankton growth in the sea around this time), triggering the onset of aridity on land. The $\delta^{13}C$ values in the peats in the palaeolakes of the Nilgiri hills in south India clearly indicate both the existence of a very wet spell 9000-5000 yrs ago, as also the existence of dry conditions during the period, 5000-2000 yr BP (Rajagopalan et al. 1997). Thus, arid conditions prevailed in the Indian sub-continent during the period, 4000-2000 yr BP. Incidentally, it was during this period, that the mighty Sarasvati river dried up, and the Thar desert of Rajasthan came into existence (apparently, coevally with the Sahara desert). Major changes in the monsoon circulation appear to have been caused not only by orbital forcing but also by episodes of tectonic uplift of the Himalayas. These caused the displacement of low-pressure area and brought about epochs of heavy rainfall in the early part of Holocene, and aridity in the later part of Holocene.

## 11.3 CLIMATE CHANGE, WATER AND HEALTH

### 11.3.1 *What is El Niño?*

In the tropics, the evaporation from the sea surface and hence the heat input to the atmosphere is particularly large. The surface temperature of oceans in the tropics has a profound influence on the rainfall pattern, and hence on the occurrence of floods and droughts in the tropical and semi-tropical regions. It has been observed that large area of warmer water appears periodically in the Pacific off the coast of South America. By preventing upwelling (i.e. cold, nutrient-rich waters from reaching the surface), the warm water has a devastating effect on fisheries. Since this process generally happens around Christmas time, it has been named El Niño ('the boy child'). The southern oscillation of El Niño is called ENSO. The 'teleconnections' linked with ENSO are lower-than-normal precipitation in western Oceania, India, southeastern Africa, and northeastern South America, on one hand, and excessive precipitation in western South America, and eastern equatorial Africa, on the other. For the first time in the history of meteorology, the El Niño episode of 1997-1998 could be predicted several months in advance. As the proverb says, 'forewarned is forearmed'. This enabled the Governments of the regions that were likely to be affected to take appropriate steps to mitigate the adverse consequences. Similarly, the floods in southern Africa and Australia, and the drought in western India in early 2000 have been attributed to teleconnections of La Niña (which apparently occurs with a lag of a couple years after El Niño ).

### 11.3.2   *El Niño episodes and incidence of diseases*

Abnormal weather events (such as, very heavy rains or drought) affect the outbreaks of illnesses, through synergies based on the availability of water. Epstein (1998) used the database of CDC (Centers for Disease Control) of USA to delineate the linkage between the availability of water and outbreaks of illnesses.

*Water-borne diseases:*
Etiological agents for various waterborne illnesses are: Vibrio cholerae, Shigella, Chemical, Hepatitis-A, Giardia, Norwalk Agent, Campylobacter, Cryptosporidium, Salmonella, typhi, and E.coli 06: H16, etc.

The heavy rains and flooding in the Horn of Africa and East Africa (Somalia, Kenya, Tanzania and Mozambique ) during the 1997-1998 El Niño episode, led to the incidence of mosquito-borne, rodent-borne and water-borne diseases. There were outbreaks of cholera, malaria and Rift Valley fever, which killed an estimated 5000 people.

In 1993, there was an outbreak of the water-borne disease, Cryptosporidium, in Milwaukee, USA, when heavy rains and flooding of the Mississippi caused the entry of sewage water into Lake Michigan, which then entered the Milwaukee's clean water system. More than 400,000 people were affected, and about 100 people died, mostly persons infected with HIV and AIDS. Enteric diseases have a devastating effect on immuno-compromised persons. It has been found that rates of incidence of diarrheal diseases among the HIV-infected persons in the developing countries are higher than the rates in the developed countries. This is attributed to the more frequent exposure of HIV-infected persons in the developing countries to enteric pathogens through contaminated food and water.

Some times the effects of weather events may accentuate the consequences of local contamination. Increased nutrient supply compounded by heavy rainfall, increased runoff, warmer temperatures, resulted in the outbreak of viral gastroentertides associated with the consumption of shellfish and cyclospora-related diarrhea.

In Latin America, flooding is associated with the outbreaks of cholera, typhoid, shigella and hepatitis. The adverse effects of drought due to ENSO (El Niño Southern Oscillation) episode accentuated the local vulnerabilities (improper ecological practices), such as extensive clearance of Amazon forests in Brazil and forest burning in Indonesia. The forest fires in Indonesia created a haze which affected the respiratory health of millions of people not only in Indonesia but in other countries in the neighborhood (such as, Malaysia and Singapore). Air services were badly affected, and tourism dwindled.

*Rodent-borne infections:*
The emergence of hantavirus (which is spread from the salina, droppings and urine of the rodents) in USA can be traced to prolonged drought in US Southwest in early 1990s. The population of animals (such as, owls, coyotes and snakes ) that prey upon the rodents, declined sharply. When the drought was followed by heavy rains in 1993, grasshoppers and piñon nuts on which the rodents feed became plentiful, so much so there was a ten-fold increase in the rodent population. The Hantavirus Pulmonary Syndrome (HPS), which remained dormant till then, became virulent. Subsequently, the predators did return and kept the rodent population in check.

The resurgence of the rodent-borne hantavirus in a number of countries (such as, Yugoslavia in Europe, and Argentina, Bolivia, Chile and Paraguay in South America) in late 1990s, is associated with droughts and heavy rains. The most worrisome aspect of HPS is that it may spread from person to person. An unusual rodent pest has been reported in Tehran, Iran, in May, 2000. When the melting of snow in the mountains flooded the qanats (see Section 4.2.2 for a description of qanats), millions of rodents emerged out from the qanats into homes and streets in Tehran, thus creating a huge public health problem.

*Mosquito-borne diseases:*
Way back in 1920, Gill showed that the incidence of malaria in India is closely related to the rainfall pattern. The distribution of mosquitoes has shifted dramatically in the 1990s – they have become resistant to drugs and pesticides, and have thus become more dangerous than ever. The large increase in the incidence of malaria in several parts of the world (notably, Costa Rica, Pakistan, northwest India, Venezuela, etc.) is coincident with ENSO event or teleconnections of ENSO. Flooding appears 12 to 18 months following a warm event. Similarly, dengue fever in South Pacific is driven by ENSO.

Malaria kills a million people world-wide annually, and 90% of the deaths due to malaria take place in Africa. The matter has become so serious that Heads of the African States met in Malaria summit in April, 2000, to work out an effective strategy for malaria control (including particularly, the development of malaria vaccine). After extremely heavy rains in southern Mozambique due to hurricanes Elaine and Felicia during February & March, 2000, there were stagnant water pools every where, and the incidence of malaria doubled.

*Food production:*
Most of the agricultural pests are cold-blooded stenotherms. They and the pathogens that they may carry, are strongly influenced by weather conditions. For in-

stance, drought favors aphids, locusts and whiteflies which can devastate the crops and vegetation, and thus have marked adverse impact on food security.

*Linkage of El Niño to childhood diarrhea – a case study of Peru:*
A joint study by epidemiologists, statisticians, climate scientists and clinical doctors conclusively established the linkage between El Niño and increase in childhood diarrhea, a leading cause of premature death (*Eos*, AGU, Feb. 15, 2000). As diseases tend to be multi-factorial, a statistical analysis is necessary to identify unequivocally the etiology of a disease.

Hospital admissions for diarrheal diseases in a clinic in downtown Lima, Peru, increased by 200% during the El Niño period of 1997-1998, compared with the pre – El Niño period of 1993-1997. The 1997-1998 El Niño episode increased the city's mean ambient temperature by 5°C. The incidence of diarrhea can only be attributed to increase in temperature. Flooding is not an issue, since Peru is arid. Higher temperatures have two consequences which increase the incidence of childhood diarrhea – they lengthen the period of survival of bacteria (such as, *E. coli* ) in contaminated food, and they increase exposure to bacterial and parasitic diarrhea. Conversely, lower temperatures increase the transmission of viral diarrhea. When ambient temperature increases, acute respiratory inflammation may decrease.

If the findings in Peru are applicable to other parts of the world, then diarrhea may increase by millions of cases per degree increase in ambient temperatures above normal. Currently, diarrheal diseases kill about two million children and cause about 900 million episodes of illnesses each year (p. 4, *Development and the Environment*, World Bank Report 1992).

## 11.4　CLIMATE CHANGE IMPACTS ON AQUATIC ECOSYSTEMS

Heavy rains are followed by increased runoff, and increased input of nutrients and pollutants to the aquatic ecosystems. Warming of coastal waters has a number of ecological consequences: the formation of stable, thermal layers inhibit vertical circulation, and increase in the rate of photosynthesis and algal growth may result in the proliferation of toxic algal species, such as cyanobacteria and dinoflagellates. Warming may adversely affect the immune system of sea mammals and corals, besides promoting the growth of harmful bacteria and viruses in their tissues.

It is not uncommon in the developing countries that municipal sewage (containing high loads of viruses, bacteria and protozoa) is discharged into the inland, estuarine and marine coastal waters untreated. When this happens, the aquatic ecosystems in these waters get stressed. Algae may facilitate the survival and increase of vibrios and other gram-negative bacteria, such as salmonella and *shigella*. There is convincing evidence to show that algal blooms and weeds are increasing in magnitude and frequency in the various types of waters. Epidemics of red, green and brown algal blooms have been reported from regions as diverse as California, North Carolina, Guatemala, Iceland, Japan, Thailand and Tasmania (Andersen 1992, as quoted by Epstein 1998). The algal blooms endanger the health of the seagrasses, corals, fish, birds, sea mammals and humans.

Climate change accentuates the stresses in aquatic systems caused by the following sources:

– Excessive mineral and organic nutrients (particularly nitrogen overload) which favor plant growth at the expense of animal growth,
– Degradation of wetlands: Wetlands serve as 'nature's kidneys' by filtering nitrogen and pollutants from the coastal environments. When wetlands are drained to provide space for various civil constructions along the coast, the aquatic systems get more stressed. Wetlands serve as nurseries for birds and fish. Mangroves provide spawning grounds of fish and shrimp, and protect the inland freshwater resources from salinization. Mangroves and corals protect the shorelines from storms. California lost 91% of its wetland area.
– Chemical pollution and UV-B radiation causing mutational changes in sea life: Lipophilic organic compounds (such as PCBs and PAHs), heavy metals (such as, Pb, Cd) and pesticides (such as, lindane and methoxychlor) bioaccumulate in the food chain and cause histopathological changes and genetic damage to the near-shore sea life.

### 11.4.1 *Harmful Algal Blooms (HABs )*

The following case studies illustrate how climate change, as manifested by warming and flooding, adversely affects the health of humans through impacts on aquatic ecosystems.

'Red tides' (toxic phytoplankton blooms) have been known since Biblical times (in those days, they were believed to be the blood of the sparring whales or menstrua). There was worldwide concern when algal blooms caused poisoning and mass mortality of fish and shellfish in several parts of the world. Consequently, the International Oceanographic Commission (IOC) and the Food and Agricultural Organization (FAO) established in 1987 an Intergovernmental Panel on Harmful Algae Blooms (HABs).

HABs have adverse effects on tourism, aquaculture and human health. For instance, *Pyrodinium* blooms in Philippines in 1983 and 1987 caused 1127 cases of illnesses, and 34 deaths. In 1988, immense toxic gelatinous blooms of *Chrysochromulina polyepsis* devastated 200 salmon farms in Norway and Sweden, and caused an estimated loss of about USD 200 million.

The strong El Niño warming of the North Atlantic in 1982/1983 brought about profound changes in zooplankton, finfish, seal and seabird communities, by favoring the growth of cyanobacteria, and toxic and unpalatable (to fishgrazers) dinoflagellates. The sea is a global thermostat – it absorbs heat for worldwide distribution. So global warming is expected to result in stronger and more frequent El Niño episodes in future.

The following case illustrates most vividly the linkage between extreme weather events and their health consequences: In 1987, along the eastern Atlantic coast of Canada, drought was followed by heavy rains and increased runoff. When this interacted with the warm eddies of the Gulf Stream, there was a bloom of pennate diatom (*Nitzschia pungens*) near Prince Edward Island. The bloom produced domoic acid which entered the mussels. There were 156 cases of amnesia (some permanent) and five deaths among the mussel consumers (Epstein et al. 1993). A similar episode of domoic poisoning occurred in Monterey Bay, California, USA, in 1991/1992, where hundreds of pelicans and cormorants died after eating the fish poisoned with domoic acid.

Sverdlow et al. (1992) and Epstein et al. (1993) gave a blow-by-blow account of how poor sanitation facilitated the spread of cholera in Peru. In January 1991, cholera first appeared in Chancay, a port city 60 km north of Lima. A day later, it appeared in Chimbote, a seaport 400 km north of Lima. Soon the epidemic spread all along the 2000 km coast of Peru. The Government advised the people in the coastal areas to take two precautions to protect themselves from cholera:

1. Boil the water before drinking, and
2. Do not eat the fish, as they are contaminated with cholera inoculum. But the poor could not observe either of the two directions. Fuel in Peru is expensive, and boiling water would have involved about a quarter of their meagre earnings. Fish happens to be the cheapest food available, and the poor could not afford any thing else. Cholera then spread to Ecuador, Colombia and Chile. Later it spread to Brazil, Venezuela and Bolivia – invariably through waterways, and aided by poor sanitation. By April 1992, there were 533,000 cases of illnesses, and 4700 deaths among the Latin American nations.

Since 1960, there have been annual epidemics of cholera in Bangladesh, but the reservoir was unknown. It has since been found that a viable but non-culturable, 'quiescent' form of *Vibrio cholerae* is hosted by water hyacinths, duckweed plants, cyanobacteria, silicated diatoms and zooplankton (primarily copepods). It appears the vibrios could hibernate by contracting 15 to 300 fold. With warming and availability of nutrients, they rapidly revert to infectious stage. DNA studies have established the genetic identity of the Bangladeshi and Peruvian strains of *Vibrio cholerae* (Epstein 1998). That the bacterium evidently travelled across the Pacific as a stow-away has been surmised by the study of the bilge of the vessels.

Large Marine Ecosystems (LMEs) along the continental coasts occupy large areas (of the order of 200,000 km$^2$) and account for 95% of the annual marine resources used by the humans. The sustainability, health and biomass yields of the LMEs need to be optimized by ecologically based strategy of managing the coastal ecosystems, and protecting the LMEs from the adverse consequences of climate change and mitigating the stresses on ecological systems of LMEs.

It is critically important to monitor the algae (including those that can harbor bacteria) in coastal LMEs. On one hand, their distribution serves as early warning signals of ecosystem stresses, and on the other they alert us about possible adverse health effects. Technology is now available for the remote sensing of the HAB species and associated pathogens (through Advanced Very High Resolution Radiometer – AVHRR, and Sea Wide Field Sensors – Sea WiFS images).

## 11.5  CLIMATE CHANGE AND THE HYDROLOGICAL CYCLE

The Intergovernmental Panel on Climate Change (IPCC 1996) found 'discernible evidence' to show that humans are having an impact on the natural climate regime by accelerating changes in the forcing factors (such as, climate relevant emissions)

It is well established that temperature rises as a consequence of the radiative forcing of climate by $CO_2$. The pre-industrial and present day concentrations of $CO_2$ in the atmosphere are 280 ppmv and 360 ppmv respectively. When the doubling of the $CO_2$ atmospheric concentrations (i.e. 560 ppmv) takes place, it is expected that the temperature will rise by 0.8°C. The temperature rise is expected to result in the rise of evaporation and precipitation globally by 7-15%, leading to intensified hydrological cycle (IPCC 1996). There is likely to be a shift in the rainfall pattern in two ways:

1. Some areas may receive more rainfall than before and some areas, less, and
2. At a particular location, there may be no change in the quantum of annual rainfall, but it may be distributed in such a manner that there may be less frequent but heavier precipitations. Statistical analysis suggests that the frequency of extreme weather events (such as, floods and droughts) is likely to increase – for instance, in USA the frequency of heavy rains ($> 5$ cm $d^{-1}$) has increased by 20% since 1900. Coastal damage from sea level rise may get accentuated by storm surges.

A warmer atmosphere holds more moisture (6% more for every 1°C). As a consequence of intensified hydrological cycle, cloudiness may increase and block the outgoing long-wave radiation (heat). The projected increase in the frequency of heavy rains and flash floods is attributable to enhanced evapotranspiration and increased cloudiness.

These developments have profound implications for agriculture and food security, water supply and sanitation and waterborne diseases.

We need to understand three critically important issues concerning the Water Cycle (*Eos*, AGU, Mar. 7, 2000):

1. What are the underlying causes of variation in the water cycle on both global and regional scales, and to what extent is this variation attributable to human activities?
2. How do the global movement of water in and between the atmosphere and land surface, and between the atmosphere and the oceans, affect the fluxes and their variability on continental and local scales?

3. How exactly does the water cycle interact with the waterborne movement of carbon, nitrogen and other nutrients, and how are the aquatic ecosystems influenced by biogeochemical fluxes of these nutrients?

## 11.6 CLIMATE CHANGE IMPACTS ON WATER SUPPLIES

This section is largely drawn from the detailed account given by Georgiyevsky (1998) on how global warming will affect the water resources.

General Circulation Models (GCMs) are used to predict the kind of climate change that could be expected in the future. The GCMs most commonly used are: Geophysical Fluid Dynamics Laboratory (GFDL), and United Kingdom Meteorological Office (UKMO). Other models that are some times used are: Goddard Institute for Space Studies (GISS) and the Canadian Climate Centre Model (CCCM). The models rarely yield the same results, and some times, the results may even be contradictory. For instance, for the Volga and Dneiper basins, the GFDL models project an increase of precipitation by 10-20%, whereas the UKMO models yield a figure of 30-40%. In the case of mean annual temperature, the GFDL scenario projects a rise of 3-5°C, whereas the UKMO model suggests a rise of 7-9°C.

It has been found that the coupled atmosphere – ocean model developed by the Max Planck Institute, Hamburg, Germany, yields good simulations of the precipitation patterns under the current climate. Murari Lal (1994) used this model to make projections about the monsoon regions of south-east Asia, assuming business-as-usual increase of greenhouse gas emissions, and assuming no significant increase of aerosols. His projections (up to 2080) indicate an average surface temperature increase over the land area of 3°C, with increased precipitation and runoff in the flood-prone areas of northeast India and south China during summer.

Aerosol particles are so small that millions can dance on a pinhead, but yet they have profound influence on the climate of the globe. Atmospheric aerosols directly influence the radiative transfer in the atmosphere. By providing condensation nuclei, they facilitate cloud formation and precipitation. The size of the aerosols may range from 0.001 to 100 μm, with the majority lying in the range of 0.01 to 100 μm. Anthropogenic particles (< 2 μm) dominate the urban aerosols, whereas natural aerosols, such as windblown dust, are coarser (> 2 μm). There is increasing recognition of the role of the aerosols in climate change forcing – for instance, precipitation has been found to be sensitive to aerosols. Murari Lal et al. (1995) showed that if a substantial increase in sulfate aerosols is assumed, precipitation in northeast India would decrease, rather than increase.

According to Houghton (1997, p.121), there may be two effects of climate change on water supplies:
1. Increasing the vulnerability of communities already facing water shortages, and
2. Large changes in water supplies in many places.

*Vulnerability*: According to United Nations Report entitled, 'Comprehensive Assessment of the Freshwater Resources of the World' (1997), about 80 countries in the world (several of them in Sub-Saharan Africa and Middle East), making up 40% of the world population, are already suffering from serious water shortages, which had become a limiting factor in their economic and social development. The Report forecasts that by 2025, as much as two-thirds of the world's population will be affected by moderate to severe scarcity, unless appropriate mitigation measures are taken (see Section 1.7, and Fig. 1.25). To meet the extra demand due to increasing populations and greater use of water per capita, groundwater is being extracted faster than it is being replenished. Consequently, the water table keeps going down. For instance, in Beijing, China, the water table is falling by two metres a year as groundwater is pumped out. Similar situation exists in several cities in India (e.g. Bangalore, Hyderabad).

Under these circumstances, communities living in arid and semi-arid lands become more vulnerable, and droughts will become far more disastrous than before (bronchitis may cause only minor discomfort to a healthy person, but it may kill an AIDS patient).

Another dimension of the vulnerability arises from the fact that many of the world's water resources are shared. The Danube passes through 12 countries which use its water, the Nile water is shared by 9 countries, and five countries are involved in the case of Ganges-Brahmaputra system. When the availability of water decreases, there is inevitably scramble for water. Water resources of rivers like Jordan and Euphrates-Tigris are so crucial to the very existence of communities that Boutros-Boutros Ghali, former Secretary General of the United Nations, once remarked, 'The next war in the Middle East will be fought over water, not politics'!

*Changes in water supplies*: Global warming will intensify the hydrological cycle, with increased precipitation than before in some regions, and decreased precipitation in others. What is not clear yet is which regions will be affected in either ways, and more importantly, to what extent. There is a general view that global warming will hit the arid and semi-arid areas hardest. According to IPCC review (Stakhiv et al. 1992), a rise in temperature by 1-2°C, and precipitation of reduction by 10%, will lead to decrease in water resources by 40-60%. For the monsoon regions (such as, India), the river runoff is expected to increase during the rainy season, and to decrease during the dry season, with droughts becoming more frequent.

*Enhanced supplies:*
Analysis of water balance data from seven stations in the East European Plain for the period, late 1940s to mid-1950s, led to the following conclusions:
– Annual precipitation increased by 35-80 mm (i.e. 5-15%), with most significant increase occurring during the summer – autumn period,
– During the cold season of the year, air temperatures rose by 1.0-1.2°C. This had a number of beneficial consequences, all along the line. There was less

freezing, and more thawing during winter. The consequent rise in soil moisture, led in due course, to increase in groundwater resources to the extent of 10-30% during summer-autumn period. This benefited the aquifers, and the water-table rose by 50-100 cm. Increased groundwater recharge enhanced the river flow. In the case of South America, there has been an increase in water supply since 1980s. The annual river runoff in Great Britain is expected to increase by 10-25% in the winter and spring periods.

*Reduced supplies:*
During the period, 1981-1990, the inflow of the African rivers to the Atlantic Ocean decreased by 17% compared to the average for the period, 1951-1990. The decrease in runoff from the arid part of Africa was even greater (27%). There was severe drought in southern California during 1987-1992, when the runoff decreased by 23% relative to the normal. The river discharges in most parts of Australia went down consequent upon reduction in precipitation. For instance, in the case of rivers originating in the Darling Range in eastern Australia, runoff decreased by 40% during the period, 1975-1994, compared with the period, 1911-1974.

In the case of USA, it is a 'mixed bag':
– Northwest: a slight increase in annual runoff and floods,
– California: a considerable increase in winter runoff and decrease in summer with insignificant increase in annual runoff,
– Great Lakes: a decrease in runoff, and
– Rest of USA: uncertain changes in precipitation (Climate Change and Water Resources Planning Criteria, quoted by Georgiyevsky 1998).

## 11.7   ADAPTATION SCENARIOS

A consensus is emerging in favor of the view that adverse environmental changes that are evident all around us are driven more by human activities (such as, damming of the rivers, water diversions and water withdrawals, land-use changes through engineering constructions, deforestation and agriculture, etc.) rather than by climate change. Global climate may, however, be the primary cause for system function and services changes in areas such as coastal zones where sea level rise could salinize the coastal freshwater resources through salt water intrusion.

The following account is case study in respect of the vulnerability and adaptation assessments in regard to River Gambia (in the Gambia) in west Africa (source: 'Report on the Vulnerability and Adaptation Assessment Study in the Gambia', 1997; with the permission of the Secretary of State for the Presidency, Fisheries and Natural Resources, Banjul, the Gambia):

The river flow is very sensitive to climate change conditions, especially to rainfall variation. Generally, a 1% rise in rainfall will result in 3% change in runoff.

The vulnerability analysis indicates that climate variables alone can cause a 50% change in the runoff. The cumulative consequence of these changes will be to increase the maximum salt water intrusion into the River Gambia by 40 km. inland.

The relationship between the average monthly position of the saline front, and the monthly mean flow of the river, can be expressed by the following equation:

$$D = bQ^a \qquad (11.1)$$

where $D$ = distance (in km ) of the position of the saline front to the mouth of the river at Banjul, $Q$ = monthly mean flow ($m^3 s^{-1}$) of the river at Gouloumbu, and $a$, $b$ are constants (which vary for different months).

| $Q\,(m^3\,s^{-1})$ | 20 | 30 | 40 | 50 | 60 | 70 | 80 | 90 | 100 | 125 | 150 |
|---|---|---|---|---|---|---|---|---|---|---|---|
| D (km) | 195 | 180 | 170 | 160 | 152 | 148 | 143 | 139 | 135 | 125 | 115 |

The adaptation measures are presented under three heads. It may be noted that their successful implementation is critically dependent upon educating the farmer.

*Demand-driven measures:*
It is critically important to maintain a minimum flow during the dry season in order to keep the salt water intrusion into the Gambia River in check. It has therefore been recommended that the daily abstraction rate be reduced from the present 1.0 to 0.7 $m^3 s^{-1}$.
A number of consequential steps need to be taken to implement this decision:
1. Administrative steps (such as, licensing for withdrawal of river water for irrigation) may have to be taken to ensure compliance with the reduced abstraction rate,
2. Early maturing rice varieties and low water need crops need to be grown. Biotechnology can be used to develop water-efficient strains of crops. Preference should be given to low-water need crops – for instance, as against 300 ha.cm of water needed for the irrigation of one ha of sugarcane, the irrigation needs per hectare of sorghum and pearl millet are as low as 20 ha.cm and 10 ha.cm respectively. Crops growing during winter should be preferred, as they would need less water, since PET (potential evapotranspiration) is much less in winter than in summer.
3. Salt-tolerant crops and varieties should be introduced. Where a three crop cycle is no longer possible, a two-crop cycle with improved varieties need to be adopted, with improved farming techniques, so as to avoid a drop in food production,
4. The efficiency of the irrigation system can be improved by increasing the canal sizes, canal lining and elimination of leakages. More efficient irrigation systems, such as sprinkler irrigation, may be adopted wherever possible (See section 6.9.2, Irrigation methods, and Section 6.12.3, Irrigation water quality and salinization ).

*Supply driven adaptation measures:*
1. Construction of anti-salinity barrage at Balingho, some 130 km. from the river mouth, will not only control the salt water intrusion, but also increase the available freshwater supplies for irrigation,
2. If a dam/reservoir is constructed in the upper basin at Kekreti, it could be used to control the saline front by releasing pre-determined flows during the dry season.
3. Tidal irrigation in the Gambia estuary sub-basin should be promoted.
4. Rainwater harvesting on the rooftops, and harvesting of surface runoff and storing them in ponds, will provide water for domestic, livestock and wildlife uses, and thus relieve the stresses on river water (see Section, 4.1, Rainwater harvesting, and Section 4.2, Harvesting of surface runoff).
5. Groundwater may be tapped through wells or boreholes, particularly during the dry season, when the availability of river water is constrained by low flow and salt water intrusion.

*Environmental aspects and conservation-driven adaptation measures:*
Environment is currently being degraded by salinization of soil and water, coastal erosion and wind and water erosion. Climate change will exacerbate the degradation. The following measures have been recommended:
1. Pending the construction of the anti-salinity barrage at Balingho, a predictive salt water intrusion model capable of simulation on a daily basis should be built and used as a planning tool, and as an early warning system,
2. The Gambia has a number of small streams in which the water flows 2-4 months a year. The water flow in such streams may be conserved through the construction of small check dams, so as to improve the soil moisture and groundwater recharge,
3. Improvements in the monitoring of the tides would facilitate tidal irrigation in a controlled manner,
4. Water discharge data should be collected systematically for all the streams. Such data will be useful in planning the location of small dikes, and control of soil erosion,
5. Greater intrusion of salt water into the Gambia river would degrade the groundwater in the densely populated western or coastal areas of the Gambia. In the case of the Kombo peninsula, where presently the groundwater abstraction is within limits, precautions have to be taken to avoid the risk of groundwater pollution from domestic and industrial wastes.

### 11.7.1 *Saline agriculture*

Saline agriculture is emerging as a viable adaptation strategy in the context of the expected increase in sea-water incursion into the coastal lowlands as a consequence of global warming (see Aswathanarayana 1999, p. 122-123; 'Saline Agriculture' 1990).

The ability of plants to tolerate salt depends upon interactions between the salinity, and various soil, water and climatic conditions. Some halophytes require fresh water for germination and early growth, but can tolerate higher levels of salt in the later stages.

Any strategy for the promotion of saline agriculture has to have the following components:

- Identification of salt-tolerant plants that could serve as food, fuel, fodder and other products, such as essential oils, pharmaceuticals, and fiber;
- Improving the salt-tolerance of the known plants, for instance, cells of rice (*Oryza sativa*) subjected to salt stress and then grown to maturity tend to have progeny with improved salt tolerance (up to 1% salt);
- Exploration for new salt-tolerant wild species that have economic importance, and ways of improving their agronomic qualities.

Some important plants that figure in saline agriculture are as follows:

*Food:*
Vegetables: asparagus (*A. officianalis*), beet root (*Beta vulgaris*), lettuce (*Lactuca sativa*),
Crops: rice (*Oryza sativa*), barley (*Hirdeum vulgare*), pearl millet or bajra (*Pennisetum typhoides*),
Foliage and leaf protein: Indian saltwort (*Suaeda maritima*).

*Fuelwood trees and shrubs:*
*Prosopis juliflora, Eucalyptus occidentalis, Casuarina obesa.*
Kallar grass (*Leptochloa fusca*) can be used for biogas production (energy yield is $15 \times 10^6$ kcal ha$^{-1}$).

*Fodder:*
Kallar grass (*Leptochloa fusca*), silt grass (*Paspalum vaginatum*), salt grass (*Distichlis spicata*), *Atriplex nummularia*
Trees: *Acacia cyclops*

*Essential oils:*
Kewda (*Pandanus fascicularis*), *Mentha arvensis*.

## REFERENCES

* Suggested reading

Alley, R. et al. 1993. Abrupt increase in the Greenland snow accumulation at the end of the Younger Dryas event. *Nature* 362: 527-529.
Alverson, K. 1999. The PAGES/CLIVAR Intersection: providing the palaeoclimate perspective needed to understand climate variability. *IGBP Newsletter*, Dec. 99, p.11-14.

Alverson, K. et al. (eds) 2000. *Past Global Changes and their significance for the future*. Quaternary Science Reviews 19: 1-5, 465 pp.

Anon. 1990. *Saline Agriculture-Salt-tolerant plants for Developing countries*. Washington, D.C., USA: Nat. Acad. Press.

*Aswathanarayana, U. 1999. *Soil Resources and the Environment*. Enfield, USA: Science Publishers.

DeMenoccal, P. et al. 2000. Abrupt onset and termination of the African humid period: Rapid climate responses to gradual insolation forcing. In Alverson, K. et al. (eds), *Past Global Changes and their significance for the future*. Quaternary Science Reviews 19(1-5): 465 pp.

Epstein, P.R., T.E. Ford & R.R. Colwell 1993. Marine Ecosystems. *Lancet* 342: 1216-1219.

*Epstein, P.R. 1998. Water, Climate Change and Health. In Zebedi, H. (ed.), *Water: A looming crisis?* Paris: UNESCO, p. 67-81.

Gerstle, B. 1992. Sea level rise and the effects on Australia. *Aquapolis* 4: 18-20.

*Georgiyevsky, V.Yu. 1998. On global climate warming effects on water resources. In Zebedi, H. (ed.), *Water: A looming crisis?* Paris: UNESCO, p. 37-46.

Holmgren, K. et al. 1999. A 3000-year old high-resolution stalagmite-based record of palaeoclimate for northeastern South Africa. *The Holocene* 9(3): 295-309.

*Houghton, J. 1997. *Global Warming*. Cambridge: Cambridge University Press.

*IPCC (Intergovernmental Panel on Climate Change) 1996. *Climate Change 1995 – Impacts, Adaptations and Mitigation of Climate Change: Scientific – Technical Analyses*. Cambridge, England: Cambridge Univ. Press.

Lal, M. 1994. Water resources of the south-east Asian region in a warmer atmosphere. *Advances in Atmospheric Sciences*, 11 (2), China: Academia Sinica.

Lal, M. et al. 1995. Effect of transient increase in greenhouse gases and sulphate aerosols on monsoon climate. *Curr. Sci.* (Ind.) 69: 752-763.

Mann, M., R.S. Bradley & M.K. Hughes 1998. Global scale temperature patterns and climate forcing over the past six centuries. *Nature* 392: 779-788.

Mann, M. et al. 1999. Northern hemisphere temperatures during the past millennium: Inferences, uncertainties and limitations. *Geophy. Res. Lett.* 26(6): p. 759.

Rajagopalan, G. et al. 1997. Late Quaternary vegetational and climatic changes from tropical peats in southern India – An extended record upto 40,000 yr BP. *Curr. Sci.* (Ind.) 73: 60-63.

Stahle, D.W. et al. 1998. Experimental dendroclimatic reconstruction of southern oscillation. *Bull. Am Met. Soc.* 79: 2137-2152.

Stakhiv, E., H.Lins & I. Shiklomanov 1992. Hydrology and Water Resources. In W. Tegard & G. Sheldon (eds), *The Supplementary Report of the IPCC Impact Assessments*.

Swerdlow, D.L. et al. 1992. Waterborne transmission of epidemic cholera in Rujillo, Peru: Lessons for a continent at risk. *Lancet* 340: 28-33.

*Valdiya, K.S. 1999. *Crucial Holocene Developments in the Indian subcontinent in the context of societal concern*. Fourth Foundation Lecture of the Indian Geological Congress, Roorkee, India.

CHAPTER 12

# Paradigms of water resources management

This is basically a synthesis chapter. The various strands of water resources management dealt with in the previous chapters are brought together for being harmonized and integrated.

## 12.1 MANAGEMENT OF RIVER BASIN UNITS

Water resources management involves the transformation of natural hydrological resources into a managed resource to be used by the society for various purposes.

We should distinguish between two terms:

1. Hydrologic *effects* – how hydrological parameters such as rainfall, evapotranspiration, infiltration, and runoff, are affected by global warming and associated hydrometeorological changes, and
2. Water resource *impacts* – how control, use and distribution of the available water supply for the use of the people and protection of the environment are impacted by changes in the natural watershed (Stakhiv 1998).

The goal of the water resources management is to increase the reliability of water-related services, and ameliorate the hydrologic extremes (such as, floods and drought). It should be a self-adapting endeavor whereby the management system responds and adjusts to various challenges, such as 'climate variability (particularly extremes), water availability, shifts in water uses and demands, demographic changes and public preferences, along with technological innovations and institutional requirements' (Stakhiv 1998). IPCC (1994) is strongly in favor of continuous adaptive management in water-dependent sectors, such as, agriculture, hydropower, municipal water, navigation, etc. Additionally, water resources management is being increasingly called upon to address water-related environmental issues, such as, aquatic ecosystems, wet lands, endangered species and waterborne diseases.

Water resources management is as much concerned with the efficient operation of the physical structures (such as reservoirs, irrigation canals and distribution system) as with institutional structures (such as, regulatory measures, economic instruments, behavioral changes, water conservation and technological innovations, etc.). By the application of such management techniques, it has been possible for USA to bring down the total freshwater withdrawals from 378 bgd (billion gallons/day) (1465 M m$^3$ d$^{-1}$) to 339 bgd (1314 M m$^3$ d$^{-1}$) in 1995.

Unfortunately, the developing countries are ill equipped to undertake adaptive management systems, for the simple reason they generally do not have either the physical structures or institutional frameworks needed for the purpose.

A consensus is emerging in favor of the view that adverse environmental changes that are evident all around us are driven more by human activities (such as, damming of the rivers, water diversions and water withdrawals, land-use change changes through engineering constructions, deforestation and agriculture, etc.) rather than by climate change. Global climate may, however, be the primary cause for system function and services changes in areas such as coastal zones where sea level rise could salinize the coastal freshwater resources through salt water intrusion.

### 12.1.1   *Water resources impact and adaptation assessments*

The following account is a summary of the US CSP (US Country Study Program, 1994, Guidance for Vulnerability and Adaptation Assessment. Version 1.0, Washington, DC, USA). Though this model was designed for the assessment of the climatic change impacts, it could be used for almost any other type of challenge that a water resources management system may have to face. Hence the model is given in some detail. The program simulates the joint impact of population growth, economic development and climate change on regional water resources. In the case of climate change impacts, 2075 is taken as the base year.

Four steps are involved:

### 1. *Impact on hydrologic sources – Supply*

Each country is considered in terms of its river basin units. A given river basin unit may cover a river basin which entirely lies within a country, or it may relate to the national portion of an international river basin. Mean monthly runoff, mean monthly temperatures and precipitation values are used for modeling. Where a country is downstream of a river basin, it is necessary to model the upstream of the river basin as well (for instance, though Mozambique is concerned with the last 820 km. length of the Zambeze River, modeling studies have to be made for the river segments in the five upstream countries as well).

A hydroclimatic model is developed on the basis of a monthly, spatially-lumped, one-dimensional water balance studies (Fig. 12.1; source: Kaczmarek 1993).

Data requirements and possible sources for the data are as follows:
– Monthly river runoff values for each river basin to be studied (obtainable from local hydro-met service or Global Runoff Data Centre, D-56068, Koblenz, Germany; fax: +49-261-1306 280; e-mail: <GRDC@BFGKO.KFG.BUND 4000.DE >),

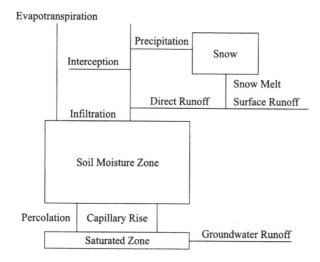

Figure 12.1. Schematic of one-dimensional soil column model (source: Kaczmarek 1993).

- Spatially arranged mean monthly values of temperature and precipitation covering the watershed of each river basin unit (Local hydro-met service or IIASA (International Institute for Applied Systems Analysis, Laxenburg, Austria) Global Climate Database),
- Total and elevation area values for each river basin unit to be modeled (local survey/mapping service or IIASA Global Climate database),
- GCM-generated, spatially averaged estimates of the changes in the mean monthly temperature and precipitation due to the doubling of the atmospheric carbon dioxide ($2 \times CO_2$).

## 2. *Impact on water uses – Demand*

Water demand is estimated for each basin on the basis of population and economic growth, according to the methodology of Kulshreshtha (1993). Climate-induced or any other changes in water demand driven directly (as in the case of agriculture) or indirectly (as in the case of coastal zones) are then estimated (Fig. 12.2; source: Kulshreshtha 1993).

Data requirements and possible sources for the data are as follows:
- Current water use for agricultural, domestic, industrial and energy-related sectors for each river basin unit (local hydro-met service),
- Current economic activity indicators for agricultural, domestic, industrial, and energy-related sectors for each river basin unit (local economic/information service or United Nations/World Bank database),
- Estimated economic growth rates for agricultural, domestic, industrial and energy-related sectors for each river basin unit (local economic/information

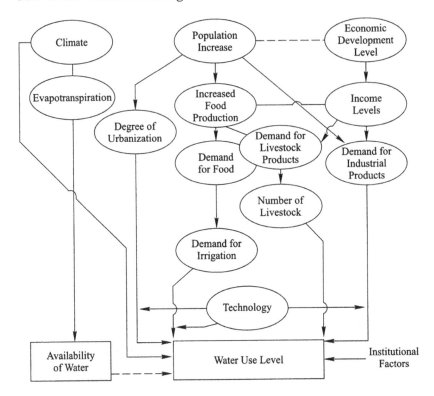

Figure 12.2. Interrelationship of water demand (source: Kulshreshtha 1993).

service or United Nations/World Bank database, and local and Country Studies Program technical experts),
– Estimates of water use coefficients due to technological changes in water use (local economic/information service or United Nations/World Bank database, and local and Country Studies Program technical experts), and
– Impacts of climate change scenarios on water use coefficients (biophysical assessments that might have been made for other country studies, and local and Country Studies Program technical experts).

### 3. Impact on supply-and-demand balance – Vulnerability

Vulnerability factor is assessed on the basis of the supply-and-demand balance for each basin unit.

The supply and demand data for each river basin unit is plotted on GIS. The vulnerability of each basin is classified as per the vulnerability index of Shuval (Fig. 12.3; source: Shuval 1987). Each basin is assigned an yield index, which may vary from 0 to 1. The natural water supply is multiplied by the yield index to give the amount of water that can be currently supplied to meet the water demand (for

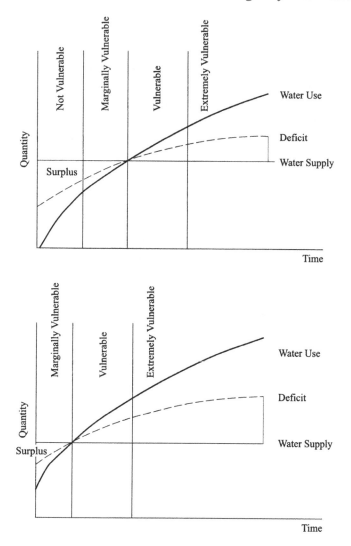

Figure 12.3. The impact of climate change on vulnerability (source: Shuval 1987).

instance if the water supply in regard to a given water basin unit is 40 km$^3$ yr$^{-1}$, and the yield index is 0.5, the amount of water that could be supplied from that basin would be 20 km$^3$ yr$^{-1}$). Possibility of transfer from surplus basins to deficit basins is considered, in order to meet the projected demand.

Data requirements and possible sources for the data are as follows:

— Estimates of the current and future water resource infrastructure development, especially reservoir storage for each river basin unit (local and Country Studies Program technical experts),

— Estimate of the current and potential interbasin transfers for each river basin unit (local and Country Studies Program technical experts).

## 4. *Alternative management strategies – Adaptation*

Alternative management strategies are considered taking into account of various possibilities, such as interbasin transfers of water, and shifts in population and economic growth. The assessment may include alternative ways of using the existing water resource infrastructure, and additional capital investments (such as building new reservoirs and canal linings, groundwater recharge, etc.).

Data requirements and possible sources for the data are as follows: Broad estimates of adaptations and costs in water use and management for a river basin unit (local and Country Studies Program technical experts).

The models for individual water basin units are integrated into a national model.

The framework for assessing the water resources vulnerability as a consequence of climate change is given in Figure 12.4 (US CSP 1994).

The following software is used for the water resource impact and adaptation assessments:

- WATBAL: Excell-based water balance model for analyzing the impacts of climate change on mean water supply,
- CLIRUN- Stochastic: A Fortran-based program for analyzing uncertainty in impacts of climate change on water supply,
- CLIRUN- TS: A Fortran-based program for modeling the impact of climate change on runoff time series.

Where a country lacks the necessary database to perform the simulations, recourse may be taken to simpler screening techniques as follows:

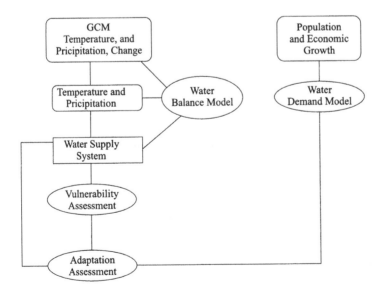

Figure 12.4. Framework of assessing water resources vulnerability (source: US CSP 1994).

| Component | Screening methodologies |
|---|---|
| Supply | Data analytic methods: Aridity Index, Climate/runoff elasticity approach |
| Demand | Linear regression techniques, Analogous country approach |
| Vulnerability | Application of Shuval Index using the screening methods above; A Delphi methodology using local experts |
| Adaptation | A workshop and roundtable meeting of water resource, agricultural and economic experts from the country and experts from countries at a comparable development level |

## 12.2   GROUNDWATER PROTECTION

This section draws heavily from the excellent account of Foster & Skinner (1995).

The importance of groundwater in the water resources management arises from the following considerations:

- The quantity of fresh groundwater within drillable depth is about 70 times greater than all the surface waters (rivers, lakes, reservoirs, etc.) in the world put together,
- The quality of groundwater is generally superior to surface water,
- Unlike surface water, groundwater hardly suffers any evaporational losses, and
- Groundwater is the main source of potable water for domestic purposes in many parts of the world.

### 12.2.1   *Vulnerability mapping*

The French hydrogeologist, Margat, pioneered the concept of groundwater vulnerability, and ways and means of addressing it through vulnerability maps (Albinet & Margat 1970).

Groundwater originates principally through the infiltration of the precipitation falling on the land surface. Thus, many activities (domestic, agricultural, industrial, etc.) taking place on land have the potential to adversely affect the quality and availability of groundwater. An important consideration to be kept in mind is the slow response of most of the groundwater systems. Once pollution has occurred, the water has to be treated at the point of abstraction. As has been explained in Section 8.8, the cleanup of an aquifer is technically complex, extremely expensive, and takes a long time to accomplish. It follows that every effort should be made to prevent the contamination of the groundwater in the first instance. An ideal situation would be to manage the land solely to protect the gathering of groundwater (a few such areas continue to exist in Germany and England since historical times). In practice, this is just not practicable. We have to accept trade-offs between economic development and aquifer protection as a fact of life – only, we should do this objectively. Groundwater protection strategies have to be designed by meshing hydrogeological understanding with the requirements of land-use policies, so as to lead to land-surface zoning (Foster & Skinner 1995).

The protection policy is based on the following elements of aquifer pollution vulnerability:
- We start with defining what activities could be permitted on land without entailing no more than acceptable risk to the groundwater,
- We take advantage of the fact that soils overlying a saturated aquifer have the capacity to attenuate the contaminants passing through the soil, and
- It may not be always possible or even necessary to give equal protection to the total available groundwater resource, such as an aquifer. Protection may be selectively given for potable and sensitive uses or specific sources, such as boreholes, wells and springs. Thus, Source Protection Areas (SPAs) may be earmarked around existing or projected sites of major potable water supply.

The vulnerability to pollution of an aquifer is a measure of its sensitivity to being adversely affected by surface-applied contaminant load. It depends upon the extent to which the strata separating the saturated aquifer from the land surface:
- Allow the penetration of the pollutants, and
- Attenuate the pollutants, as a consequence of the adsorption on solid aquifer material or other chemical reactions (for instance, some pesticides characterized by high solubility, may be immobilized because of their high affinity for absorption on organic carbon – see Section 8.7.4).

Groundwater pollution is defined as 'the probability that groundwater in the uppermost part of an aquifer will become contaminated to an unacceptable level by activities on the immediately-overlying land surface, and will be the result of interaction between the aquifer vulnerability and contaminant loading applied to the location concerned' (Foster & Skinner 1995).

When Integrated Vulnerability maps are prepared, special efforts should be made to spell out the health risks involved at a given level of vulnerability. The three 'laws' of NRC (1993) should be borne in mind: 'All groundwater is to some degree vulnerable', 'Uncertainty is inherent in all vulnerability assessments', and 'There is a risk that the obvious may be obscured by the subtle and indistinguishable'.

The conceptual basis of the British approach about vulnerability is schematically shown in Figure 12.5 (source: Foster & Skinner 1995):
- The strata above the aquifer are grouped into four classes on the basis of lithology and thickness. Experience has shown that the crucial aspect of lithology is the extent of presence of well-developed fracturing (rather than the recharge time lag which would be a measure of when, as opposed to if and which, pollutants reach the aquifer). The greater the fracturing in the overlying strata, the more vulnerable the aquifer is.
- The soil cover is grouped into three classes, based on the extent of their ability to attenuate the pollutants. The less the ability of the soil to attenuate, the more vulnerable the aquifer is. In some urban areas, soil may be deemed to be absent.

Table 12.1 (source: Foster & Skinner 1995) describes the practical significance of different classes of vulnerability.

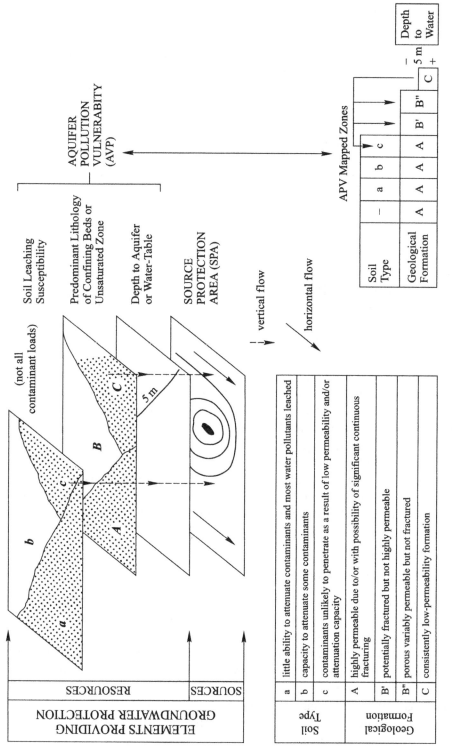

Figure 12.5. Elements of groundwater protection and classification of aquifer vulnerability (source: Foster & Skinner 1995).

Table 12.1. Practical significance of different classes of vulnerability.

| Vulnerability Class | Definition |
| --- | --- |
| Extreme | Vulnerable to most water pollutants with relatively rapid impact in many pollution scenarios. |
| High | Vulnerable to many pollutants, except those highly absorbed or rapidly transformed, in many pollutant scenarios |
| Moderate | Vulnerable to some pollutants but only when continuously discharged or leached |
| Low | Only vulnerable to conservative pollutants in long-term when continuously and widely discharged/leached |
| Negligible | Confined beds present with no significant groundwater flow |

### 12.2.2 *DRASTIC methodology*

In contrast with the British methodology which is essentially qualitative, the DRASTIC methodology developed by Aller et al. (1987) attempts to quantify the vulnerability assessment through the use of weighted indices (given in parentheses) for the various parameters, as follows:

D – Depths to groundwater ($\times$ 5), R – natural Recharge rates ($\times$ 4), A – Aquifer media ($\times$ 3), S – Soil media ($\times$ 2); T – Topographic impact ($\times$ 1), I – Impact (effect) of the vadose zone ($\times$ 5); C – hydraulic Conductivity ($\times$ 3).

In seismology, the earlier qualitative earthquake scale of Mercalli has been completely replaced by the quantitative Richter magnitude scale which is computed from the seismograph readings. Richter magnitude which is a number on logarithmic scale, has also a physical significance and scientific meaning.

While the quantification of a process is always desirable, the scientific significance of a given DRASTIC Index number is often unclear. It has been found that similar DRASTIC indices can be obtained by a very different combination of circumstances, which makes it unsafe to use the number in the decision-making process.

### 12.2.3 *Source protection areas*

Source protection (or wellhead protection as it is called in USA) is an essential part of groundwater protection. The objective of demarcating source protection areas or zones, is to protect the groundwater from two kinds of pollutants:

– Contaminants which decay with time – the longer they reside in the subsurface, the greater will be their attenuation, and the less will be their ability to contaminate the groundwater,
– Non-degradable contaminants, whose adverse impact on the groundwater can be mitigated through flowpath-dependent dilution (see Section 8.7, Transport of contaminant solutes in aquifers).

Three source protection zones are prescribed according to the current British practice (Foster & Skinner 1995):

1. Zone III (catchment area): Catchment of the source, whose area is demarcated on the basis of water balance considerations, and whose geometry, on groundwater flowpath considerations,

2. Zone I (Microbiological protection): This is defined by the mean 50-day saturated zone travel time, based on pathogen decay criteria. As mentioned in Section 8.4.2, the pathogens (say, in human faeces) have persistence times ranging from weeks to months, and the length of travel of the biological pollutants from the source is of the order of 20-30 m. To be on the safe side, a zone of 50 m radius is recommended, as high-storage aquifers are prone to fissure flow.

3. Zone II (Outer Protection): The purpose of this zone to provide delay, dilution and attenuation of slowly-degrading contaminants, and to allow gradational land-use controls between zones I and III. The extent of this zone is arbitrarily set at 400 day saturated travel time.

Mention may be made of some complexities that may be encountered. In the case of karstic aquifers, the annual variation of the catchment area may be very large. In that case, the maximum (rather than average) catchment area should be taken into consideration. Heavy pumping could create complications. For instance, there may be interference between municipal well (continuous pumping for potable supplies) and irrigation wells (seasonal heavy pumping for less sensitive uses, such as agriculture). Hence, SPAs need to be demarcated for different combinations of pumping.

There may be disagreements about the areal extent of zones I, II and III – some may say that they are unduly large and are needlessly restricting land-use, whereas some others may argue that they are too small and will not be able to fulfill their purpose. The uncertainties are inherent in the nature of things, and trade-offs are inevitable. Final decisions have to be taken on the basis of objective discussions.

Uncertainties in delimiting Zones I, II and III are real, and have to be addressed. The National River Authority (NRA) of Great Britain developed the concept of 'envelopes of protection 'to take care of the uncertainties. Best estimates and credible limiting values for the critical variables (such as, recharge rate, hydraulic conductivity and effective porosity) are made use of to compile the envelopes of protection, as follows:

1. Best-Estimate Catchment Area: which is the officially recognized zone for purposes of public policy (such as, compensation paid to the farmers for restrictions on nitrate use, so as to control the soil nitrate leaching to the groundwater) and which satisfies the groundwater balance considerations,

2. Zone of Confidence (ZOC): demarcated by the overlap of all plausible combinations,

3. Zone of Uncertainty (ZOU): the outer envelope formed from the boundaries of all plausible combinations.

The confidence index (area ZOC/area ZOU) is a useful indicator of relative uncertainties. It can indicate where additional resources could be usefully deployed.

Decision-making process in the context of uncertainties has been described in Section 9.6.3.

Most of the hydrogeological environments can be modeled using the computer program, FLOWPATH. Where the hydrological complexity is greater, recourse may be taken to the use of the programs, MODFLOW – MODPATH – MODEL-CAD.

Manual techniques of demarcating the SPAs may be employed where there is lack of data and where the aquifers (e.g. karstic aquifers) are highly heterogeneous. In any event, any zoning, however imperfect, is better than no zoning. Even a simple circular zone around a water supply source (or a semi-circular zone around a spring) does serve the useful purpose as a precautionary marker.

In sum, the vulnerability maps and demarcation of source protection areas are useful in enabling the water users and regulatory agencies to take informed decisions about ways and means of protecting the groundwater.

## 12.3  GROUNDWATER MONITORING SYSTEMS

Groundwater monitoring has been defined as 'a scientifically designed program of continuing surveillance, including direct sampling and remote quantity and quality measurements, inventory of existing and potential causes of change, and analysis and prediction of future changes' (Meyer 1973).

Groundwater monitoring is concerned with the following activities:
1. Gathering information on the physical, chemical and biological parameters of the hydrologic system,
2. An understanding of the vulnerability of the hydrological system,
3. Evaluation of the groundwater resources,
4. Forecasting the possible effects of the natural processes and human impact on the groundwater, and
5. Planning of the groundwater resource protection and conservation, etc. (Sytchev 1988, p. 55).

The parameters of water that are to be routinely measured are the following:

*Physical parameters*:
Turbidity, Suspended Solids, Electrical Conductivity, Color, Odor

*Chemical parameters*:
– pH, carbonate, chloride, sulfate, nitrate, Total Dissolved Solids, ammonia, alkalinity, hardness, BOD, COD, etc.
– Fifteen trace elements (as recommended by IMEP – 9 of the European Commission): B, Ca, Cd, Cr, Cu, Fe, K, Li, Mg, Ni, Pb, Rb, Sr, U, Zn, and F.

*Biological parameters*:

Total coliforms/100 ml. Also as many pathogens as possible from among the following: *Campylobacter* spp., *Giardia lamblia, Entamoeba histolytica, Shigella* spp., *Vibrio cholera, Salmonella typhi, Salmonella* spp., *Escherichia coli* (path.), Enteroviruses, Hepatitis A virus, *Ancyclostoma duodenale, Trichuris trichuiria, Taenia saginata, Ascaria Lumbricoides*, etc.

Groundwater monitoring cannot be done in isolation. It has to be carried out necessarily in conjunction with study of the surface water, precipitation, evaporation, soil, vegetation, etc. The mode of integration and coordination of the groundwater monitoring is shown schematically in Table 12.2 (source: Sytchev 1988, p. 55).

Though the monitoring programs are expensive, they are cost-effective. The cost of monitoring is small compared to the cost of rehabilitation of a groundwater system which has been polluted or depleted. Prevention is always better than cure!

The process of groundwater monitoring has four components: strategy, objectives, program and methods. The International GEMS (Global Environmental Monitoring System) has one station per 10,000 to 100, 000 $km^2$, national or state monitoring networks have a density of one station per 100 to 100, 000 $km^2$, and regional and provincial networks have one station for 10 to 100 $km^2$, and local or site monitoring programs have the highest density one station for several $m^2$ to 1 $km^2$. GEMS project gives priority to aquifers which are spread across several countries, such as the Nubian Sandstone of Africa and the Cretaceous aquifer of central Europe. This kind of study would help in anticipating and solving competition for water by different countries for the same aquifer. GEMS programs are also concerned with long-term trends of water pollution by certain persistent and hazardous substances like DDT, PCBs, etc.

Most countries have monitoring systems of some kind, though in the case of the Third World countries, lack of trained manpower and physical facilities constrain their effectiveness. The Groundwater Quality Monitoring network in the

Table 12.2. Scheme of integration of groundwater monitoring with other networks.

| Groundwater quality program | Integration | Coordination recommended | Coordination possible |
|---|---|---|---|
| *Separate* | | | |
| Local | Groundwater quantity MP | Precipitation MP | |
| *Interconnected* | | | |
| Regional | Groundwater quantity MP | Surface water & Precipitation MP | Climate-soil MP |
| National | Groundwater quantity MP | Surface water MP | Precipitation-climate-soil MP |
| International | Groundwater quantity MP | Surface water MP | Precipitation-climate MP |

MP: Monitoring Program.

Netherlands is a good example of an effective network. Its features are summarized as follows: Density of monitoring stations: one per 100 km$^2$; Kind of aquifer and tested segments: shallow aquifer; 10, 15, 25 m segments; Monitoring frequency: one test per year; Monitoring variables: 19 parameters related to drinking water standards, and concentration of some organic pollutants; Construction cost per monitoring well, 25 m deep: USD 3500.

### 12.3.1  *Groundwater monitoring methods*

Figure 12.6 (source: Sytchev 1988, p. 62) shows a schematic arrangement of a monitoring network. The following preliminary information has to be gathered and steps have to be taken before a monitoring system is brought into existence:
1. Objectives of monitoring, area to be covered, and financial resources available,
2. Technical data regarding the permeability and thickness of the unsaturated zone, and the geometry and hydraulic behavior of the aquifer/aquiclude,
3. Toxicity, mobility and persistence of the existing or potential pollutants which could contaminate the groundwater, and

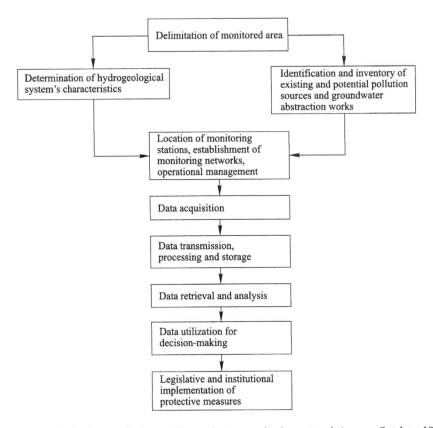

Figure 12.6. Simplified scheme of groundwater monitoring network (source: Sytchev, 1988, p. 62).

4. Analysis of the records of the groundwater abstraction in the area, etc.

There is no prescribed standard for the density of monitoring stations per unit area. More stations will be needed if the piezometric head is highly variable or if the aquifer system happens to be of crucial importance or it is highly vulnerable (see Section 12.2.1 for the explanation of vulnerability). Generally, pollution is more likely in the case of shallow, unconfined aquifers with a high groundwater table, and hence such aquifers need more monitoring stations. Monitoring wells should be carefully drilled so as not to accidentally provide interconnection between the pollution source and the aquifer via the well (Sytchev 1988, p. 62).

If the objective of a monitoring well is to keep track of the movement of pollutants in rocks characterized by fracture porosity, the orientation, geometry and the permeability of the fracture system need to be measured and made use of. Some monitoring wells are located in undepleted and/or unpolluted areas to serve as baseline stations.

Groundwater monitoring involves the measurement of piezometric head, magnitude of groundwater withdrawals and the yields of springs. The data is acquired, transmitted to the central station, processed and stored. In situations where the groundwater receives contribution from more than one aquifer, it is necessary that the monitoring well is provided with a screening arrangement which would permit the measurement of the piezometric head and quality of groundwater from individual aquifers. In the absence of such an arrangement, it should be difficult to identify which particular aquifer is more vulnerable or has been polluted.

In order to arrive at the full picture regarding the dynamics of the groundwater, it is necessary to study the vertical movement of water and the pollutant in the unsaturated zone as well. This can be accomplished in the following ways:

1. Use of lysimeters to monitor the top part of the unsaturated zone,
2. Centrifugal extraction of interstitial water from the core samples, and
3. Use of compressed gas injection or suction to obtain samples of water from different depths (Sytchev 1988, p. 63).

In the case of saturated zones, monitoring is particularly concentrated in the shallow, vulnerable aquifers. The parameters measured are the horizontal and the vertical distribution of the groundwater and possible contaminant, and the direction and the velocity of movement of such contaminant (see Section 8.7, Transport of contaminant solutes in aquifers, for explanation). Multistage groundwater samplers to monitor multilayered groundwater systems can be used for the purpose (Riha 1981). A simple but costlier method would be to have several wells of different depths at the same place, in such a way that groundwater and possible contaminant from a particular aquifer can be sampled separately.

The frequency of sampling and parameters to be measured depend upon the objectives of monitoring and the financial resources available. In the pilot stations, groundwater levels are monitored continuously. Usually, the regional stations monitor the groundwater levels once a day, and the national networks once a week. In the case of monitoring of groundwater quality, some parameters (e.g. temperature, pH, conductivity and Dissolved Oxygen) are measured continuously. The sampling frequency for other parameters may be once a week. National groundwater quality sampling is carried out biannually. Portable units (such as, Hach Radiometer hydro-chemical field kit) are available for making on-site measurements of a number of hydrochemical parameters and pollutants. However, pollutants, such as heavy metals, pesticides and organochlorines are usually studied in the laboratory as soon as possible after sampling. Toxicity of a metal is highly dependent on its speciation (see Chapter 10), which can be investigated only in the laboratory.

The recent advances in information technology have made it possible to design an effective but relatively inexpensive system of monitoring the data on-line and transmitting it to the central database station. Even in remote locations, samples of water can be analyzed in situ with portable kits. The data is recorded on a laptop computer, and then transmitted to the central station by e-mail coupled to a cell-phone. The whole system (portable analytical kit, laptop computer, modem, cell-phone, etc.) costs no more than USD 25,000.

The usual practice is for the central laboratory to plot the monitoring data received from the various stations on GIS (Geographic Information System). GIS a powerful tool for planning purposes, and presentation of spatial information. It normally involves a spatially referenced database and appropriate applications software. A series of maps with various attributes can be overlaid on one another. The outcome is a map with the desired combination of attributes consolidated together (Fig. 12.7; source: Linden, Koudstaal & Nyongesa 1989). For instance, GIS can be used to figure out which kind of pollution is coming from which source, and temporal and spatial variations in water quality along a river, etc. so that appropriate corrective decisions can be taken.

It is possible that the earlier data available in the databank may not be of the same quality and dependability as the present day data. In the past, there may have been errors in sampling and measurement either because the personnel were not well trained or the instrumental set-up available was not of sufficient accuracy, etc. It may be necessary to weed out what is evidently 'bad data'. The recommended practice is to standardize the process of sampling, measurement, reporting, storage, retrieval, etc., so that the data collected is of uniform quality, and meaningful conclusions can be drawn by the comparison of data (vide Environmental Monitoring Methods Index – EMMI – of US EPA, referred to under Section 3.4).

### 12.3.2 *Use of statistical analysis in monitoring*

The purpose of the statistical analysis of groundwater is to quantify the dynamics of the groundwater and pollutant movement, and to elucidate the anthropogenic impact on the hydrological system. The information may be presented in a numerical or graphical form. The statistical (particularly, the stochastic) methods that are generally used in the groundwater studies are summarized as follows (Sytchev 1988, p. 72):

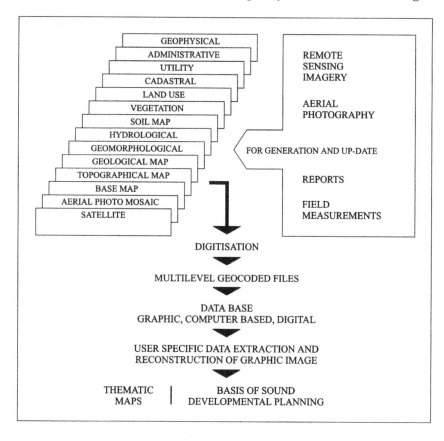

Figure 12.7. Production of thematic maps using GIS (source: Van der Linden et al. 1989).

- Correlation and regression analysis for the physico-chemical parameters and components;
- Trend analysis in regard to physico-chemical components of water in space and time;
- Spectral analysis of the periodic changes in time of the physical and chemical composition of water;
- Factor analysis to study the relationship between the chemical and hydraulic parameters of the hydrogeological subsystem within the soil-water-vegetation system.

The longer the time period over which the data is available and the more the number of data sets, the more dependable the conclusions would be. Statistical techniques are now available to draw useful conclusions even when the data sets available are limited.

One important application of the statistical methods is to determine the optimal sampling frequency. If it has been found that (say) samples taken monthly could lead to a conclusion at about the same confidence level as the data arising from the sample frequency of once a week, it would evidently be cheaper to have the monthly frequency as nothing is lost by doing so. Generally, the analysis of the sample is only a small part of the expense – most of the expense arises from the need to move men and equipment to the sampling site.

Feedback is an integral part of the monitoring system. The monitoring methodology is revised continuously on the basis of the feedback. If some thing works, it is continued; if some other thing does not work, it is either discontinued or modified suitably. We should always look for cheaper, faster and simpler ways of doing things. Any strategy for the utilization and protection of groundwater resources, has to take into consideration the techno-economic and socio-political, etc. implications of various options open in that regard.

Monitoring activity is useful in the planning and management of the water-supply systems, in the following ways:
1. Site selection, and size of the water-works,
2. Projection of the impact of water -works on the moisture, soil, biota and micro-climate of the area,
3. Cost-benefit study of the water-works,
4. Socio-economic criteria for the allocation of water among the competing users (drinking and domestic purposes, sanitation, irrigation, industries, recreation, etc.).

## 12.4   CONJUNCTIVE USE OF WATER RESOURCES

The purpose of the conjunctive use of water resources is simple: surface water surplus during the wet periods is temporarily stored in groundwater reservoirs, to be withdrawn for use during the dry periods. Conjunctive use is particularly necessary in the monsoon regions. During the rainy period lasting for a few months, the rivers are full and there is more water than could be used. During the dry period when water is desperately needed, the rivers are dry or have very little flow. Conjunctive use of water is a sensible strategy to address the mismatch in time and space between water demand and water availability (for details, see IAHS publication no. 156, 'Conjunctive Water Use').

Buras & Hall (1961) used dynamic programming and economic considerations for optimizing the conjunctive use of water. The latter involves the determination of the necessary volume of the groundwater reservoirs to be used conjunctively with the surface reservoirs, in the long term.

### 12.4.1   *Design and management of groundwater reservoirs*

A groundwater reservoir needs to have two facilities: areas where water can be recharged naturally or artificially, and pumping installations. The availability of surface water in a given project is first determined on the basis of runoff records for the last 50 years. That data is then used to estimate the volume of the groundwater reservoir needed for conjunctive use. Geological considerations determine the maximum storage capacity that a groundwater reservoir could have. The extent to which the stored water could be withdrawn depends upon the availability of finances (e.g. pumping and recharge costs), and environmental considerations (e.g. saline intrusion, subsidence of land). Models should therefore simulate parameters such as storage coefficient (S), transmissivity (T) and the contour conditions of the aquifers.

The water balance of the groundwater reservoir is given by the following formula (Correa 1987):

$$V_{(t+1)} = V_{(t)} + I_{(t)} - O_{(t)} + R_{(t)} - P_{(t)} \qquad (12.1)$$

where $V_{(t)}$ = groundwater volume at the end of period, $t$, $I_{(t)}$ = underground inflow to groundwater reservoir in period of time, $t$, $O_{(t)}$ = Underground outflow from groundwater reservoir in period of time, $t$, $R_{(t)}$ = Groundwater recharge (natural and artificial) in period of time, $t$, $P_{(t)}$ = pumping volume from groundwater reservoir in period of time, $t$.

The variables, $I_{(t)}$ and $O_{(t)}$, are dependent upon the piezometric surfaces, but their yearly mean values can be made use of in the calculations. The volume of water that needs to be pumped will be the deficit between the water demand and the quantity of surface water available. Evidently, the maximum pumping capacity of the system limits the quantity of groundwater that could be extracted.

The groundwater recharge involves the following terms:
– Groundwater recharge from precipitation: the rate of infiltration is controlled by the nature and intensity of precipitation on one hand, and the permeability of the rocks and soils in the zone of aeration, on the other. Groundwater recharge from precipitation may be insignificant in arid areas,
– Artificial recharge: the maximum possible volume of artificial recharge would depend upon the geological conditions (recharge area), physical characteristics (permeability) and economic considerations (costs),
– Channel seepage, deep percolation of unconsumed irrigation water on unconfined aquifers: this may be taken as constant for every year, if the water demand is constant,
– Percolation from streams and ponds: this depends upon the surface water surplus that flows over the recharge area (unconfined aquifer).

The long-term equilibrium of the groundwater reservoirs or safe yield is given by the following equation. It corresponds to the condition that the total inputs and outputs to the groundwater reservoir are equal (Correa 1987).

$$\sum_{t=1}^{N} (<F_{(t)}> + Rj_{(t)} + I_{(t)} - P_{(t)} - O_{(t)}) = 0 \qquad (12.2)$$

where $N$ = Years of observed runoff sums, $<F_{(t)}>$ = function of artificial recharge + percolation from streams and ponds, $Rj_{(t)}$ = percolation from irrigation losses, $I_{(t)}$ = underground inflow to groundwater reservoir in period of time, $t$, $P_{(t)}$ = pumping volume from groundwater reservoir in period of time, $t$, $O_{(t)}$ = Underground outflow from groundwater reservoir in period of time, $t$.

The above equation has a solution if the mean value of water demand in $N$ years ($Wd_m$) falls within the range of $WD_{min}$ and $WD_{max}$, corresponding to the coefficient of recharge ($k$) being within the range of $k_{min}$ and $k_{max}$. The system would evidently not be sustainable if $Wd_m > WD_{max}$. This point can be illustrated with a simple example. If the monthly earnings of a person vary from (say) a minimum of USD 3000 per month to a maximum USD 6000/m, with an annual average of USD 48,000, the 'safe' amount that such a person could spend (without getting into problems) is USD 4000 per month. If such a person starts spending USD 5000 per month, he will land himself in troubles.

The coefficient of recharge ($k$) can also be used to determine the recharge policy. For water demand with $k_{max} > k > 1$, it is necessary to have a recharge policy to ensure the long-term sustainability of the reservoir system. On the other hand, for water demand with $k_{min} < k < 1$, no artificial recharge policy would be needed as the system can be sustained through controlling the natural percolation from the river.

The objective function to be applied for the optimization of the system is based on the difference between the groundwater actually stored and the target groundwater volume at the beginning and at the end of the hydrological year.

### 12.4.2 *Case history of conjunctive use through over-irrigation of paddy fields*

Tsao (1986) gave a case history from Taiwan of an innovative conjunctive use through over-irrigation of paddy fields.

Wetlands serve as 'kidneys' (purifiers) of the contaminated water, besides playing a useful role in the hydrological cycle. The Cho-Shui river basin in Taiwan with an area of 5335 km², has fertile soils, average annual temperature of 22°C, and average annual precipitation of 2017 mm. It is thus ideally suited for paddy cultivation round the year. But the catch is, that 80% of the precipitation occurs during May to September. It is dry during October to April. Since the river is heavily loaded with sediment, and since a single-purpose hydropower reservoir already exists on the river, it has not been found to be feasible to construct surface reservoirs. Under the circumstances, recourse has been taken to the withdrawal of groundwater for irrigation during the dry period. Consequently, more than 100,000 wells have been dug. To complicate the matters further, large quantities of groundwater have been withdrawn to raise eels, shrimp, fish and clams. As is to be expected, the groundwater levels and yields declined sharply, land subsided, and incursions of seawater occurred (in some places, the zero groundwater level is 25 m below the mean sea level). The tidal wave due to typhoon Wyne in 1980 overflowed the sea dykes erected near the coast, and caused great damage to the paddy fields.

Two types of conjunctive use of water through artificial recharge were attempted. Injection recharge was technically feasible, but as the water is heavily loaded with sediment, the experiment proved to be too expensive, and was therefore abandoned.

Artificial recharge through the over-irrigation of the paddy fields proved to be techno-economically feasible. Since the Cho-Shui river basin is a granary of Taiwan, any modification of the irrigation system has profound socio-economic implications. For this reasons, detailed simulation studies were made before decisions were taken, and elaborate monitoring systems have been set up to keep track of the environmental and socio-economic implications of over-irrigation.

An orientation study was initially made with a 195 ha plot, where the shallow top soil was underlain by a 30 to 50 m deposit of alluvial coarse sand and gravel which serves as a natural recharge area. Since the average intake rate of paddy fields is 150 mm $d^{-1}$, the groundwater responded well to over-irrigation. A salt balance study was also made in order to determine the optimal rate of fertilizer application, so as to minimize the loss of fertilizers due to over-irrigation. Weirs were constructed to provide ponds for the supply of irrigation water. As the drainage water from the paddy fields is invariably contaminated, it could not be used for aquaculture which is extremely sensitive to contamination.

Experience has shown that an over-irrigation of 50 mm $d^{-1}$ is entirely feasible. About 0.77 Bt (billion tonnes) of groundwater per year can be safely extracted from the system, while satisfying all the constraints. Mathematical models predicted that with an artificial recharge of 86.8 Mt (million tonnes) per year by over-irrigation, it is possible to prevent the intrusion of saline water inland.

## 12.5   WATER MANAGEMENT UNDER DROUGHT CONDITIONS

Statistical analysis of climate data suggests that the frequency of extreme weather events (such as, floods and droughts) are likely to increase as a consequence of global warming and El Niño phenomena (see Section 11.5 for details). Water managers are expected to come up with strategies to cope with both drought (too little water) and floods (too much water).

There is a well-established link between precipitation and ecosystems, to wit:

| Ecosystem | Precipitation (mm $yr^{-1}$) |
|---|---|
| Desert | < 100 |
| Arid lands | < 250 |
| Semiarid lands (range lands) | > 250 |

Semiarid tropical lands are characterized by long dry seasons, low and unpredictable rainfall and poor soils.

Droughts are manifestations of significant shortages in all domains of the hydrological cycle. They have adverse impacts on the environment, water quality and water availability, water supply system, hydropower generation, navigation, vegetation cover, dilution capacity of the rivers, groundwater balances, deposition of sediments in lakes and reservoirs, etc.

Falkenmark et al. (1990) emphasize that droughts and famines should not be regarded as disaster *events* but *processes*. Drought occurs when the available moisture in the root zone falls below 30% of the water-holding capacity of the soil. They distinguish four different kinds of water scarcity (Fig. 12.8; source: Falkenmark et al. 1990).

*Natural*:
– Aridity, reflected in the short length of the growing season, and
– Intermittent droughts, reflected in recurrent drought years with risk of crop failure;

*Man-made*:
– Landscape desiccation caused by soil degradation, resulting in water not reaching the plant roots, and
– Water stress, whereby the demand for water by the growing population exceeds the regenerative capacity of the system.

The significance of El Niño, ENSO and La Niña episodes have been explained under Section 11.3.1.

In 1991-1992, Southern Africa experienced the worst drought of the century. The cereal harvest in the Southern Africa Development Community (SADC) region halved, causing immense hardship to about 100 million people in Southern

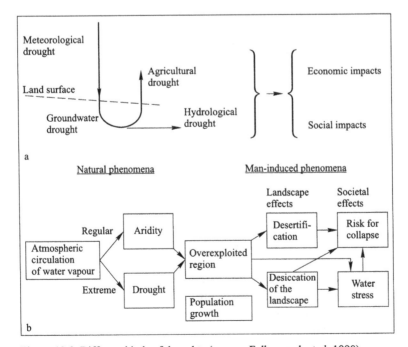

Figure 12.8. Different kinds of droughts (source: Falkenmark et al. 1990).

Africa. The scientific observation linking El Niño warm events in the Pacific Ocean to below average rainfall in southern Africa, holds a great promise for predicting drought a year ahead (*Nature*, 21 July 1994, v. 370, p. 175-176, p. 204-205). This would make it possible for the countries concerned to be prepared for drought, through measures such as drought-tolerant varieties of crops and species, improved conservation of moisture, storage of food and fodder, etc. Mark Cane and Gideon Eshel of Lamont – Doherty Observatories of Columbia University, USA, and Roger Buckland of SADC Food Security Unit in Harare, Zimbabwe, found a strong correlation between El Niño index and both rainfall and yield of maize (largely rainfed) in Zimbabwe. Surprisingly, the correlation coefficient with maize yield (0.78) is stronger than that with the rainfall (0.64). The yield of maize in Zimbabwe, a high of 2.4 t ha$^{-1}$ in 1986, and the low of 0.4 t ha$^{-1}$ in 1992, are accounted for by El Niño.

According to the estimates of the Organization of African Unity (OAU, Addis Ababa), the number of people affected by disasters in Africa during the period, 1980-1989, are as follows:

|  | No. of people dead | No. of people rendered homeless |
| --- | --- | --- |
| Drought | 450,000 | 200,000 |
| Floods | 1,000 | 500,000 |

This shows clearly that drought is the most serious killer in Africa.

### 12.5.1 *Modeling of drought*

Frequency analysis of drought characteristics is the traditional method of evaluation of drought impact. This takes into consideration only one parameter, namely, drought magnitude. Madsden & Rosbjerg (1995) developed a truncation level approach which takes into consideration not only the magnitude but also duration of the drought simultaneously. In this method, droughts are defined as periods during which the hydrological determinant (e.g. stream flow) is below a certain threshold. The parameters to be considered are:
- The drought duration ($D$) which is the distance between a downcross and the following upcross of the truncation level,
- The deficit volume $S$ (often termed, drought severity) which is the sum of the deficits within the dry spell period.

Drought magnitude ($M$) = $S/D$.

The observed distributions are compared with the estimated distributions.

Drought can occur in temperate climates too, but they hardly pose a threat to life, as happens to be the case in arid lands (say, of Africa). In UK, drought contingency plans form a continuum with normal operations (Walker & Smithers 1995) and this is an eminently sensible principle. Droughts may be divided into two categories in the context of their impact:

– Environmental drought, caused by diminished precipitation, as manifested by reduced flow in rivers and having impacts on agriculture and fisheries, and
– Water supply drought, which affects the domestic and industrial consumers of water. Often, a drought may have both environmental and water supply impacts. The National Rivers Authority (NRA), a public agency, has the responsibility to ensure that the utilities companies which are privately owned, do not cause environmental damage in their quest to cut costs.

There are two management options to mitigate the drought:

1. Those that will have the effect of reducing the demand on the water resources system. Three levels of demand reduction could be considered: 5% (equivalent to a call on voluntary savings), 10% (equivalent to ban on hosepipes), and 20% (equivalent to a ban on non-essential uses).
2. Changes in the management of the water supply networks.

By using a combination of stochastic dynamic programming and simulation to optimize the operation policies, it is possible to investigate the benefits and trade-offs between the different demand reduction and supply enhancement measures. A contingency plan is drawn up incorporating the various measures and implemented (Fig. 12.9; source: Walker & Smithers 1995).

### 12.5.2   *Remote sensing of drought*

According to the US National weather service, a precipitation deficiency of 15% or more during a six-month period can be considered as the onset of drought. The decrease of soil moisture during drought will have a significant impact on vegetation. Remote sensing images by AVHRR (Advanced Very High Resolution Radiometer) aboard NOAA's Polar Orbiting Environmental Satellite, are capable of providing a detailed picture about the temporal and spatial distribution of the impacts on vegetation due to drought.

Green leaf foliage is characterized by a strong absorption in the red region, and a strong reflectance in the near IR (Infrared) region due to scattering. The Normalized Difference Vegetation Index (NDVI given by the ratio: NIR – Red/NIR + Red) has been found to correlate well with the green Leaf Area Index (LAI). A decrease in NDVI is indicative of reduced photosynthetic activity and green biomass. Thus, with the help of AVHRR, it is possible to quantify and monitor the impact of drought on vegetation.

Lamblin & Ehrlich (1996) employed AVHRR images to study land cover changes in Africa. AVHRR channel 3 (3.5-3.9 μm) delineates the boundary between forest and non-forest, and land surface skin temperatures are derived from channels 4 and 5.

Figure 12.10 (source: Lamblin & Ehrlich 1996) elegantly illustrates the relationship between the surface temperature (°C) and vegetation index. Over bare soil, there is a good correlation between the radiant surface temperature and soil water content. Point A in the figure refers to dry bare soil, and point B, moist bare soil. As

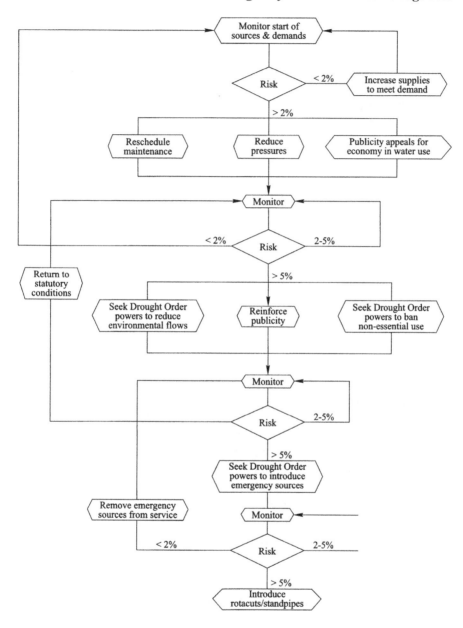

Figure 12.9. Contingency plans depending upon the assessed risk of failure of the source (source: Walker & Smithers 1995).

the vegetation cover increases, surface temperature decreases (see trapezium, ACDB). Point C corresponds to continuous vegetation canopies with high resistance to evapotranspiration (due to low-soil water availability). Point D corresponds to continuous vegetation canopies with low resistance to evapotranspiration (be-

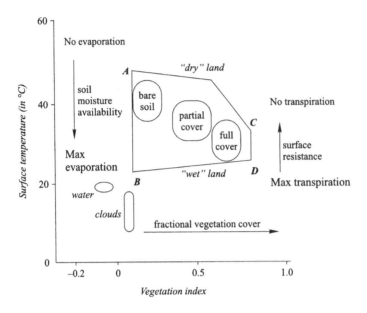

Figure 12.10. Relationship between the surface temperature and vegetation cover (source: Lamblin & Ehrlich 1996).

cause the surface is well-watered). The upper envelope line A-C corresponds to low evapotranspiration line (dry conditions), whereas the lower envelope line B-D represents the potential evapotranspiration line (wet conditions).

In the arid lands of Sub-Saharan Africa, enhancing the access of water to the roots of the crops is the key to increased crop yields. This can be achieved by

– Soil conservation to secure maximum infiltration,
– Increase in the water holding capacity of the soil and nutrients by increasing the content of organic matter and clays, and
– Increase in the availability of water, through rainwater harvesting, water storage, etc. (see Section 6.8.3, Strategies for improving the availability and the use of moisture).

## 12.6  WATER MANAGEMENT UNDER FLOOD CONDITIONS

According to WMO, during the period, 1947-1980, tropical cyclones, hurricanes and typhoons killed an estimated 499,000 people worldwide, and floods (unrelated to these) killed 194,000 people. Extremely strong winds and high tides can cause heavy surge of water which can reach far inland. Low-lying Bangladesh is particularly vulnerable to such disasters – over 250,000 people died in such storm surges in 1970. In another similar event in 1988, 80% of the entire Bangladesh was affected. The disastrous floods in Yangtze (China) in 1991 and in the Missouri and Mississippi rivers (USA) in 1993 are well known. The floods in southern Mozambique during Feb.-April, 2000 give an indication of the extent of damage that they could cause. The floods left an estimated one million people homeless and hungry. The death toll has been put at about 700 (among the survivors who have been rescued is a child born to her mother on a tree-top where the mother has taken refuge!). The floods have been attributed to La Niña, which is probably linked to the strong El Niño event of 1997-1998. The torrential rains in Jan. 2000, were followed by cyclone Connie. The cyclone Elaine in mid-February and tropical depression Gloria in early March resulted in the enormous swelling of the rivers, Limpopo, Save and Zambeze in southern and central Mozambique. At the height of the flooding, the Limpopo river widened from its normal 100 m to 3.5 km.

The flood hazard itself cannot be prevented, but through an understanding of the land conditions which are prone to a given hazard and the processes which could culminate in the damage to life and property, it is possible to minimize the damage through preparedness for a particular eventuality.

Flooding takes place when the river channels are unable to contain the discharge. In the tropical countries, floods are caused by various factors:
- Climatological (rain),
- Part climatological (coastal storm surges, estuarine interactions between stream flow and tidal conditions), and
- Others (failure of dams and other control works, excessive release from dams).
  Floods could get intensified because of basin characteristics, network characteristics, and channel characteristics, each of which have both stable (unvarying) and variable components (Table 12.3).

The improper land-use practices accentuated the flood devastation. There are hardly any forests left in the catchment area of the rivers. It is well known that the forest areas are characterized by high infiltration capacity and transmissibility. The infiltration capacity of the forest areas is 2-3 times greater than in the open fields. The surface runoff in the forested areas may be as little as one-tenth of that of the open fields. There is hardly any protective vegetation on the banks of the Save and Limpopo rivers. Crops are grown right to the edges of the rivers, even

Table 12.3. Unvarying and variable characteristics.

|  | Stable (unvarying) | Variable |
|---|---|---|
| Basin characteristics | Slope, attitude, shape, etc. of the basin | Arising from the interactions between climate, geology, soil type, vegetation cover, etc. which are manifested through the storage capacity of soil and bedrock, extent of infiltration, and transmissibility of soil and bedrock, that are affected by anthropogenic activities |
| Network characteristics | Pattern | Surface storage, under-drainage, channel length, contributing or source area, etc. |
| Channel characteristics | Slope, flood control and river regulation works | Roughness, load, shape, storage |

on river slopes. Thus, the combination of absence of forest cover on one hand, and inappropriate farming practices on the other, intensified the floods.

Floods can be mitigated by structural, water control and non-structural measures. The structural methods include dams, reservoirs, and retarding basins, channel management and embankments. The water control methods include flood proofing and catchment modifications. Schemes of drainage and flood protection, flood forecasting, flood warning and emergency preparedness systems, flood insurance, public information and education, and flood relief constitute the non-structural methods.

*Early warning systems*: By developing the infrastructure to link with GOOS (Global Ocean Observing System) of WMO, and combining it with GIS, it is possible to simulate the potential spread of floods as a consequence of extreme weather and climate events, with sufficient lead time to allow action to save lives, minimize property damage and protect the environment.

### 12.6.1 *Flood management*

From times immemorial, flood plains have been the preferred locations for human settlements and agricultural development. Unfortunately, the very same rivers periodically overflow their banks, and cause considerable loss of life and property. Most of the flood management methodologies address the economic issues, whereas both economic and social criteria should be included in the decision-making process. Simonovicic (1998) described a decision-making framework for a multi-objective approach to flood management.

There are three stages in the flood management process:
1. Planning: techno-socio-economic evaluation of various alternative measures to minimize flood damages,
2. Flood emergency management: appraisal of the current flood situation and daily analysis of how the flood control units are working, potential events that could affect the current flood situation (such as, breaches of dikes and bunds, winds, heavy rainfall, releases from dams upstream, etc.) and urgent capital works that may be needed, etc.
3. Post flood recovery: evaluation of flood damage, rehabilitation of damaged properties, and assistance to flood victims to return to normal life.

There are two types of flood mitigation measures:
1. Structural methods which include local construction measures as also flood protection methods upstream. They involve the construction of levees or walls to prevent inundation from floods, drainage and pumping facilities:
   – diversion structures to divert the flow from the protected region,
   – channel modification to increase the hydraulic capacity of the river, and
   – construction of storage reservoirs upstream.
2. Non-structural methods: zoning (to permit such land uses as may not be severely damaged by floods, such as pasture land), protection of individual prop-

erties (such as, waterproofing of lower floors), construction of flood shelters (where people can take refuge during the duration of the floods), flood warning systems (to evacuate the residents), and flood insurance (compensation payable to individual families and companies), etc.

### 12.6.2 *Flood Forecasting and Regulating Decision Support System (FFRDSS)*

Qi et al. (1995) developed a computer software program for comprehensive flood control decision support system for the Dongtiao river in east China. It includes a data acquisition sub-system, a forecasting and regulating sub-system, and flood control decision making sub-system. It can be applied for regulating the reservoirs, lakes, detention basins, flood ways, intervening area inflows, tributaries, backwater influences, etc. It makes use of the following models quoted by Qi et al. (1995).

NRIHM (Nanjing Institute of Hydrology Runoff Generation Model) – Wen et al. 1982; GIUH (Geomorphological Instantaneous Unit Hydrograph) – Rodriguez-Iturbe et al 1979; TLR (Total Linear Response model) – Nash et al. 1980; DWOPER (Water Balance Model and Dynamic Wave Operation Model) – Fread 1978.

Every program has a module for graphic visualization of results.

FFRDSS is a completely interactive and menu-driven. It allows the user to simulate various possible future situations, and natural and human interventions. A techno-socio-economically viable operational management decision can be taken in the context of flood control measures contemplated.

The main menus and the corresponding sub-menus are as follows (Qi et al. 1995):

a) Introduction of the river system:
 – river system map, and
 – route of flow.
b) Forecasting and regulating:
 – access to data library,
 – forecasting of water stage or discharge for predefined sites, and
 – decision support.
c) Display of regime flow:
 – map of distribution of the amount of rainfall,
 – map and table for regime flow,
 – printing of hydrograph, and
 – cross-section of predefined sites.
d) Display of hydrograph:
 – reservoir inflow flood,
 – water stage of reservoir,
 – stage hydrograph of predefined sites,
 – intervening area rainfall chart, and

  – intervening area discharge hydrograph.
e) Display of regime of disaster:
  – historical regime of disaster,
  – assessment of flood damage, and
  – space distribution of flood disaster.
f) Hydrological simulation:
  – reservoir water stage and inflow,
  – intervening area inflow, and
  – water stage and discharge of predefined sites.
It has been reported (*Eos*, AGU, Mar. 28, 2000) that a number of American agencies (US AID, NOAA, USGS, etc.) are collaborating to develop a Flood Risk Monitoring Model in the context of southern Africa. It is said to be a physically-based, catchment-scale hydrologic model which includes a GIS-based module and rainfall-runoff simulation model.


## 12.7   PLANNING OF SUSTAINABLE WATER SUPPLY SYSTEMS

Sustainable development has been defined as that kind of development that 'meets the needs of the present without compromising the ability of the future generations to meet their own needs'. In the context of water resources, this could probably be worded as follows: Sustainable development of water resources encompasses those patterns of water resources utilization that will enhance social and economic benefits for the present and the future generations, without impairing the eco-hydrological processes. This implies that any use of water resources which either leads to or has the potentiality to lead to, the degradation of the eco-hydrological system is *ipso facto* unsustainable and hence unacceptable.

  Schultz & Hornbogen (1995) point out that simply forecasting an attribute of water resources management, such as water demand, and designing projects to meet the future water demand, is inadequate. They propose an 'anticipatory approach', whereby our predictions should involve not only parameters directly concerned with water, but also in terms all relevant variables, such as, hydrological and ecological conditions, as well as socio-economic conditions, including a potential change in the society's value system. Using the concepts of fuzzy-sets, and probability functions, they quantify a 'possibility function' on the basis of a large number of possible future development paths.

  Figure 12.11 (source: Schultz & Hornbogen 1995) illustrates the assessment of alternative strategies for possible future conditions in regard to water availability and water demand on the basis of the possibility functions for likely changes in hydrological and socio-economic conditions. Figure 12.12 (source: Hornbogen & Schultz 1998) depicts the driving forces of both water supply and water demand. By combining two one-dimensional possibility functions of (a) and (b), we can derive the two dimensional graph shown in (c), which takes care of the areas of

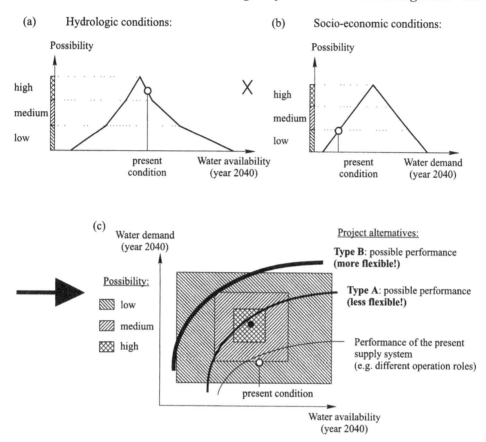

Figure 12.11. Assessment of alternate strategies based on possibility functions (source: Schultz & Hornbogen 1995).

low, medium and high water demand as well as water availability. Suppose the realization of sustainable development needs a new water project (say, a water supply dam) which is capable of addressing all possibilities. This would correspond to the outer edge of the rectangle specified as low possibility. Three curves are shown in (c): performance of the present supply system, type A project alternative which covers the core conditions of high possibility, but which does not allow modifications, and type B which is close to the optimum solution but is flexible. If the project alternative B does not cost much more than A, it should be preferred as it allows greater flexibility. However, if alternative B is much more expensive than A, then one has to decide whether the extra expense is worth the flexibility.

The point made above may be illustrated with a common experience. If an à la carte meal in a restaurant does not cost much more than a table d'hote (fixed menu) meal, then the preference would naturally be for an à la carte meal. However, if the table d'hote meal contains most of your favorite dishes but costs much less

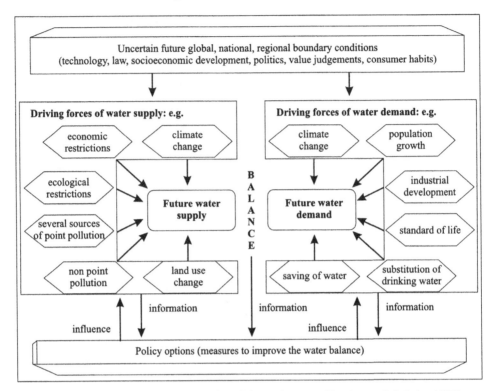

Figure 12.12. Driving forces of water supply and demand (source: Hornbogen & Schultz, 1998).

than à la carte meal, it would obviously be sensible to go in for a table d'hote meal (unless you feel a strong craving for a particular dish which is not included in the table d'hote meal and what is more, can afford to pay for an à la carte meal!).

### 12.7.1 *Stakeholder involvement in Integrated land and water management*

It is now widely realized that mitigation methodologies developed on the basis of biophysical studies alone have very little chance of success, unless they have been drawn up in consultation with the stakeholders right from the beginning. It is for this reason that IGBP (International Geosphere – Biosphere Program) promoted close linkage between two relevant sub-programs, namely, LUCC (Land Use/Land Cover Change) and CGTE (Global Change and Terrestrial Ecosystems).

Climate change impacts on freshwater resources appear in sub-Saharan Africa in their starkest form. The high rate of population growth with associated changes in land use, extreme variability of precipitation, both in time and space, and high potential evapotranspiration in dry lands – these issues can only be addressed by an imaginative integration of biophysical approaches with socioeconomic approaches. As Coleen Vogel (1999) puts it elegantly, 'In short, there is need to be able to identify landscape patterns as products of both distinctive physical and social interconnections, and which in turn feed back into the Earth system, rather than focusing, singularly, on modulating and shaping the roles of the biophysical drivers of change'. Table 12.4 (source: Coleen Vogel 1999) gives the framework of how the big issues are to be addressed through specific tasks and activities.

The poor people in the South have to cope not only with socio-economic realities (such as HIV/AIDS, armed conflict, population dislocation and programs of economic reform), but also environmental degradation and extreme weather events (such as, currently floods in Mozambique, and drought in Ethiopia). Research is needed to find out how these socioeconomic factors are contributing to the global change, and how the global change is feeding back to the system and affecting the communities. In the case of Mozambique, poverty drove the farmers to cut down vegetation (for fuelwood) and do farming right up to the edges of the streams. There is little doubt that these improper land-use practices exacerbated the devastating floods in Mozambique during Feb.-April, 2000. Almost certainly, the floods will be followed by droughts in Mozambique in the coming years, and the people are as unprepared for droughts as they were for floods. Thus, with each disaster, the people become less able to cope with extreme weather events. This vicious cycle would only stop when mitigation measures are taken and preparedness systems are put in place.

Figure 12.13 (source: IGBP Newsletter, Dec. 1999) illustrates a systems approach for integrating the biophysical and socioeconomic approaches in regard to range lands as they affect the rural livelihoods.

IGBP jointly with its human resources analogue, IHDP, came up with an Integrated Land and Water Management paradigm for Africa (IGBP Newsletter, Dec. 1999), which is valid for most of the developing countries. The paradigm has the following components:

1. Horizontal integration: integration among adjacent land users and land uses within catchments; between upstream and downstream users; among domestic, industrial, urban and other users, and among governments sharing river systems,

2. Vertical integration: integration among the range of organizations and institutions functioning at different scales and strives to achieve (a) maintenance of adequate amounts and quality of water to all water users, (b) prevention of soil degradation, (c) food security, and (d) prevention and resolution of conflicts between water users'.

IGBP – IHDP point out that, in order to be really effective on the ground, freshwater resources experts in the developing countries need to be trained not only in the scientific and technological methodologies but also in institutional management and skills in community involvement. They identified six constraints or deficiencies:

1. Deficiency of information and understanding: recognition that water has a value in all its competing uses, and incorporation of this concept in the planning process; collation of existing data,

Table 12.4. 'Socializing the pixel' – Integration of biophysical and socioeconomic aspects of climate change – The big issues and some examples of specific tasks and activities (source: Coleen Vogel 1999).

| Big issues | Focus 1 | Focus 2 | Focus 3 |
|---|---|---|---|
| Transition to a Sustainable World | Activity 1: Understanding land-use decisions<br>Task 3.2: Simulation modeling of land-use/land – cover change to identify sustainable future scenarios | Task 1.2: Definition of land-cover change indicators | Activity 2: Major issues in methodologies of regional land-use/land-cover change models<br>Activity 3: land-use/land-cover change and the dynamics of interrelated systems<br>Activity 4: Scenario development and assessments of critical environmental themes |
| Biogeochemical cycles and Biodiversity | Activity 2: From process to pattern: linking local land-use decisions to regional and global processes<br>Task 3.1: Identify key biogeochemical and climate variables associated with changes in land cover over a long period | Task 1.1: Monitoring bio-physical and socio-economic variables<br>Activity 2: Socializing the pixel | Activity 2: Major issues in methodologies of regional land-use/land-cover change models<br>Activity 3: land-use/land-cover change and the dynamics of interrelated systems<br>Task 3.2: Water issues in regional and-use/land-cover change |
| Critical regions and vulnerable places | Task 3.3: Simulation models that identify key interactions associated with degradation and vulnerability | Activity 1: Land-cover change, hot- spots and critical regions<br>Task 3.32: Definition of risk zones and potential impacts | Task 2.2: Improving the environment – economy linkages<br>Task 3.2 Water issues in regional land-use/land-cover change<br>Task 3.3: Expanding the global food and fibre production<br>Activity 4: Scenario development and assessments of critical environmental themes |

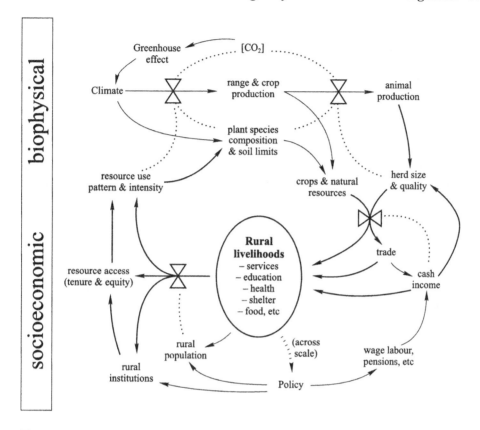

Figure 12.13. Integration of biophysical and socioeconomic approaches (IGBP Newsletter, Dec. 99).

2. Deficiencies of governments – existence of networks of institutions, policy makers, managers, stakeholders and scientists,
3. Deficiency in capacity: availability of necessary technical and managerial skills,
4. Deficiencies in land management options: development of coping and adaptation mechanisms,
5. Deficiencies in water management options: ways and means of addressing water storage, water quality, water treatment and use, and equitable distribution, and
6. Deficiencies in stakeholder involvement: involvement of the stakeholders from the outset in planning and development.

## 12.8  GLOBAL FRESHWATER – SUPPLY CRISIS

Scarcity of freshwater has different connotations in different regions. In the case of humid regions, the problem is not that water is scarce *per se*, but that there has been degradation in the quality of aquatic environment. In the case of arid lands or lands characterized by monsoon climate, freshwater is scarce during the dry season.

The problem of degradation of water quality is described from the following perspectives (Peters et al. 1998):

*Causes*: Pathogens, organic pollution, nitrate pollution, heavy metals, salinity from changed land use, pesticides and other xenobiotic chemicals, nuisance and toxin-producing species, introduction of exotic species.

*Effects*: Scarcity, sustainability and economics of water users, human health, ecosystem health, global decline in the microbial safety of water, biodiversity and species extinction.

*Ecosystem stresses*: Economic development, financial and human costs of waterborne diseases.

The major water quality issues and their controlling factors are summarized in Table 12.5 (source: Peters et al. 1998).

Earth system processes are closely interlinked. Every land use decision we make has an impact on water resource. Every time we use fuelwood for cooking, we are not only affecting the air we breathe, but we are having an impact on water

Table 12.5. Major water quality issues.

| Major causes/issues | Major related issues | Major controlling factors |
|---|---|---|
| Natural ecological conditions | Parasites * | Climate and hydrology |
| Natural geochemical conditions | Salts, Fluoride ** | Climate and lithology |
| | Arsenic **, metals ** | |
| Population | Pathogens | Population density and |
| | Eutrophication * | Waste treatment |
| | Micropollutants | Miscellaneous |
| Water management # | Eutrophication * | Flow velocity |
| | Salinization | Water balance |
| | Parasites | Hydrology |
| Land use | Pesticides | Agrochemicals |
| | Nutrients ($NO_3^-$) | Fertilizer |
| | Suspended solids * | Construction/clearing |
| | Hydrological changes # | Cultivation, damming |
| Long-range atmospheric transport | Acidification * | Cities, Smelting and |
| | Micropollutants | Fossil fuel emissions |
| Hot spots | | |
| Megacities | Pathogens, Micropollutants | Population and waste treatment |
| Mines | Salinization, Metals | Types of mines |
| Nuclear industry | Radionuclides | Waste management |
| Global climate variability and climate change | Salinization | Fossil fuel and CFC emissions; |
| | Sea salt intrusion ** | Temperature and precipitation, Sea level change |

\* Is relevant primarily to surface water, ** Is relevant primarily to groundwater, # As drains, diversion and over-pumping of aquifers.

resource as well. Natural and anthropogenic processes affecting water quality have to be traced as water runs downhill. Hence, a drainage basin is the most natural framework for monitoring the water quality.

Figure 12.14 (source: Peters et al. 1998) shows schematically how Policy and management, Scientific understanding of the processes and assessments, and Monitoring are to be integrated on a drainage basin scale ($< 1\text{-}10 \text{ km}^2$). As it is too expensive and time-consuming to investigate all drainage basins equally intensively, Peters et al. (1998) recommend that a representative drainage basin be investigated in close association with stakeholders (bottom-up component) but on a globally consistent design (top-down component). The methodology developed in this manner is then applied to other drainage basins, and the results can then be integrated with regional and global networks. Thus, though each basin is investigated individually on a bottom-up basis, greater benefits accrue by the adherence to a common, standard typology (Fig. 12.15: source: Peters et al. 1998).

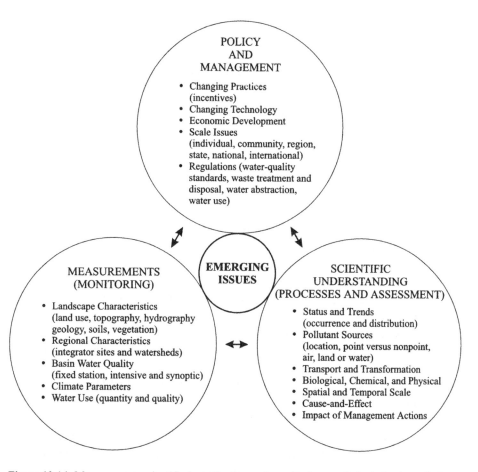

Figure 12.14. Management, scientific investigation and monitoring on drainage basin scale (source: Peters et al. 1998).

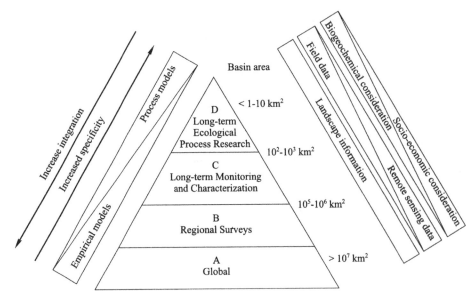

A framework for investigating water quality:

Figure 12.15. A framework of investigating water quality (source: Peters et al. 1998).

A template for integrated basin-scale studies has been made use of in the case of Murray-Darling basin of Australia.

## 12.9   DECISION – SUPPORT SYSTEM FOR WATER RESOURCES MANAGEMENT

The water resource issues have so many dimensions and complexities that decision-support is necessary for water managers. Kaden et al. (1989) describe a simple computer-aided water management system, to assist operators, managers and planners (Fig. 12.16). It has three components:

1. Measuring and information systems (data acquisition, transmission and storage),
2. Software system (user software),
3. Organizational system (organizational structure, legal and economic regulations).

Haagsma (1995) describes a more sophisticated, Internet-linked Decision Support System (Fig. 12.17). We should realize that decisions in regard to water resources are not necessarily taken on the basis of objective information alone, and that political considerations do come into the picture. Large amounts of data have to be transformed into useful information. The aim of the decision support system is to make available the relevant information to both the experts and water authorities, to enable them to have informed consultation.

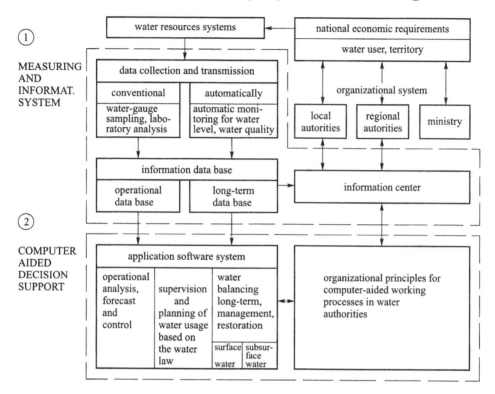

Figure 12.16. Computer-aided water management system (source: Kaden et al. 1989).

We should distinguish between coordination and integration. In the case of co-ordination, there is hierarchy whereby one model delivers the data to another in a specified manner, but there may be no feed-back (for instance, a water technician may make hundreds of piezometric measurements and then hand over the information to his boss who may be a local water engineer, who in his turn may send a summary of it to his superior, and so on; the important point is that there is no feed-back to the technician).

On the other hand, integration has no such built-in hierarchy. Models, information (including non-machine-readable information) and databases can be integrated into a decision-support system. Integration is thus analogous to different ministries setting up a joint task force to solve a problem. Different ministries bring in their own view-point to bear on the issue. There is no hierarchy here. Haagsma (1995) describes a recursively linked approach whereby different models can be used for different parts of the system (e.g. study of groundwater – surface water interaction, according to the time scales of the underlying hydrological processes) and run simultaneously, communicating through a network. Communication is facilitated by a communications server.

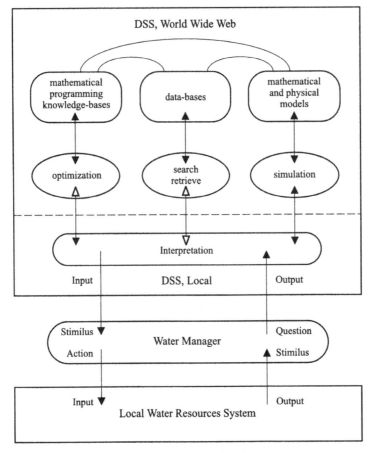

Figure 12.17. Decision Support System for water resources management (source: Haagsma 1995).

## 12.10 EPILOGUE

There is little doubt that in the twenty-first century, water issues are going to emerge as the most serious problem facing humankind.

The visionary paradigm for water resources management that L'vovich proposed way back in 1979 continues to be valid to this day. It presupposes:
– Cessation of the discharge of sewage into rivers and lakes,
– Reuse of wastewater,
– Closed circuit recycling of industrial water,
– Underground reservoirs of the same order of storage capacity as surface reservoirs,
– Protection of forest cover to reduce runoff, and increase percolation and soil moisture,
– Regulation of evaporation,
– Water transfers, etc. (Fig. 12.18; source: L'vovich 1979, p. 373).

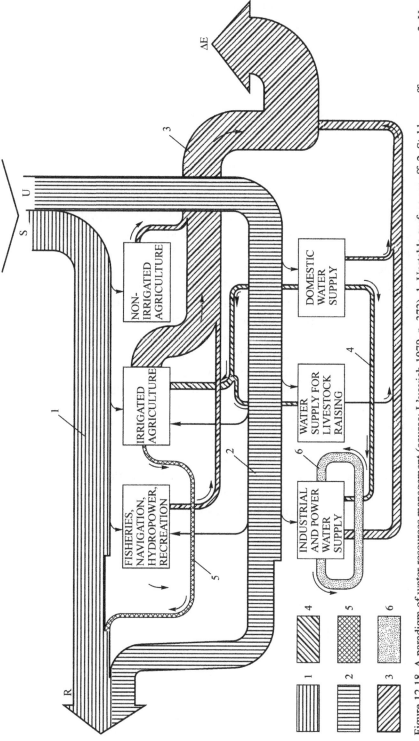

Figure 12.18. A paradigm of water resources management (source: L'vovich 1979, p. 373). 1. Unstable surface runoff, 2. Stable runoff resources, 3. Unrecoverable water consumption, 4. Sewage and polluted river water, 5. Water returning following irrigation, 6. Closed water recycling system.

The recommendations made by Shiklomanov (1998) to address the situation, and the way the issues raised by him have been covered in the present volume, are explained as follows:

1. Protection of water resources: Chapter 3 deals with all aspects of protection of water quality. Groundwater is sought to be protected through an understanding of its vulnerability (Section 12.3).

2. Drastic decrease in specific water consumption, particularly in irrigation and industry: Figure 12.19 (source: Shiklomanov 1998, p. 25) gives a forecast of the increase of population and increase in irrigated land (in M ha) upto the year 2025. In 1995, the agriculture sector accounted for 67% of the total water withdrawal and 86% of its consumption. This is not sustainable. In future, the efficiency of irrigation has to be improved greatly (through measures such drip irrigation and cutting down of conveyance losses), so that with lesser total withdrawal (about 60%), more food could be grown (Chapter 6). Ways and means of reducing the water consumption in industries by extensive recycling have been described in Chapter 7, and Chapter 6 explains how to reuse various kinds of waste water.

3. Complete cessation of the practice of discharging waste water into the hydrological systems (such as, rivers and lakes): As pointed out by Shiklomanov (1998), every cubic metre of contaminated water discharged into rivers and lakes spoils up to 8-10 m$^3$ of good water. This is a monstrous situation, and every effort should be made to stop this practice.

4. Harvesting of precipitation and making a more efficient use of runoff: Chapter 4 describes in detail ways and means of rainwater harvesting from rooftops, harvesting of surface runoff, groundwater recharge, etc.

5. Greater use of salt and brackish waters: Saline agriculture using brackish water or sea water has been described in Section 11.7.1.

6. Influencing the precipitation-forming processes.

7. Use of water stored in lakes, underground aquifers and glaciers: The conjunctive use of water resources, and the design and management of groundwater reservoirs, have been described under Section 12.4.

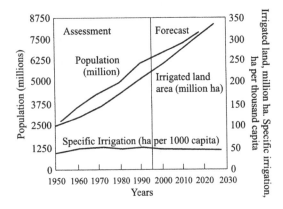

Figure 12.19. Projected increase of population and irrigated land (source: Shiklomanov, 1998).

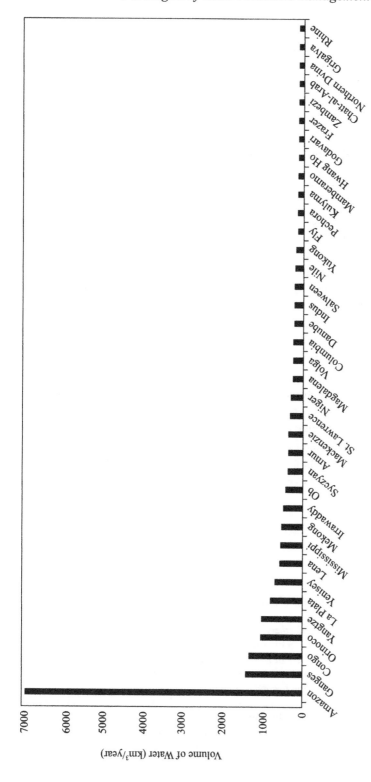

Figure 12.20. Mean annual runoff of the major rivers of the world (source: Shiklomanov 1998).

8. Spatial and temporal redistribution of water resources: Section 6.2 deals with the techno-socio-economic issues involved in water transfers.

The total global runoff is more than adequate to meet the demands for water for the humankind for many decades to come. But the catch in it is that the distribution of freshwater resources in the world is extremely uneven. Anthropogenic activities tend to degrade the freshwater resources, thus complicating the problem even further. Under the circumstances, the transfer of waters between one country to another and between different parts of a country are unavoidable, and every effort should be made to minimize the problem.

The total runoff from the earth's rivers is estimated to be 42,700 $km^3$ $yr^{-1}$. The Amazon alone accounts for about one-sixth of the total runoff. The total sustainable yield of freshwater in the world has been estimated to be 12,500 $km^3$. According to Shiklomanov (1998), the total global withdrawal by 2025 is projected to be 5100 $km^3$ $yr^{-1}$, with a consumption of 2860 $km^3$ $yr^{-1}$.

The world's major rivers (i.e. those whose long-term runoff is more than 100 $km^3$ $yr^{-1}$) are shown in Figure 12.20 (source: Shiklomanov 1998, p.14). Though some rivers like the Danube and the Rhine in Europe, Nile in Egypt, Colorado in USA and Cauvery in India, are intensively used, most of the rivers in the world are used only to a limited extent. Despite enormous techno-socio-economic problems, massive transfer of waters (e.g. Three Gorges dam on River Yangtze in China which transfers about 70 $km^3$ $yr^{-1}$ of water from south to north) has to be undertaken in some situations. For example, transfer of water from the Congo river could provide water supplies to several parts of Africa. Similarly, transfer of waters from the Ganges-Brahmaputra system within India can virtually alleviate water supply problems in good part of the country. It should be emphasized that high dams are not the only means of bringing about water transfers – a number of other techniques are available (e.g. groundwater recharge).

## REFERENCES

* Suggested reading

Albinet, M. & J. Margat 1970. Mapping of groundwater pollution vulnerability. *Bull. BRGM 2ème Serie* 3(4): 13-22.
Aller, L. et al. 1987. *DRASTIC: A standardized system for evaluating the groundwater pollution potential using hydrographic settings.* US EPA Report 600/2-87-035.
Buras, N. & W.A. Hall 1961. An analysis of reservoir capacity requirements for conjunctive use of surface and groundwater storage. In *Groundwater in Arid zones.* IAHS Publ. no. 56.
Correa, N.R. 1987. Determination of the necessary volume of the groundwater reservoirs for optimal conjunctive water use for irrigation in an arid region. In T.H. Anstey & U. Shamir (eds), *Irrigation and Water Allocation,* IAHS Publ. 169: 221-228.
*Falkenmark, M., J. Lundquist & C. Widstrand 1990. Water scarcity – An ultimate *constraint in Third World Development.* Tema V, Rept. 14. Univ. of Linköping, Sweden.

Foster, S.S.D. & A.C. Skinner 1995. Groundwater protection: the science and practice of land surface zoning. In K.Kovar & J. Krásný (eds), *Groundwater quality: Remediation and Protection.* IAHS Publ. 225: 471-482.

Haagsma, U.G. 1995. The integration of computer models and databases into a decision support system for water resources management. In S.P. Simonovic et al. (ed.), *Modelling and Management of sustainable basin-scale Water Resources Systems.* IAHS Publ. 231: 253–261.

Hornbogen, M. & G.A. Schultz 1998. A quantitative technique for identification of a potential future water crisis in a water supply area. In H. Zebidi (ed.), *Water: A looming crisis?* Paris: UNESCO, p. 357-362.

IPCC (Intergovernmental Panel on Climate Change) 1994. *IPCC Technical Guidelines for assessing Climate Change Impacts and Adaptation.* Cambridge, England: Cambridge Univ. Press.

Kaczmarek, Z. 1993. Water balance model for climate change impacts. *Acta Geophysica Polonica* 41(4): 1-1.

Kaden, S., A. Becker & A. Gnauck 1989. Decision-support system for water management. In D.P. Loucks & U. Shamir (eds), *Closing the Gap between Theory and Practice.* IAHS Publ. 180: 11-21.

Kulshreshtha, S.N. 1993. *World Water Resources and Regional Vulnerability: Impact of future changes*, RR-93-10, Internat. Inst. for Appl. Systems Analysis, Laxenburg, Austria.

Lamblin, F.F. & D. Ehrlich 1996. The surface temperature – vegetation index space for land cover and land-cover change analysis. *Int. J. Remote Sensing,* 17(3): 463-487.

L'vovich, M.L. 1979. *World Water Resources and their future.* Washington, D.C.: Am. Geophy.Uni.

Madsden, H. & D. Rosbjerg 1995. On the modelling of extreme droughts. In S.P. Simonovic et al. (ed.), *Modelling and Management of sustainable basin-scale Water Resources Systems.* IAHS Publ. 231: 377-385.

Meyer, C.F. 1973. *Polluted Groundwater.* US EPA Rept. no. 600/4 – 73- 0016.

NRC (National Research Council) 1993. *Groundwater vulnerability Assessment: Contamination potential under conditions of uncertainty.* Washington, DC, USA: Nat. Acad. Press

*Peters, N.E. et al. 1998. Water quality degradation and freshwater availability – Need for a global initiative. In H. Zebidi (ed.), *Water: A looming crisis?* Paris: UNESCO, p. 195-201.

*Qi, Li et al. 1995. Flood control decision support system for Dongtiao River in China. In S.P. Simonovic et al. (ed.), *Modelling and Management of sustainable basin-scale Water Resources Systems.* IAHS Publ. 231: 125-130.

Riha, M. 1981. Multistage sampling and testing in groundwater – A prerequisite for maximum utilization of aquifer systems. *Groundwater* 17: 423-427.

Schultz, G.A. & M. Hornbogen 1995. Sustainable development of water resources systems with regard to long-term changes of design variables. In S.P. Simonovic et al. (ed.), *Modelling and Management of sustainable basin-scale Water Resources Systems.* IAHS Publ. 231: 329-338.

*Shiklomanov, I. 1998. *World Water Resources: A new appraisal and assessment for the 21st. century.* Paris: UNESCO.

Shuval, H.I. 1987. The development of water reuse in Israel. *Ambio,* 16, 186-192.

Simonovic, S.P. 1998. Criteria for social evaluation of flood management. In H. Zebidi (ed.), *Water: A looming crisis?* Paris: UNESCO, p. 337-342.

Stakhiv, E.Z. 1998. Induced climate change impacts on water resources. In H. Zebidi (ed.), *Water: A looming crisis?* Paris: UNESCO, p. 249-258.

*Sytchev, K.I. 1988. *Water Management and Geoenvironment.* Paris – Nairobi: UNESCO-UNEP.

*Tsao, Y-S. 1987. Over-irrigation of paddy fields for the purpose of artificially recharging groundwater in the conjunctive use scheme of the Cho-Shui River basin. In T.H. Anstey & U. Shamir (eds), *Irrigation and Water Allocation.* IAHS Publ. 169: 99-109.

Van der Linden, J.Ph., R. Koudstaal & L.M. Nyongesa 1989. Water Resources Information Systems for regional planning. In D.P. Loucks & U. Shamir (eds), *Closing the Gap between Theory and Practice.* IAHS Publ. 180: 85-94.

Vogel, C. 1999. Facing the challenges of the new millennium: a LUCC/IGBP perspective. *IGBP Newsletter*, June 1999, p. 3-4.

Walker, S. & H.A. Smithers 1995. Recent advances in drought management with particular reference to northwest England. In S.P. Simonovic et al. (ed.), *Modelling and Management of sustainable basin-scale Water Resources Systems*. IAHS Publ. 231: 107-116.

# Exercises

(The assistance of Dr I. Radhakrishna in the preparation of some Exercises, is gratefully acknowledged).

1. It is planned to dispose off wastewater from a fertilizer plant at the rate of 2000 $m^3 d^{-1}$ for the next 20 years in a deep sandstone, which is 40 m. thick and has an average effective porosity of 0.20. Calculate the minimum distance (in m) of waste migration from the disposal well.

   According to the equation of Warner (1975),

   $$r = \sqrt{(V/\pi b n_e)}$$

   Where $r$ = Radial distance of the wastewater front from the injection well,
   $V = Qt$ = Cumulative volume of wastewater injected, ($Q$ = Rate of injection ($m^3 d^{-1}$); $t$ = time period of injection), $b$ = effective thickness of the bed (in m), $n_e$ = effective porosity (%).

   $$V = Q t = 2000 \ m^3 \ d^{-1} \times 20 \ y \times 365 \ d = 14.6 \times 10^{-6} \ m^3$$
   $$r = \sqrt{(14.6 \times 10^6 / 3.14 \times 40 \times 0.2)} = 760 \ m$$

2. Calculate the volume (in $m^3 \ ha^{-1}$) of irrigation water to be applied, given the following conditions:
   1. Depth of the root zone = 1.3 m;
   2. Field capacity (in volumetric terms) = 20%;
   3. Decrease in wetness since the last irrigation (in volumetric terms) = 12%.
   Equivalent depth of water contained in the root zone = 1300 mm × 0.20 = 260 mm.
   Volume of the root zone in one ha of land = 10,000 $m^2$ × 1.3 m = 13,000 $m^3$.
   Total water content of the root zone = 13,000 m3 × 0.20 = 2600 m3.
   Fraction of deficit in the field capacity that needs to be replenished = 0.20 − 0.12 = 0.08.
   Volume of the deficit in the root zone that needs to be replenished = 1.3 m × 0.08 = 0.104 m.
   So the amount of water that needs to be applied per ha = 0.104 m × 10,000 $m^2$ = 1040 $m^3$.

3. Calculate the water requirements for a crop with four months growing season, and express it as a percent of total PET, given the following particulars:

   |       | May | June | July | August |
   |-------|-----|------|------|--------|
   | KC    | 0.6 | 0.8  | 1.1  | 0.7    |
   | PET   | 5   | 7    | 9    | 9      |

   Water requirement for the crop for the month of May =

   | | | |
   |---|---|---|
   | KC × PET × no. of days in the month = 0.6 × 5 × 31 = | | 93 mm. |
   | Water requirement for June = 0.8 × 7 × 30 | = | 168 mm. |

| Water requirement for July = 1.1 × 9 × 31 | = | 307 mm. |
| Water requirement for August = 0.7 × 9 × 31 | = | 195 mm. |
| Total water requirement for the season | = | 763 mm. |

Total PET for the season = $(5 \times 31) + (7 \times 30) + (9 \times 31) + (9 \times 31) = 923$ mm.
Total water requirement as a percent of total PET = $(763/923) \times 100 = 83\%$.

4. A groundwater system with piezometric surface as a regional slope (i) of the order of 10 m km$^{-1}$. Estimate the groundwater specific discharge through a confined aquifer with transmissivity $T = 0.002$ m$^2$ s$^{-1}$.

   Specific discharge $(\phi/A) = T.\ i = 0.002 \times (10\ \text{m}/1000\ \text{m}) = 0.2 \times 10^{-4}$ m$^2$ s$^{-1}$.

5. Estimate specific retention value $(S_r)$ of an unconfined aquifer, given the specific yield to be 20%, and porosity to be 30%.

   $\phi = S_y + S_r$
   $S_r = \phi - S_y = 0.3 - 0.2 = 0.1 = 10\%$

6. A basin with an area of 600 km$^2$, has a rainfall input of 700 mm. Assuming the evaporation – evapotranspiration rate to be 60%, and runoff rate to be 20%, calculate the water infiltration input to the subsurface system.

   Total precipitation in the basin = $600 \times 10^6$ m$^2 \times 0.7$ m = $420 \times 10^6$ m$^3$ = 420 M m$^3$.
   Volume of evaporation and evapotranspiration = 420 M m$^3 \times 0.60 = 252$ M m$^3$.
   Volume of runoff = 420 M m$^3 \times 0.20 = 84$ M m$^3$.
   Thus, the infiltration input = $420 - (252 + 84) = 84$ M m$^3$.

7. Estimate the apparent velocity of the groundwater system of an aquifer, given that the length of the aquifer to be 20 km, head difference to be 40 m, and permeability to be 20 m d$^{-1}$. Also, calculate the actual velocity of the system assuming a porosity of 30%, and compare it with the apparent/ darcy velocity.
   Apparent velocity,

   $$V = -k(dh/L) = -20 \times \frac{40}{20 \times 10^3} = -0.04\ \text{m d}^{-1} = 4\ \text{cm d}^{-1}\ (\text{numerically})$$

   Actual velocity

   $$V_a = V/\alpha,$$

   where $\alpha$ is the porosity $= 0.04$ m d$^{-1}$/0.3 = 0.13 m d$^{-1}$ approximately.
   Thus, $V_a = 3\ V$

8. Estimate the interface depth in an unconfined aquifer near the coastal boundary, taking fresh water density $(\rho_f)$ as 1.00 g cm$^{-3}$, saline water density $(\rho_s)$ as 1.025 g cm$^{-3}$, and height of the water table to be 2 m above the mean sea level.

   Interface depth, $Z$ (in m) is given by $= \dfrac{\rho_f}{\rho_s - \rho_f} \times h_f = \dfrac{1000}{1 \cdot 025 - 1 \cdot 000} \times 2\ \text{m} = 80\ \text{m}$

9. The total runoff volume during a 6 hr storm with a uniform intensity of 1.5 cm h$^{-1}$, is $21.6 \times 10^6$ m$^3$. If the area of the basin is 300 km$^2$, calculate the average infiltration rate for the basin.

Total rainfall = Intensity of rainfall × duration = 1.5 cm × 6 hr = 9.0 cm.

Depth of the runoff = Volume of runoff/area of the basin = $21.6 \times 10^6/300 \times 10^6 = 0.072$ m = 7.2 cm

Thus, the total infiltration = Total rainfall – Depth of runoff = 9.0 – 7.2 = 1.8 cm

Average infiltration rate = 1.8 cm/6 hr = 3 mm h$^{-1}$.

10. Estimate the combined subsurface fresh water flow into the oceanic basin of a coastal aquifer extending for a length of 400 km along the coast, given the permeability value of 40 m d$^{-1}$, average thickness 50 m, and piezometric gradient of 1 m km$^{-1}$.

   The flow rate $\phi = -K.A.i$ or $\phi = K.A.i$ (numerically), where $K$ = permeability, $A$ = cross section of the basin; $i$ = piezometric gradient

   $$\phi = 40 \text{ m d}^{-1} \times (400 \times 10^3 \text{ m} \times 50 \text{ m}) \times 0.001 = 8 \times 10^5 \text{ m}^3 \text{ d}^{-1}$$

11. A 5-m deep canal is constructed in an area of 1 km$^2$, with top surface of the unconfined aquifer matching the bottom basement of the canal. Given the hydraulic gradient of 10 m km$^{-1}$, permeability of 20 m d$^{-1}$, and porosity of 30%, calculate the amount of time required to saturate the unconfined aquifer.

   Total volume of pore space available = porosity × area × thickness = $0.30 \times (1 \times 10^6 \text{ m}^2) \times$ × 5 m = $1.5 \times 10^6 \text{ m}^3$

   Flow rate = $K.A.i = 20 \text{ m d}^{-1} \times (1 \times 10^6 \text{ m}^2) \times (10/1000) = 0.2 \times 10^6 \text{ m}^3$

   Time required to saturate the aquifer = $1.5 \times 10^6 \text{ m}^3/0.2 \times 10^6 \text{ m}^3 = 7.5$ days

12. The chemical analysis of a water sample, obtained from a medium-depth bore-well, yielded the following data:

| Constituent | Ca | Mg | Na | HCO$_3$ | CO$_3$ | SO$_4$ | Cl |
|---|---|---|---|---|---|---|---|
| (mg l$^{-1}$) | 56 | 16 | 85 | 256 | 24 | 43 | 82 |

Calculate SAR (Sodium Absorption Ratio) and the Total Dissolved Solids (TDS)

|  | mg l$^{-1}$ | eq.wt. | meq l$^{-1}$ |
|---|---|---|---|
| Ca | 56 | 20 | 2.75 |
| Mg | 16 | 12.2 | 1.32 |
| Na | 85 | 23 | 3.70 |

   SAR = Na/[(Ca + Mg)/2]$^{1/2}$ = 3.70/[(2.75 + 1.32)/2]$^{1/2}$ = 2.58

   TDS = 56 + 16 + 85 + 256 (0.49) + 24 + 43 + 82 = 431.5 mg l$^{-1}$

13. A groundwater system with an areal extent of 10 km$^2$, and a thickness of 5 m, is being charged with water at the rate of 10 m$^3$ s$^{-1}$. Estimate the time needed for the system to get saturated.

   Total volume of the groundwater system = $10 \times 10^6 \text{ m}^2 \times 5 \text{ m} = 50 \times 10^6 \text{ m}^3$

   Flow rate = 10 m$^3$ s$^{-1}$ = $0.864 \times 10^6 \text{ m}^3 \text{ d}^{-1}$

   Time required to saturate the system = $50 \times 10^6 \text{ m}^3/0.864 \times 10^6 \text{ m}^3 \text{ d}^{-1} = 58$ days (approximately)

# Appendices

## APPENDIX I

## SYMBOLS AND CONVERSION COEFFICIENTS

Prefix names of units of multiples and submultiples

| Prefix | Symbol | Factor by which unit is multiplied |
|--------|--------|-----------------------------------|
| Exa | E | $10^{18}$ |
| Peta | P | $10^{15}$ |
| Tera | T | $10^{12}$ |
| Giga | G | $10^{9}$ |
| Mega | M | $10^{6}$ |
| Kilo | k | $10^{3}$ |
| Hecto | h | $10^{2}$ |
| Deka | dk | $10^{1}$ |
| Deci | d | $10^{-1}$ |
| Centi | c | $10^{-2}$ |
| Milli | m | $10^{-3}$ |
| Micro | μ | $10^{-6}$ |
| Nano | n | $10^{-9}$ |
| Pico | p | $10^{-12}$ |
| Femto | f | $10^{-15}$ |
| Atto | a | $10^{-18}$ |

Base units in Systéme International (SI).

| Property | SI unit | Symbol |
|----------|---------|--------|
| Length | Meter | m |
| Mass | Kilogram | kg |
| Time | Second | s |
| Electric current | Ampere | A |
| Temperature | Kelvin | K |
| Amount of substance | Mole | mol |

In this book, $t$ means ton ($10^6$ g = $10^3$ kg). When m is used as a prefix (as in mg = milligram, or mmol = millimole), it means milli ($10^3$). When m is used as a suffix (as in 97.8 m), it means meter. M means million.

| Physical quantity | Name of SI unit | Symbol for SI unit | Definition of unit |
|---|---|---|---|
| Force | Newton | N | $kg \, m \, s^{-2}$ |
| Pressure | Pascal | Pa | $kg \, m^{-1} \, s^{-2} \, (= N \, m^{-2})$ |
| Energy | Joule | J | $kg \, m^2 \, s^{-2}$ |
| Power | Watt | W | $kg \, m^2 \, s^{-3} \, (= J \, s^{-1})$ |
| Frequency | Hertz | Hz | $s^{-1}$ (cycles per second) |

Conversion of older units into SI units.

| Quantity | SI unit | Old unit | Value of old unit in SI unit |
|---|---|---|---|
| Force | Newton (N) | Dyne | $10^{-5} \, N$ |
| Pressure | Pascal (Pa) | Atmosphere | 101.325 kPa |
| Energy | Joule (J) | Calorie | 4.184 J |

1 bar = $10^5$ Pa = $10^6$ dynes/cm$^2$ = 750 Torr = 0.98692 atm = 14.504 lb/in$^2$ (psi: pounds per square inch), 1 MN/m$^2$ = 1 N/mm$^2$ = 1 MPa = approx. 145 psi; 1 Mg m$^{-3}$ = 62.4 pcf (pounds per cubic foot).

Some commonly used units (in relation to SI base units).

| Property | Unit | Symbol | SI relation |
|---|---|---|---|
| Charge concentration | moles of charge per m$^3$ | $mol_c \, m^{-3}$ | |
| Concentration | moles per m$^3$ | $mol \, m^{-3}$ | |
| Electric capacitance | Farad | F | $M^{-2} \, kg^{-1} \, s^4 \, A^2$ |
| Electric charge | Couloumb | C | $A \, s$ |
| Electric potential difference | Volt | V | $M^2 \, kg \, s^{-3} \, A^{-1}$ |
| Electrolytic conductivity | Siemens per meter | $S \, m^{-1}$ | $M^{-3} \, kg^{-1} \, s^3 \, A^2$ |
| Energy | Joule | J | $M^2 \, kg \, s^{-2}$ |
| Force | Newton | N | $m \, kg \, s^{-2}$ |
| Mass density | Kilogram per cubic meter | $kg \, m^{-3}$ | |
| Molality | Moles per kilogram of solvent | $mol \, kg^{-1}$ | |
| Pressure | Pascal | Pa | $M^{-1} \, kg \, s^{-2}$ |
| Specific adsorbed charge | Moles of charge per kilogram of adsorbent | $mol_c \, kg^{-1}$ | |
| Specific surface area | Hectare per kilogram | $ha \, kg^{-1}$ | $10^4 \, m^2 \, kg^{-1}$ |
| Viscosity | Newton-second per m$^2$ | $N \, s \, m^{-2}$ | |

Values of some important physical constants.

| Constant | Symbol | Value |
|---|---|---|
| Atomic mass unit | u | $1.6606 \times 10^{-27}$ kg |
| Avogadro constant | $N_A$ | $6.022 \times 10^{23}$ mol$^{-1}$ |
| Boltzmann constant | $k_B$ | $1.3807 \times 10^{-23}$ J K$^{-1}$ |
| Diffuse double-layer parameter(at 298.15 K) | b | $1.084 \times 10^{16}$ m mol$^{-1}$ |
| Faraday constant | F | $9.6485 \times 10^4$ C mol$^{-1}$ |
| Molar gas constant | R | $8.3144$ J K$^{-1}$ mol$^{-1}$ |

*Some useful equations*

*Exchangeable – sodium percentage(ESP)*

$$= \frac{\text{Exchangeable sodium (mol}_c \text{ kg}^{-1})}{\text{Cation Exchange Capacity (mol}_c \text{ kg}^{-1})} \times 100$$

Sodium absorption ratio (SAR) = $Na/[(Ca + Mg)/2]^{1/2}$

(Concentrations of Na, Ca and Mg expressed in m mol l$^{-1}$)

*Soluble – sodium percentage(SSP)*

$$= \frac{\text{Soluble} - \text{sodium concentration (meq l}^{-1})}{\text{Total cation concentration (meq l}^{-1})} \times 100$$

$$ESP = \frac{100 - ESR}{1 + ESR} = \frac{100/2K_G}{1 + (SAR/2K_G)} SAR$$

where ESR = Exchangeable sodium ratio, and $K_G$ is Gapon selectivity coefficient $(0.015 \text{ m mol}^{-1})^{-1/2}$

*Useful conversion coefficients*:

1 BTU (British Thermal Unit) = $1.055 \times 10^3$ Joules (J)

1 erg = 1 dyne /cm = $2.39 \times 10^{-8}$ calorie = $1 \times 10^{-7}$ Joule = $9.4805 \times 10^{-11}$ BTU

Fuel value of 1 m$^3$ of fuelwood = 9.4 gigajoules (GJ); 1 t of coal = 28.9 GJ

1 ton of oil = 41.7 GJ = 1.44 t of bituminous coal

1 micron (μm) = $10^{-6}$ m = $10^{-4}$ cm = $10^{-3}$ mm = $10^4$ Å

1 Ångstrom (Å) = $10^{-4}$ μm = $10^{-8}$ cm = $10^{-10}$ m; 1 nm = $10^{-9}$ m =10 Å

1 metre = 100 cm = 1000 mm = 3.2808 ft = 1.0936 yd

1 sq. metre (m$^2$) = 10.764 sq. ft = 1.196 sq. yd.

1 cubic metre = 1 m$^3$ = $10^6$ cm$^3$ = 35.31 cu. ft = 1.308 cu.yd

1 hectare (ha) = 100 m × 100 m = $10^4$ m$^2$ = 2.47 acres

1 sq. km (km$^2$) = 100 ha = 247 acres; 1 acre= 4840 sq. yd = 4046.8 m$^2$

1 cu. km (km$^3$) = $10^5$ ha. m.; 1 M ha m = 10 km$^3$; 1 ha.m = 8.1 acre-ft

1 acre-ft = 0.1235 ha.m = 1235 m$^3$; 1 Maf (million acre-ft) = 1.235 km$^3$

1 l = 1 dm$^3$ = $10^{-3}$ m$^3$; 1 m$^3$ = $10^3$ L = $10^6$ ml

1 US gallon = 3.875 l; 1 Imperial gallon = 4.546 l

1 barrel (crude oil) = 42 US gallons = 35.80 Imp. Gallons = 162.75 l

1 acre-ft = 326,000 gallons

$1 \text{ m}^3 \text{ s}^{-1} = 0.03156 \text{ km}^3 \text{ yr}^{-1}$; $1 \text{ km}^3 \text{ yr}^{-1} = 31.68 \text{ m}^3 \text{ s}^{-1}$

$1 \text{ l d}^{-1} = 0.365 \text{ m}^3 \text{ yr}^{-1}$; $1 \text{ m}^3 \text{ yr}^{-1} = 2.74 \text{ l d}^{-1}$

$1 \text{ l s}^{-1} = 15.48 \text{ gpm (gallons per minute)}$; $1 \text{ gpm} = 0.0646 \text{ l s}^{-1}$

$1 \text{ tonne (t)} = 10^3 \text{ kg} = 10^6 \text{ g}$; $1 \text{ kg} = 2.2046 \text{ lb} = 32.150 \text{ oz}$

$1 \text{ part per million (ppm)} = 10^{-6} \text{ g g}^{-1} = 1 \text{ g t}^{-1} = 0.032 \text{ oz t}^{-1} = 0.644 \text{ dwt t}^{-1}$

$1 \text{ part per billion (ppb)} = 10^{-9} \text{ g g}^{-1} = 1 \text{ mg t}^{-1}$

$K = T°C + 273.15$

$1 \text{ year} = 365.25 \text{ days} = 8,766 \text{ hours} = 5.26 \times 10^5 \text{ min} = 3.156 \times 10^7 \text{ sec}$

$1 \text{ day} = 24 \text{ hours} = 8.64 \times 10^5 \text{ sec.}$

## APPENDIX II

Water resources endowment, and annual water withdrawal of countries

| Country | 1 | 2 | 3 | 4 |
|---|---|---|---|---|
| Mozambique | 0.8 | 1 | 13.0 | 13 + 40 = 53 |
| Tanzania | 0.5 | 1 | 3.0 | 8 + 28 = 36 |
| Nepal | 2.7 | 2 | 7.8 | 6 + 149 = 155 |
| Bangladesh | 22.5 | 1 | 19.6 | 6 + 205 = 211 |
| Madagascar | 16.3 | 41 | 22.8 | 17 + 1658 = 1675 |
| Nigeria | 3.6 | 1 | 2.5 | 14 + 30 = 44 |
| India | 380.0 | 18 | 2.2 | 18 + 594 = 612 |
| China | 460.0 | 16 | 2.3 | 28 + 434 = 462 |
| Kenya | 1.1 | 7 | 1.1 | 13 + 35 = 48 |
| Pakistan | 153.4 | 33 | 3.3 | 21 + 2032 = 2053 |
| Indonesia | 82.0 | 3 | 12.8 | 9 + 443 = 452 |
| Egypt | 56.4 | 97 | 0.9 | 84 + 1118 = 1202 |
| Zimbabwe | 1.2 | 5 | 1.8 | 18 + 111 = 129 |
| Thailand | 31.9 | 18 | 3.0 | 24 + 575 = 599 |
| Tunisia | 2.3 | 53 | 0.4 | 42 + 283 = 325 |
| Malaysia | 9.4 | 2 | 22.6 | 176 + 589 = 765 |
| Iran | 45.4 | 39 | 1.7 | 54 + 1308 = 1362 |
| Mexico | 54.2 | 15 | 3.8 | 54 + 847 = 901 |
| South Africa | 9.2 | 18 | 1.2 | 65 + 339 = 404 |
| Brazil | 35.0 | 1 | 43.0 | 91 + 121 = 212 |
| Ireland | 0.8 | 2 | 14.1 | 43 + 224 = 267 |
| Israel | 1.9 | 88 | 0.4 | 72 + 375 = 447 |
| Spain | 45.3 | 41 | 2.8 | 141 + 1033 = 1174 |
| Belgium | 9.0 | 72 | 1.2 | 101 + 816 = 917 |
| UK | 28.4 | 24 | 1.2 | 101 + 406 = 507 |
| Australia | 17.8 | 5 | 19.0 | 849 + 457 = 1306 |
| France | 40.0 | 22 | 3.4 | 116 + 612 = 728 |
| Canada | 42.2 | 1 | 98.5 | 193 + 1559 = 1752 |
| USA | 467.0 | 19 | 9.4 | 259 + 1903 = 2162 |
| Germany* | 41.2 | 26 | 2.1 | 67 + 601 = 668 |
| Sweden | 4.0 | 2 | 20.5 | 172 + 307 = 479 |
| Japan | 107.8 | 20 | 4.4 | 157 + 766 = 923 |
| Switzerland | 3.2 | 6 | 6.9 | 115 + 387 = 502 |

* Before reunification. 1. Total internal renewable water resources ($km^3$), 2. Percentage of annual withdrawal out of total water resources (%), 3. Internal renewable water resources ($\times$ 1000 $m^3$ per capita per year), 4. Annual per capital withdrawal of water in $m^3$ (domestic + industrial and agricultural = total). The countries are arranged in order of increasing income. (source of 1, 3 and 4. World Development Bank; source of 3. Human Development Report, 1997, of UNDP).

# Author index

# Subject index

Printed and bound by CPI Group (UK) Ltd, Croydon, CR0 4YY

23/10/2024

01777678-0018